Chemistry in Microelectronics

Chemistry in Microelectronics

Edited by
Yannick Le Tiec

Series Editor
Robert Baptist

iSTE

WILEY

First published 2013 in Great Britain and the United States by ISTE Ltd and John Wiley & Sons, Inc.

ISTE Ltd
27-37 St George's Road
London SW19 4EU
UK

www.iste.co.uk

John Wiley & Sons, Inc.
111 River Street
Hoboken, NJ 07030
USA

www.wiley.com

Library of Congress Control Number: 2012955113

British Library Cataloguing-in-Publication Data
A CIP record for this book is available from the British Library
ISBN: 978-1-84821-436-1

Printed and bound in Great Britain by CPI Group (UK) Ltd., Croydon, Surrey CR0 4YY

Table of Contents

Preface

Over the last few decades, the size of transistors has been drastically shrunk although functionality of their circuits has increased. The physics models have been widely investigated and corrected to take into account physical phenomena happening for very small features. The chemistries used for surface preparation have long been considered as black boxes. Materials consumption has seriously been impacting on transistor performance pushing the engineering teams to develop new alternative chemical mixtures and sequences. At the same time over recent years, the metrology made huge progress, allowing a better characterization and understanding of the phenomena.

This book is divided into five chapters to analyze the role of chemistry in the fabrication of advanced transistors: the chemicals and sequences used in processing, from cleaning to etching, the role and impact of the purity of the sources and materials used in "front end of the line (FEOL)" which correspond to the "heart" and performance of individual transistors, moving further to the "back end of the line" (BEOL) which is related to the interconnection of all the transistors designed earlier during the FEOL processing steps. Finally, the need of specific functionalization also requires key knowledge in surface treatments and chemical management, allowing applications like biosensors to develop new nanosystems. A new era is widely opening with the convergence of electronics and biology.

Chapter 1 allows the reader to enter the fascinating world of the FEOL, which can be considered as the place where the heart of the transistor is manufactured. Many domains are pretty important to properly optimize the transistor performance: from doping level and profile to capacitance and junction consideration, optimized integration of new materials in terms of nature, properties and thicknesses. In this chapter, the focus is on three key modules: the gate stack definition, the crystalline materials involved in source–drain areas and the contacts definition. These key elements will give a transistor its final performance either being on the "high performance" side or being on the "low power" side. The gate stack optimization is mandatory and material choice is very impacting in this area: a careful choice of the best precursors for the deposited materials is necessary. Material science is also much involved not only in the gate area but also in the two main parts of the transistor that are source and drain areas. Optimized undoped (or more recently *in situ* doped) monocrystalline materials have to be deposited on these areas and a few steps later, specific contacts have then to be realized.

Chapter 2 brings the reader to the next step, which is the understanding of the interconnection between transistors themselves, which is called the BEOL. This chapter covers the aspects of optimizing the material properties so that a given transistor will be able to switch as quickly as requested, even if the signal needs to move through many metal layers of interconnection. Today, in most advanced technologies, there are up to 13 metal levels, which allow connection between a few billions of transistors on "first – ground – floor or level 0". The chip can be assimilated to a building (like a tower) with an ever smaller surface as a base and ever higher height adding new floors, up to 13 today. Engineering teams have then to design all the interconnections inside such a tower building which could be seen as putting in place the right stairs, elevators, rooms, etc., so that carriers (electrons or holes) can move as fast as possible from one point to other of that building.

Chapter 3 covers the main aspects of the surface preparation that are used in the current industry and research laboratories. Since 1970, wet cleaning is a key driver when advanced technologies request new answers: alternate materials have been introduced to address the following nodes. Over the last 10 years, the specification list given to the cleaning experts is getting new criteria and, today, there are key focuses concerning material consumption or loss, as well as selectivity between materials during chemical exposure. Basic chemistries are revisited to address new competing materials because the "standard" semiconductor silicon surface/substrate is now challenged by germanium, silicon-germanium, and III–V compounds, among others. The main chemical reaction is still about oxidation and reduction or acidic and basic, then there comes the solution of multiple equation systems. This is nothing new to chemical science; only the theoretical approach of the basic knowledge used for years has to be carefully applied to address the future challenges in transistor technology. In successive sections, the chapter describes the cleaning, the etching and the final rinsing-drying sequence: mechanisms, chemical reactions, impact of concentration and temperature are reported.

Chapter 4 describes the use and management of chemical fluids in microelectronics. The production of semiconductor devices is extremely sensitive toward any presence of traces of organic or metallic contamination. Undesired impurities must be kept at very low levels. Hence, it is important for all production steps, starting already with the resourcing of chemicals, for example ultrapure water, gases and liquid chemicals. As device line widths continue to shrink, the requirement for ever higher water purities in semiconductor applications is expected to increase. Techniques to clean these chemicals, control them to the desired concentration and, finally, the treatment of waste water and gas abatement are covered. The chapter also presents unique problems for liquid and gas production as well as quality analysis with the outlook of reducing the overall consumption.

Chapter 5 provides information regarding the surface functionalization for micro- and nanosystems, which are natural extensions of microelectronics and are now found in many convergent applications. Different materials are being investigated, from metal oxides, semiconductors, carbon nanotubes; silicon nanowires, metals or polymers. The functionalization processes are described: activation, cleaning, plasma treatments, chemical vapor deposition, sol-gel and polymer deposition techniques. Localization processes are also discussed: spotting, photolithography, scanning beam lithography, microcontact printing, etc. Mechanisms are explained, chemical reactions and related applications are detailed. This final chapter is a fascinating one to complete the book with, describing some interactions between chemistry and microelectronics.

After this technical outline, let me finally express my thanks for the commitment and considerable work of each coauthor who spent many hours carefully describing their expertise and picking up the most relevant references to his domain. The reader will find many relevant references at the end of each chapter (around 700 references in total throughout the book) including the most recent ones. I also would like to thank the editor for giving us the opportunity to edit this book and for his great support during the review periods making possible this first edition. On behalf of all coauthors, follow us into the inside of the nanochemistry world.

Yannick LE TIEC
January 2013

Chapter 1

Chemistry in the "Front End of the Line" (FEOL): Deposits, Gate Stacks, Epitaxy and Contacts

1.1. Introduction

The manufacturing of a transistor involves several hundred production and control stages and is a process that takes several weeks, generally between four and twelve depending on the complexity of the electronic component. Production is considered to be either "front end" or "back end".

A "front end" plant will produce a component consisting of between hundreds of millions and several billion transistors, all interconnected via different stages. Today the most advanced technologies with 20 nm nodes use between eleven and thirteen metal levels to interconnect several billion transistors by means of a specific map that is determined by the feature of the component. The component is tested and the functional chips are identified and selected in order to calculate what is known as the yield, that is the number of chips made to specifications versus the total number made. A "back end" plant will receive the "front end" components and finalize the packaging through a second pathway and specific processes. Tests then

---

Chapter written by François MARTIN, Jean-Michel HARTMANN, Véronique CARRON and Yannick LE TIEC.

guarantee the reliability of the component, generally over a period of 10–20 years.

In the "front end" plant, the manufacturing of the component is itself divided into two main sectors: first, the "front end of the line" (FEOL), which could be considered as the technological heart of the operation. It involves the extreme optimization of each individual transistor: dimensioning, type (N or P), arrangement, characteristic (high performance or low consumption, for example) and application (logic, analog, etc.). The second area of manufacturing is known as the "back-end of the line" (BEOL) and will be explained in Chapter 2 of this book, simplified, its aim is to optimize the interconnection of the individual transistors made by the FEOL sector.

This chapter focuses on the FEOL sector: it can be divided into several subsectors of activity; through processes, we will successively list zones for deposition, mechanical–chemical polishing, dry and wet etching, surface preparation/cleaning, doping or lithography. For each technological node, there are groups of experts aiming to define the specifications required for these different processes, in order to obtain the desired features for various applications. This is done by the international group, International Technology Roadmap for Semiconductors (ITRS) [ITR 12].

A modern transistor must address the various technological aspects that are shown in Figure 1.1. The challenges concern the starting material "A" (generally silicon, eventually modified to a version called silicon on insulator (SOI) corresponding to an ingenious $Si/SiO_2/Si$ stack where the highest active layer of silicon is isolated from the substrate by an embedded layer of oxide), the insulation between the transistors (B), the determination of the dopant profiles (target addressed by points C, E and G), the gate stack (D and F), the contact zones of the gate as well as the "Source" and "Drain" zones of the transistor (H and I), etc. [ITR 11].

This chapter describes three important elements. In Figure 1.1, these are described mainly by points F, H and I, and somewhat by D and E as well. First, the chemistry involved in the gate stack is described as this is directly related to a fine tuning of the transistor: the optimization of the gate insulator and the gate itself. The choice of the chemical precursors is of key importance. Next, crystalline heterostructures (based on SiGe in particular) are explained and the involvement of the precursors, temperature, pressure and surface preparation is detailed. Finally, the chemistry involved to address efficient contacts is also precisely described.

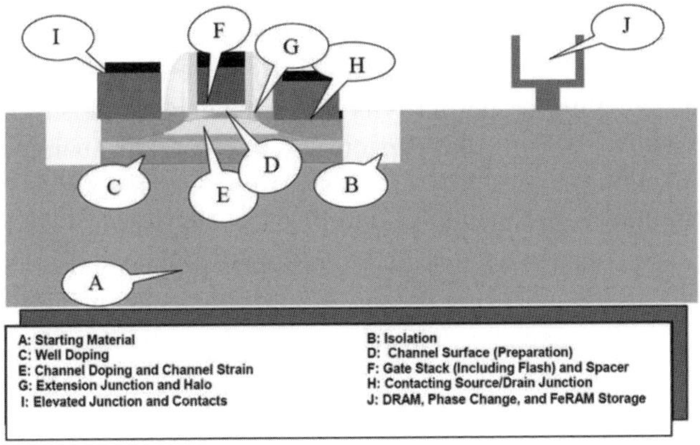

Figure 1.1. *The challenges of the "front end of the line" as defined by the ITRS [ITR 11]*

1.2. Arrangement of the gate

1.2.1. *Generalities*

After the discovery of the planar structure in the 1960s, for a long time the gate stack of the metal oxide semiconductor (MOS) transistors was made of a silicon (SiO_2) insulator covered by a gate for the channel control (itself made of N and P doped polycrystalline silicon). Looking at the chemical processes used in the production of this stack, first a thermal oxidation of the silicon takes place in the heat treatment furnace containing several tens of wafers (reaction [1.1]). Then the silicon gate is deposited by chemical deposition in the vapor phase low pressure chemical vapor deposition (LPCVD) from silane in a furnace under vacuum (reaction [1.2]).

$$Si + (O_2 \text{ or } 2\,H_2O) \rightarrow SiO_2\ (+2\,H_2) \qquad [1.1]$$

$$SiH_4 \rightarrow Si + 2\,H_2 \qquad [1.2]$$

For dimensions of about 0.18 µm and smaller, additional constraints arose: the continuation of transistor size scaling fixed the thickness of the insulation gate (Moore's Law) and while the current tunnel through the oxide increased drastically, the boron doping diffused from the P-type metal-oxide semiconductor (PMOS) side through the gate insulator to the transistor's channel. This chapter will explain how scientists first improved the properties of the silicon oxide, before introducing into the 45 to 28 nm nodes an innovation that would replace a large part of SiO_2 by a high-permittivity insulator, known as "High K": by depositing a greater thickness of a material with a higher permittivity, the leakage current is indeed reduced without adversely affecting the capacitive coupling. To guarantee a greater physicochemical compatibility with the High K material as well as an optimal capacitive coupling, the second major innovation was to deposit a metal gate, rather than a gate made from doped polycrystalline silicon. The latter had the disadvantage of having a slightly depleted interface doping, adding a parasitic capacity, which adversely affected the transistor's performance. In this section, the chemical mechanisms that come into play inside the fine layers of the gate stacks will be defined as well as the precursors used for depositing the films in the manufacturing environment. The motivation behind the choice of the High K material and the interaction with its environment will only be described briefly (for more information, the reader may refer to review papers about this topic) [WIL 01, LOC 06, CHO 11].

1.2.2. *Silicon nitriding processes*

1.2.2.1. *Thermal nitridation*

In the history of the technological breakthroughs in this field, the first requirement was to increase the resistance of the silicon oxide to electrical tension (its reliability) and against the diffusion of the boron through the insulator. Before the 90 nm nodes, boron diffusion was controlled by an optimization of the implantation conditions as well as by the activation annealing of the polycrystalline silicon gate and junctions. Processes used to increase the reliability of the Si/SiO_2 interface were investigated by introducing into the 2–3 nm oxide layers a very specific quantity of nitrogen. Experiments were first conducted by annealing silicon oxide in the presence of NH_3, thus nitriding the oxide surface by optimizing the temperature of the process; however, the presence of reducing species resulting from the thermal decomposition of NH_3 at high temperatures caused the apparition of electrical traps in the silica. This disturbance was resolved by carrying out complex reoxidizing steps on the nitrided oxide, thus increasing the difficulties for controlling the process.

Next, scientists introduced nitriding molecules that contained no hydrogen. The first attempts were carried out with N_2O; however, it appeared that its complex decomposition involves the nitridation/oxidation of the interface, giving a Si–N–O bonding state, which is difficult to control uniformly over a batch of wafers [HEG 97]. By identifying that the main product of N_2O decomposition was, in fact, nitrous oxide (NO), it seemed easier to introduce it directly into the annealing furnace. As the diffusivity of the NO molecule is dependent on the insulator's growth conditions (and its stress) [SAG 96], the annealing process was optimized in order to obtain the necessary nitrogen concentration in the internal interface, depending on the silicon's thickness. In addition, the bonds formed at the interface were of the Si–N type [HEG 97] and so by lightly reoxidizing it, the nitrogen was then in an environment of the same type as that formed by treatment with N_2O, as shown by X-ray photoelectron spectroscopy (XPS) [HAL 98]. With this sort of process, approximately 0.1% nitrogen may be obtained in an optimal chemical environment [WIL 01], that is when conditions lead to the improvement of the interface's characteristics without degrading any other properties of the transistor, such as mobility and leakage current.

1.2.2.2. *Plasma nitridation*

From the 90 nm technological nodes up to the advent of High K materials, a more efficient nitridation process was developed, which granted an increase in permittivity by nitriding the silica to the core. It improved the balance in the compromise between the electric thickness of the insulator (Equivalent Oxide Thickness (EOT)) and its leakage current, since permittivity increases with nitrogen incorporation, while providing a more efficient barrier to the diffusion of the boron. As the nitridation process in the presence of NO only allowed a reduced amount of nitrogen, approximately 1×10^{13} atm/cm^2 at the Si/SiO_2 interface, a new insulator-nitriding technology was required. It needed to be able to introduce a given percentage of nitrogen (between six to 15% depending on the devices) while preserving the silicon interface from nitridation for the reasons explained previously (degradation of transistor properties). The new process uses a low-energy nitrogen plasma (in the order of 1 eV), allowing nitrogen implantation in the surface and the volume of the 1.5 to 2.5 nm-thick oxide layers. The experimental procedure for this process is extensively documented [CUB 04]: the nitriding plasma first incorporates the nitrogen into the silicon oxide in the form of three Si–N bonds and unstable Si–N=O complexes. The latter are then removed by lightly oxidizing annealing, which reorganizes the Si/SiO_2 interface. This must be achieved without exposure to air and so is carried out in a single-wafer treatment system that contains the nitriding plasma reactor as well as the annealing reactor. Otherwise, non-reproducible exposure to the atmosphere gives way to a reaction of non-stabilized nitrogen atoms to form N_2O or NO [CUB 04]. This technology did not evolve until 40 nm nodes for the low-power technologies. To scale further electrical oxide thickness while keeping a low gate leakage current, High K/metal stacks were introduced from the 45 nm nodes for microprocessor technologies (high performance) and from the 32 nm nodes for the "low-power" technologies.

1.2.3. *The introduction of the High K/metal gate stacks*

1.2.3.1. *The choice of High K materials in a conventional process*

The selection of the best High K material for use as an insulation gate was the subject of intensive sampling by the end of the 1990s. The principal criteria of success being, besides a sufficiently high permittivity (at least four times that of SiO_2), a thermodynamic stability in relation to the silicon substrate [LOC 06], a sufficiently large forbidden band and the positioning of the bands in relation to the silicon (known as the band "overlap" of the insulator), at least 1.5 eV, respectively, above or below silicon conduction band energy (Ec) and silicon valence band energy (Ev) to limit the leakage current [ROB 00]. The material presenting the best compromise in relation to all these properties is hafnium oxide, HfO_2, mainly due to its thermal stability with underlaying silicon that sets it apart from ZrO_2 (Zr is in the same column as Hf in Mendeleyev's periodic table but is less stable on silicon/thin silicon dioxide couple). However, following an extensive study of the properties of the material, scientists decided in favor of hafnium silicate $Hf_xSi_{1-x}O_4$, a compromise between SiO_2 and HfO_2. Indeed, the aim was first and foremost to maintain the compatibility of the High K material with a polysilicon gate, which is not performed by HfO_2, due to the interaction between reducing species (resulting from the decomposition of the silane) with Hf–O bonds. Silicon in hafnium silicate also increases the crystallization temperature of the insulator compared with that of HfO_2; since crystallization does not interact with the leakage current [KIM 04a] in very thin films, researchers quickly discovered that it is generally responsible for the creation of defects that affect the mobility of the channel carriers [ZHA 06, KIR 06a]. Adding even small quantities of silicon is favorable to suppress crystallization [KIR 06a]. The same also applies, more generally, to the decrease in thickness of the High K, stopping the nucleation of the crystallites [ZHA 06]. Thermal stability of the hafnium silicate is improved by plasma nitridation. Later this also increases the permittivity of hafnium silicate as well as its chemical compatibility with the metal gate. However, unlike plasma nitridation of silicon dioxide described earlier, very unstable Hf–N bonds are also produced and must be eliminated during the annealing process. This is known as "post-nitridation anneal (PNA), and is carried out to leave only nitrogen–silicon bonds [INU 05]. The control of the nitridation plasma parameters is critical, particularly the pressure; indeed the nitrogen atoms replace oxygen atoms which diffuse into vacuum at a low working pressure. On the contrary, at higher pressures, it diffuses till Si/SiO_2 interface and contributes to an increase in the thickness of the interfacial silicon oxide layer, thus diminishing the global electric capacitance of the insulator [INU 05, KAM 08a]. The optimization of a nitrided hafnium silicate increases the final gain on the leakage current by about two orders of magnitude in relation to the equivalent SiON electrical thickness, in the 1.2 to 1.3 nm equivalent oxide thickness (EOT) range (corresponding to a 28 nm "low-power" node), in spite of a reduction of the HfSiO band-gap on the valence band side due to the addition of nitrogen [BAR 06]. A metal gate compatible with the insulator must also be introduced, providing the same work function as N^+ and P^+ doped polycrystalline silicon (respectively ~4.1 and 5.15 eV).

1.2.3.2. *Constraints and possibility of High K/metal integration*

1.2.3.2.1. "Gate first" assembly

The assembly that consists of maintaining polysilicon on silicate HfSiON was indeed only attempted in transistors whose principal objective was limiting the leakage current (application of the dynamic random access memory transistors (DRAM), rather than improving overall performance. In complementary metal oxide semiconductor (CMOS) logic, scientists first researched the insertion of the metal gate on the High K insulator in the standard way where the gate is made before the activation of the junction dopants (a rapid thermal treatment carried out above 1,050°C). It seemed that controlling the potential of the gate at the edge of the silicon band could not be done by simply using metals with corresponding theoretical potentials. This impossibility is due to the interaction of the heated metal with the High K insulator, namely its chemical affinity for the oxygen, which induces a complex phenomenon: the creation of charged vacancies in the High K [UED 06] and/or the modification of the interfacial SiO_2 layer [BER 10]. These phenomena lead to a value of the final work function close to the silicon mid-gap [AKI 07], which makes the fabrication of the metal for the PMOS transistor in the range of fine electrical thicknesses (about 1 nm) impossible. This problem was partially solved for the N-type metal-oxide semiconductor (NMOS) transistor by introducing a dipole at the IL/HfSiON interface [KIR 08]. The most frequently used elements are lanthanum for the NMOS transistor [KIR 08] and aluminum for the PMOS transistor [JUN 05] on the condition that the latter operates with an FDSOI-type substrate (see Figure 1.2, [WEB 10]): the channel is made from undoped silicon that releases the specifications to 4.8 eV for PMOS and 4.4 eV for NMOS transistors. On bulk silicon, where the channel is highly doped, it is impossible, even with a dipole, to pilot the required work function near 5.1 eV; the channel's potential is then modified using a SiGe alloy [KRI 11] (see section 1.3). Moreover nitrogen incorporated into the insulator (voluntarily and mainly by diffusion from the grid after high-temperature annealing) plays an essential role in the mobility control and the reliability of the transistor [GAR 08]. The integration of a High K/metal stack in a conventional "gate first" scenario is then actually feasible but quite complex for nodes beyond 28 nm. Then major manufacturers of integrated circuits chose to modify the transistor assembly completely in order to avoid these high-temperature interactions between the metal gate and the insulator.

1.2.3.2.2. "Gate last" assembly

A "gate last" assembly uses a "dummy" polysilicon gate during the high-temperature sequencing of the process before being replaced by a gate made from High K/metal. It was used by Intel® (see Figure 1.3) for 32 nm "high-performance" transistors [PAC 09]. In these stacks, the gate potential of the NMOS transistor is set by a TiAl alloy deposited on thin TiN [VEL 11]. The lack of high-temperature interaction between the metal and the insulator, HfO_2, means it is easier to obtain lower electrical thicknesses [RAG 05] and to adjust the PMOS gate potential with titanium nitride [PAC 09]. Looking at the subsequent technological nodes (20 nm followed by 14 nm), it is obvious that manufacturing a "planar" gate stack insulator (which has existed for nearly half a century) equivalent to 0.6 nm silica is difficult. The solution used nowadays involves a three-dimensional deployment of the transistor to increase the capacitive coupling surface [WU 10]. Others may prefer solutions where the SiO_2 interface layer (IL) is reduced to the very minimum due to the "scavenging" phenomenon [KIM 04b], based on a "suction" of the

High K's oxygen through a metal gate with a high affinity for oxygen: the High K material, which is thermodynamically more stable than silicon, compensates for the loss of oxygen by reducing the underlying SiO_2 IL and by releasing silicon into the substrate. We also mention the attempts aiming to modify the crystalline structure of HfO_2 in order to increase the value of its permittivity, either by stabilizing in a cubic or quadratic phase, rather than monoclinic. The apparition of this phase may be obtained by adding another compound, such as a small quantity of silicon [MIG 08], or by forming an HfZrO alloy [TRI 08]. Let us note that phase stabilization is a complex, stress-dependent phenomenon, and therefore depends on the gate structure and the thermal budget of the device [TRI 06].

Figure 1.2. *A "gate first" type assembly for fully depleted silicon on insulator technology (from [WEB 10])– the dark color: the High K/metal bilayer on the interfacial layer (IL) is in a lighter color*

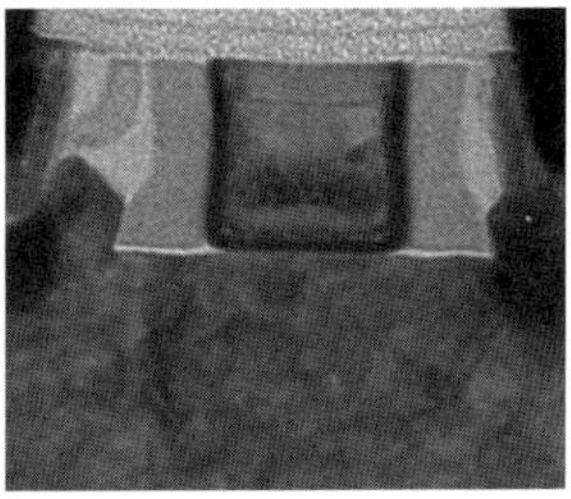

Figure 1.3. *An NMOS transistor from "Gate Last" technology showing the High K/metal bilayer inside the grid's cavity, after the removal of the dummy polysilicon grid (from [PAC 09])*

1.2.3.3. *Chemistry of High K deposition processes*

After having described the evolution of the gate integration sequences as well as the role played by solid chemistry in silicon manufacturing, this section will describe the chemical precursors used for deposition of the layers mentioned previously.

1.2.3.3.1. Deposition by CVD

Metal organic chemical vapor deposition (MOCVD) allowed the deposition of hafnium silicate from Si_2H_6 together with a hafnium alkoxy, the Tetrakis(tert-butoxy)Hafnium ($Hf(OtBu)_4$) (TBH) diluted in octane. This method involves the sectioning of the alcohol ligands by the silicon hydride, leading to the formation of an oxide with a 60% Hf/Hf+Si ratio, without any addition of oxygen in a "batch" furnace at low temperature (260°C), and so without any risk of regrowth of the SiO_2 interface sublayer [INU 05]. However, it is critical to control the mechanisms of deposition over a large surface in the furnace, which is why the single wafer systems have become more efficient for High K oxide deposition during the last decade. The first criterion for hafnium precursor selection is the capability to perform deposition at a controlled rate, which will allow for a sufficient uniformity in the thickness on the wafer. It must have a high enough vapor pressure, a property that is generally obtained from the liquid phase, either heated or not, to deliver a sufficient quantity of the precursor into the reactor. The second criterion is the thermal stability of the precursor, which guarantees the absence of a particle producing solid residue [GIR 07]. Based on the latter, hafnium's most frequently used precursor is an ethylamine, tetrakis diethylamino hafnium (see Table 1.1) only used for HfO_2 deposition [VAN 04a] or with amino precursors of silicon such as tris(dimethylamino)silane (TDMAS) or bis(dimethylamino)silane (BDMAS) [GRO 12, KAT 08] in the presence of oxygen in a range of deposition temperatures greater than 500°C to avoid carbonated contamination [VAN 04a]. Octane is always used as a solvent as it distributes small quantities of the chemical precursor into the reactor. More sophisticated processes also allow the injection of less-volatile precursors and so improve the CVD deposition recipe by injecting pulses of reactant into the oxygen with an optimal frequency [COS 04]. It is still a CVD process but the deposition is controlled to the monolayer (known as "AVD™") and is close to the way of deposition known as ALD (atomic layer deposition), which will be detailed in the next section.

Metal	Amines (*Main ligand*)			Cyclopetadienyl	Alkoxy
	methyl	*mixte*	*ethyl*		
Hf	TDMAH	TEMAH	TDEAH	$HfCp_2Cl_2$	HTB
Si	*n*-DMAS (n = 2–4)				
Zr		TEMAZ		$ZrCp_2Cl_2$	

Table 1.1. *The main organometallic precursors described in this chapter for gate stack deposition; the others are chlorides: MCl_4 with M=Ti, Zr, Hf; the latter two are solids, TiCl4 is liquid*

Ti	TDMAT	TEMAT	TDEAT	$TiCp_2Cl_2$	
Ta	PDMAT	TAIMATA			
		TBTDET			
Mg				$Mg(EtCp)_2$	
La				$La(i\text{-}PrCp)_3$	

Table 1.1. *(Continued) The main organometallic precursors described in this chapter for gate stack deposition; the others are chlorides: MCl_4 with M=Ti, Zr, Hf; the latter two are solids, TiCl4 is liquid*

1.2.3.3.2. ALD: atomic layer deposition

For gate application in microelectronic technology, a specific method was introduced to deposit High K insulator: ALD in which deposited gate oxide films typically have a thickness of 1.5–2.5 nm, plus or minus one atomic layer. Unlike CVD where the reactants are simultaneously injected into the reactor, the metallic precursors and the oxidizing reactants are separately injected by pulses in ALD, with an inert gas purge between the two (see Figure 1.4). Dedicated reactor designs have been developed to minimize the time taken by the pulse and purge while avoiding any possibility of a mixture of reactants and oxidizing reagents, leading to a CVD parasitic growth. Here, the deposition kinetic is controlled by precursor and reactant chemisorption mechanisms on the surface sites. In the case of an oxide deposition, hydroxyl groups are used whose surface density decreases by recombination when the temperature is increased. Then, unlike CVD deposition, the ALD deposition rate will decrease with any increase in reaction temperature. The total quantity of precursor (or reactant) injected over the course of each cycle must be optimized in order to cover all the reactive sites in as short a time as possible (typically between a few tens or hundredths of a millisecond).

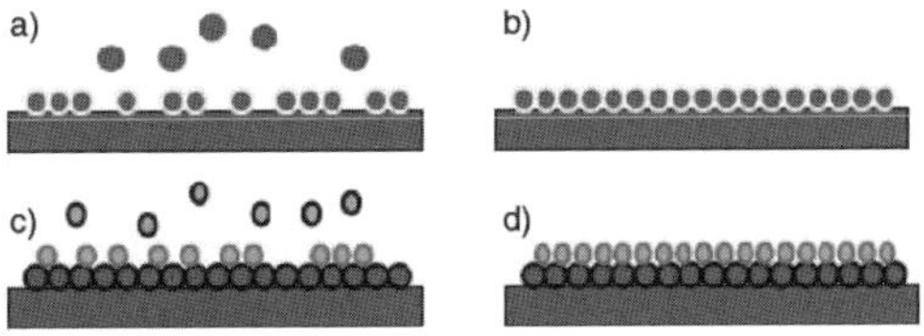

Figure 1.4. *Schematic of the ALD deposition process: a) pulse and chemisorption of the metal precursor, b) purge of the inert gases, c) pulse and chemisorption of the reactant and d) reactant purge*

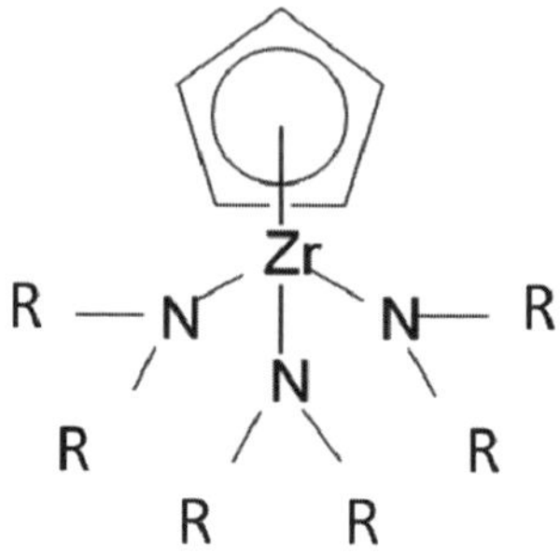

Figure 1.5. *An amine molecule with a cyclopentadienyl ligand used for zirconium oxide deposition (R being an organic radical), from [GIR 07]*

The optimal temperature range for chemisorption and desorption reactions is between 200 and 350°C as it allows the use of suitable metallic precursors: they must be stable inside this range and optimized in terms of solid residue, such as cyclopentadienyl amine derivatives, called MyALD (with M = Hf or Zr, see Figure 1.5), or ozone, O_3, as an oxidative reactant [GIR07]. These precursor pairs have been developed specifically to manufacture metal-insulator-metal (MIM)-type capacitive structures that are not, however, inside the scope of this chapter: applying ozone to the gate insulator deposition indeed helps to start the reaction on a deoxidized surface [KIR 06b], however, it leads finally to a thicker SiO_2 interface and so water is the preferred reactant even if it does require the presence of an initial silicon dioxide layer to start the reaction. There are many studies covering the advantages of different reactors and precursors, but it is acknowledged that if using water as the oxidative reactant it is simpler to use hafnium tetrachloride, $HfCl_4$, as the hafnium precursor. Others such as tetrakis dimethylamino hafnium (TDMAH) and tetrakis ethylmethylamino hafnium (TEMAH) react with H_2O, but have a much lower thermal stability (leading to an increase in dry residue [GIR 07]), meaning the reactions must be carried out below 300°C. Then carbonated residues will form in the layers, leading to a decreased film density [TRI 05].

Hafnium chloride, which is a by-product of zirconium purification and is relatively of low cost, is a solid precursor (unlike amines) and has a vapor pressure of approximately 1 torr at 160°C [PAS 63], compatible with the doses injected during each ALD cycle. Dedicated sources working in this temperature range have been developed to deposit hafnium oxide at a higher reaction chamber temperature, in order to avoid any possible recondensation of the chloride. Their design takes into account the extremely high reactivity of $HfCl_4$ in humid conditions, guarantees a sufficient amount of precursor entering the reactor for a continued delivery of reactant, especially in the case of deposition on larger areas than the chip itself. Indeed, if the gates are present as cavities (see Figure 1.3), the total amount of precursor added must be increased depending on the actual developed surface of the chip, and must also take into account the time required for the diffusion of the precursor/reactant pair inside the cavities during the pulses. Models of this process have been investigated in detail [STO 06]. Due to its present use in industry, specific details of the $HfCl_4/H_2O$ chemistry will be examined here, as well as the behavior of the residues of chemical precursors inside the films during the following process steps. Indeed, chlorine was identified early as a contaminant and then mixed chlorinated cyclopentadienyl (Cp_2HfCl_2) molecules were developed to reduce its effects. Results, however, were disappointing as the gain achieved was countered by an increase in carbonated contaminants [NIN 06].

1.2.3.3.3. HfO_2 ALD deposition with $HfCl_4$

The effect of the deposition parameters. The deposit mechanism based on $HfCl_4/H_2O$ on silica was established in an experimental reactor [AAR 99] and globally corresponds to the chemisorption on surface hydroxyls, given in reaction [1.3]:

$$HfCl_4 + n\,(\text{–}OH)_s \rightarrow n\,(\text{–}O\text{–}HfCl_{4-n}) + n\,HCl \quad \text{with } n = 1 \text{ or } 2 \qquad [1.3]$$

Next is the hydrolysis of the Hf–Cl bonds by water, producing once more HCl as well as an Hf–OH-type surface, which may then host another cycle. The actual mechanisms, which are more complicated, are modeled in terms of growth per cycle (GPC) [PUU 04]. The chemisorption mechanisms on the substrate during the first cycle have led to much research into the potentially crucial role of the heterointerface in the electrical properties of the part. Researchers also found that the first deposition cycle leaves a greater quantity of chemisorbed chlorine [BLI 03, TAL 08] than a permanent deposition regime (approximately ten times more, see Figure 1.6). This is even more pronounced on a chemical (rather than thermal) oxide, due to readsorption of HCl by the silica's hydroxyl groups. The chemisorption of chlorine increases with the decrease in reaction temperature and may affect the charge density of the oxide interface [SRE 06]; by increasing both the time taken by the water pulses and the deposition temperature, the surface and volume concentration of chlorine may be reduced [BLI 03, DEL 06] (see Figure 1.7). These steps are necessary as it is then harder to eliminate chlorine afterwards by annealing, especially at the SiO_2/HfO_2 interface [SRE 06, FER 02] where a silicate-like compound is identified [REN 02] before the appearance, after a few cycles, of the permanent deposition regime, corresponding to a speed of approximately 0.055 nm of HfO_2 deposed per cycle at 300°C. The chlorine and hydroxyl residues have a direct effect on the properties of the layers: they have an unstable structure, as shown by the strong capacity of the amorphous HfO2 to exchange oxygen during the annealing process at low-temperature as shown by O^{18} annealing experiments [GAN 04]. They will also increase the crystallization temperature [VAN 04b] as well as the regrowth of the SiO_2 oxide IL during the following annealing processes at 650°C [FER 06] (see also Figure 1.8). The undesirable regrowth at high temperature of the interfacial layer is reduced when annealing is carried out under vacuum rather than at atmospheric pressure [BAR 04]. Recent studies have developed a more specific, two-stage, desorption of the hydroxyls during the vacuum annealing of HfO_2: first the surface groups at 350°C, then those bonded to the core of the layer at 500°C [RAG 09]. This technique controls the thickness of the SiO_2/HfO_2 bilayer with an EOT of less than 1 nm.

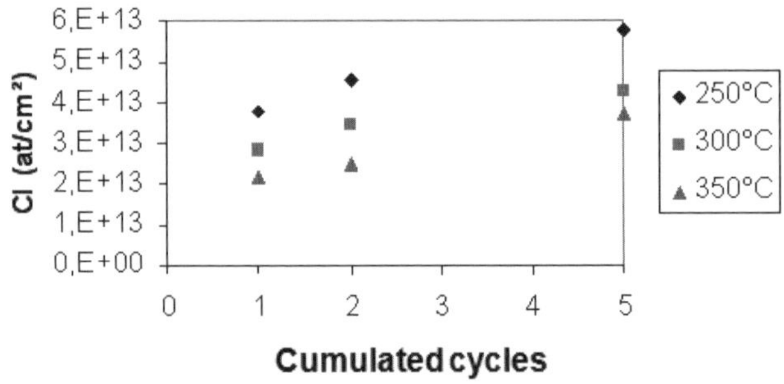

Figure 1.6. *TXRF measurement of the total amount of chlorine adsorbed versus temperature and the number of ALD cycles (HfCl4/H2O) on a chemical silica surface, from [BLI 03]*

Precursors for alloy and functional layer deposition. In section 1.2.3.2, the possibility of modifying the structural properties of the High K layer (the crystallinity), or fixing the work function of the gate stack with an La- (Mg) or Al-based dipole was mentioned. A group of precursors that allow the deposition of these alloys and functional layers exists (see Table 1.1): for example $HfZrO_4$, first obtained from organometallic precursors [TRI 08] is also deposited with chlorinated Hf and Zr precursors and water reactant, while the hydrolysis of $SiCl_4$ permits the addition of Si in HfO_2 using ALD [FED 07]. Aluminum oxide is deposed using trimethylaluminum due to its high vapor pressure, which allows the liquid source to be maintained at ambient temperature. As for the possibility of deposing a fine layer of lanthanum by ALD, a first approach based on diketonate/ozone [NIE 01] was discarded (due to the solid source with a low vapor pressure and the formation of carbonates in the deposit) in favor of cyclopentadienyl compounds, typically $La(i\text{-}PrCp)_3$. Similarly, thin layers of MgO have been deposited using $Mg(EtCp)_2$ [KAM 08b].

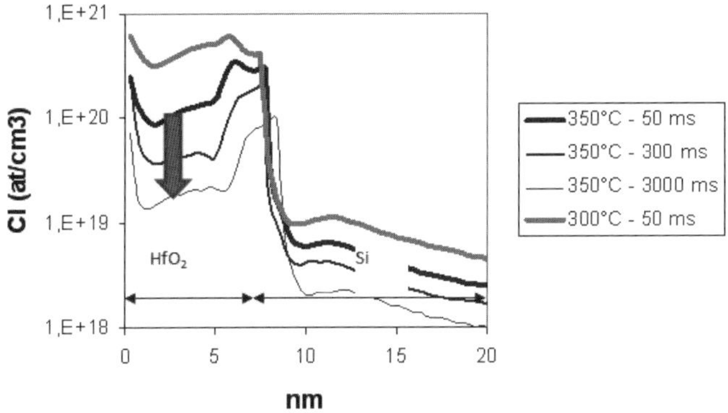

Figure 1.7. *Chlorine's profile in HfO_2 and at the interface with silicon measured by secondary ion mass spectroscopy (SIMS) versus the deposition temperature and water pulse time, from [BLI 03]; the effect of increasing the water pulse time at 350°C is shown by the black arrow*

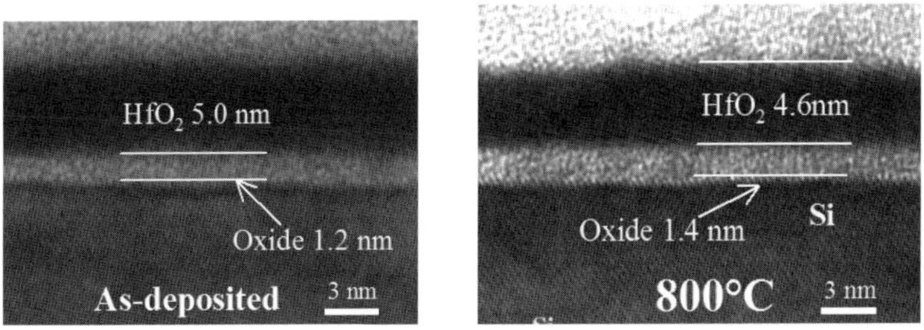

Figure 1.8. *View by transmission electron microscopy of the HfO_2 bilayer and the interfacial layer: on the left hand side, before annealing, there is a slightly abrupt interface corresponding to a silicate; on the right hand side, after annealing in the presence of N_2 at 800°C, the crystalline HfO_2 and the interfacial SiO_2 layer are perfectly distinct*

1.2.3.4. *Deposition of the metallic gate and associated precursors*

The most innovative gate last stacks (see Figure 1.3) require physical vapor deposition (PVD)-type deposition, especially in cases of non-conformal layers (e.g. no deposition on the walls of a cavity) or non-stoechiometric deposition. These kinds of deposits are outside the scope of this chapter that will mainly explain the chemistry of the processes. In this framework, CVD or ALD deposits of conventional materials for gate stacks will be described, namely titanium and tantalum compounds or nitrides. The most frequently used precursors for metal compounds are typically the same structures of amines and chlorides molecules as used for oxide deposition, but with an nitridizing reactant molecule (see Table 1.1). The nitride deposition temperature could be less critical, unlike the insulator deposit, where the oxidizing compound may directly influence the regrowth of the interfacial SiO_2 layer. However, the diffusion of the nitridizing species and its interaction with the insulator may damage certain properties [GAR 08]; so depositions at moderate (<500°C) temperatures will be favored. In the next section, the molecules that may be used in industrial depositions of metallic gates will be described.

1.2.3.4.1. Chlorinated precursors

Similarly to the chlorinated solids used for hafnium and zirconium oxide deposition, liquid titanium chloride is used to deposit TiN by the CVD and ALD processes. The latter method is preferred for reasons already described, that is a deposition temperature less than 500°C. Unlike with chloride hydrolysis, their nitridation with ammonia is less energetically favorable [RIT 99], so the deposition temperature must be above 400°C and the pulse time is increased until saturation of the Ti–Cl bond nitridation reaction occurs to facilitate the desorption of the less-volatile compounds (which may saturate the ulterior chemisorption sites of $TiCl_4$); the two half-reactions, instead of [1.4] and [1.5], involved are:

Chloride chemisorption (only one site):

$$TiCl_4\,(g) + NH\,(s) \rightarrow N\text{-}TiCl_3\,(s) + HCl\,(g) \qquad [1.4]$$

Nitridation of the Ti–Cl bonds (simplified):

$$N\text{-}TiCl_3 + 3\,NH_3 \rightarrow N\text{-}Ti\text{-}N_3 + 3\,HCl \qquad [1.5]$$

We finally end up with a low deposition speed (~0.02 nm/cycle) and relatively long cycles [RIT 99], which limit the application of $TiCl_4$ to thin films (batch treatment of wafers are in this case more favorable for the throughput). Thanks to their higher deposition rate, the organometallic precursors were more rapidly integrated into manufacturing for deposition of metal compounds and were introduced first into interconnection levels such as barrier materials or electrodes for capacitors.

1.2.3.4.2. Organometallic precursors

MOCVD and CVD metallic nitride deposits use amino compounds from Table 1.1, which, in the case of tetravalent titanium, have the same structure as those used for hafnium, namely the tetrakis-dimethylamino titanium (TDMAT), tetrakis-ethylmethylamino titanium (TEMAT) and tetrakis(diethylamino) titanium (TDEAT) (tetrakis amino compounds). Pentavalent tantalum also generates a family of amino compounds: the metal is linked either to five nitrogen atoms in the pentakis dimethyaminotantalum molecule (PDMAT) ($Ta[N(CH_3)_2]_5$) [MUS 09] or to four nitrogen atoms including one double bond in terbutylimido-tris(diethylamino) tantalum (TBTDET) $(NEt_2)_3Ta{=}NBu$ [PAR 02]; a methyled variation $Ta(N\text{-}t\text{-}C_5H_{11})[N(CH_3)_3]$ is still known by its more commercial name, TAIMATA [PAR 03] (see Figures 1.9 and 1.10).

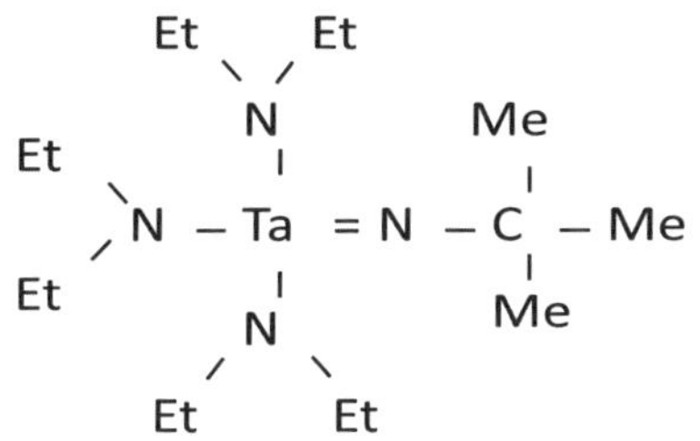

Figure 1.9. *TBTDET molecule, tantalum's precursor*

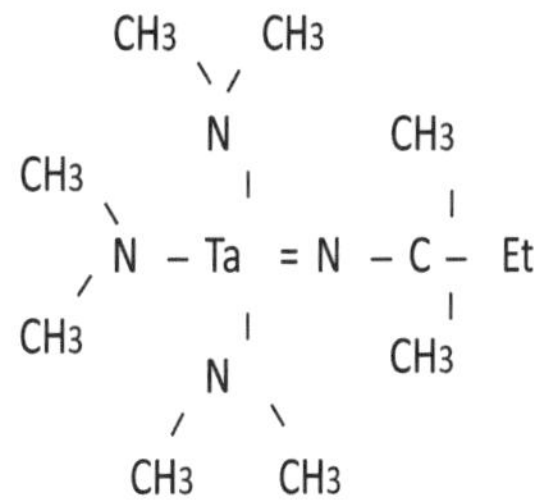

Figure 1.10. *TAIMATA molecule*

However, CVD- and ALD-made deposits (with these organometallic precursors) contain carbonated residues and have a lower density than would be expected for the material. This gives way to oxygen diffusion, for example when using TDMAT and NH_3 [ELA 03, FIL 05]. Their specific resistance is high, particularly in the case of tantalum nitride layers. In fact, the carbonated contamination is decreased by increasing the length of the titanium ligands, that is by using TDEAT rather than TDMAT [KIM 03], or by a section of tantalum precursor like TAIMATA ligands by hydrogenated silicon compounds, but leading to the formation of TaSiN [NAR 04]. To reduce the carbonated contamination of the layers, a plasma treatment can be applied after the deposit. Finally, the films are now formed directly by plasma-enhanced atomic layer deposition (PEALD), similarly to ALD, but with a reactive cycle including hydrogenated and/or nitriding reducing plasma [MUS 09]. Using this method, the mechanisms leading to the formation of a layer of TiC_xN_y are described in [CAU 08] and in the following reactions, [1.6] and [1.7]:

Adsorption of TDMAT on amino sites:

$$2\,NH + Ti[N(CH_3)_2]_4 \rightarrow N_2Ti[N(CH_3)_2]_2 + 2\,HN(CH_3)_2 \qquad [1.6]$$

Ligand division by hydrogenated plasma:

$$N_2\,Ti[N(CH_3)_2]_2 + H_2\ \text{plasma} \rightarrow TiCN + CH_3NH_2 \qquad [1.7]$$

The titanium, initially bonded to four nitrogen atoms, in its final state can be bonded to a carbon atom via a transportation mechanism where a Ti–N bond is substituted by a carbon from a methyl ligand [CAU 08]. The plasma power conditions allow the control of the concentrations of Ti_2C and TiN. These mechanisms give similar results with tantalum precursors such as TBTDET [PAR 02], leading to a mixture of Ta–C and Ta–N, again with the plasma controlling the proportions. Equally, pure metal deposits are possible but only in the presence of hydrogenated plasma from chlorinated metals and at very low deposition speeds [KIM 02]. For an application to gate stacks, it is crucial to remember that the gate insulator is a sensitive material, which may be structurally modified in the presence of active hydrogen plasma [GAR 05]. Effects on HfO_2 have been proven after PEALD of a TaCN gate, combined with the apparition of interfacial states (depending on the type of plasma used) [PAR 08]. The manufacturing of structures that include PEALD deposits in contact with the gate insulator is probably rather difficult to achieve, except perhaps in the top part of the stack, when the insulator is already protected from the active species generated during the metal deposition.

1.2.4. *Conclusion*

This section non-exhaustively describes the manufacturing challenges of an advanced CMOS gate stack, and identifies the role played by chemistry in the growth and transformations of the gate insulator and metal electrode. Compared with processes used since the invention of planar technology, a rise in material innovation has taken place on gate stacks in these last few years. The control of the film properties of the nanomaterial thickness' critical functions was done with much intense research and development: first identifying the molecules and processes that were most favorable and then optimizing the deposition equipment for the treatment or modification of layers. The development of stable chemical precursors is crucial, at the lowest possible costs and with a high level of purity. They illustrate how, in the last decade, chemistry has taken an important role in the critical stages of chip manufacturing, which are key elements in our communication tools (telephony, computers). This continuous innovation in process chemistry allows controlling the manufacturing of billions of transistors on a single chip: the thickness has a precision of approximately 1 **Å** (a tenth of an nm) on the thinnest layers such as the gate insulator of these transistors. Compared with processes used in the previous decades, critical deposits are being carried out at lower temperatures by ALD, a deposition method that may be extended to other applications not mentioned in this chapter (for instance MIM capacitive insulation or insulation on emerging semiconductor materials). The reduction of the thermal budgets shows how the energy consumption of the manufacturing process is being better managed for integrated circuits, an approach that will certainly take more and more importance in the next coming years.

1.3. Chemistry of crystalline materials

1.3.1. *Generalities*

In [BAU 11], Chapter 4, entitled "Epitaxy of Strained $Si/Si_{1-x}Ge_x$ Heterostructures", describes in great detail the reduced pressure – chemical vapor deposition (RP-CVD) of Si/SiGe(C) heterostructures for nano- and optoelectronics [BAU 11]. That chapter consists of the following parts: section 4.1 describes: (1) the evolution of Si and SiGe industrial epitaxy by CVD, (2) one of the two epitaxy tools used by CEA-LETI for research purposes and (3) a few general notions of epitaxy, which may be useful later on. The work performed on $Si/Si_{1-x}Ge_x$ and $Si/Si_{1-y}C_y$ strained heterostructure epitaxy for use as MOSFET conduction channels is detailed in sections 4.2 and 4.3. The optimization of the selective epitaxial growth of Si and SiGe:B raised Sources and Drains (RSD) on SOI substrates is described in sections 4.4 and 4.5. In section 4.6, the structural properties of SiGe virtual substrates used as templates for the fabrication of tensily strained SOI (or sSOI) substrates are described alongside the electrical gains they lead to (namely because of the dual channels of these structures). The electrical and optical properties of devices based on thick layers of pure Ge grown on blanket and patterned wafers are described in section 4.7. In section 4.8, the architectural possibilities offered by the lateral selective etching of SiGe versus Si are detailed. Finally, the most important findings are summarized in section 4.9 and several perspectives are presented.

The aim of this section here is to focus on two points that have so far scarcely been mentioned in [BAU 11]: (1) the preliminary Si (or SiGe) surface preparation used prior to epitaxy of monocrystalline layers and (2) the various advantages and drawbacks of gaseous (or liquid) precursors used to deposit high-quality $Si/Si_{1-x}Ge_x$ heterostructures.

1.3.2. *A few basic ideas about epitaxy*

Rather than proceeding directly to the results, this section will give some basic definitions and facts about epitaxy. To carry out an *epitaxy*, a *monocrystalline* layer is deposited that will adopt the same crystalline structure as the *monocrystalline* substrate on which it is deposited (which is the case of the semiconductors from the III–V or IV–IV columns, which crystallize in the diamond or sphalerite structure such as GaAs, Si and Ge), that is a structure dictated by the underlying substrate, sometimes with rotations in the crystal's direction (such as for metals on insulating substrates, etc.). This is known as *homoepitaxy* when the substrate and the layer are of the same type (e.g. Si on Si) or *heteroepitaxy* when they are not (e.g. SiGe on Si).

Intrinsic or *in situ* doped Si/SiGeC stacks (known as *heterostructures*) are most of the time grown on Si substrates with a (100) surface orientation.

Those substrates can be bulk-type or not (presence of a buried oxide layer), blanket or patterned, etc. It may be useful at this stage to mention the main characteristics of the elementary compounds of a SiGeC alloy: Si, Ge and C crystallize in the diamond phase, that is two overlapping sphalerite structures shifted one from the other by a quarter of the large diagonal (see Figure 1.11). The lattice parameter of C (i.e. the length of one of the sides of the cube in Figure 1.11) is much smaller than that of Si, which in itself is much smaller than that of Ge (a_C = 3.567 Å ⇔ a_{Si} = 5.43105 Å ⇔ a_{Ge} = 5.65785 Å). The lattice parameter increase is accompanied by a sharp decrease in the energy bandgap: E_g (C) = 5.48 eV ⇔ E_g (Si) = 1.11 eV ⇔ E_g (Ge) = 0.66 eV. Mixing Si, Ge and C in a SiGeC alloy will induce strong band-gap modifications; built-in stress will also play a role. Incorporating C by substitution into a Si or Ge matrix is otherwise quite difficult. The miscibility of C in Si is only approximately 10^{-4}% at thermodynamic equilibrium (i.e. at 1,400°C), with an unfortunate tendency toward SiC precipitates formation at higher concentrations. Si and Ge, however, are miscible in all proportions. During heteroepitaxy, there will be discrepancies between the lattice parameters of the deposited layer and the substrate, which will lead to an accumulation of elastic energy in the layer. For thin layers, the substrate's atomic columns will be extended into the *pseudomorphic* epitaxial layer. A layer is described as being in *tension* (or *compression*) when its lattice parameter a_L is lower (or higher) than that of the substrate a_S. For a pseudomorphic layer, the in-plane lattice parameter of the layer is equal to that of the substrate (i.e. $a_L^{//} = a_S$). For a layer in tension, we then have $a_L^{\perp} < a_L < a_L^{//} = a_S$; for a layer in compression, $a_L^{\perp} > a_L > a_L^{//} = a_S$. The perpendicular lattice parameter is given by $a_L^{\perp} = a_L + 2c_{12}/c_{11}(a_L - a_S)$, with $2c_{12}/c_{11}(Si)$ = 0.77, $2c_{12}/c_{11}(Ge)$ = 0.74 and $2c_{12}/c_{11}(C)$ = 0.61. However, when the thickness of the layer increases, it can become energetically favorable to introduce misfit dislocations to minimize the elastic energy that has accumulated and accommodate the lattice parameter mismatch between the layer and the substrate. The layer is then said to be *plastically relaxed.* The thickness at which the transition from pseudomorphic to plastically relaxed layer occurs is called *critical thickness for plastic relaxation.* Schematic diagrams in Figure 1.12 show the different configurations of the {epitaxial SiGe layer on Si substrate} system.

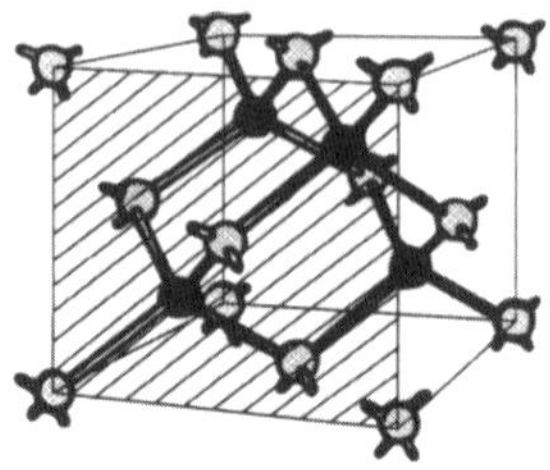

Figure 1.11. *Diamond's structure and the normal (100) plane growth*

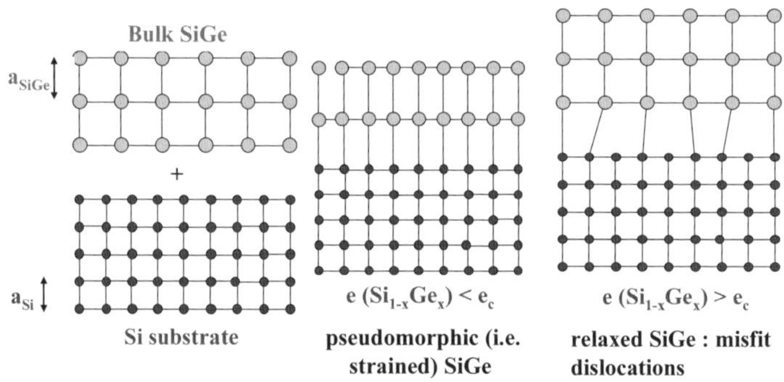

Figure 1.12. *Schematic diagrams of the atomic arrangement when growing a SiGe layer epitaxially on a Si substrate*

Through the 1980s, researchers from several institutes proved that the lower the epitaxial temperature, the higher the critical thickness for plastic relaxation of SiGe on Si(001) was (kinetic barrier to the formation and propagation of dislocations in metastable layers). Values 10 times larger than those predicted by equilibrium theories were obtained [VAN 75, HOU 91]. Solid source molecular beam epitaxy (SS-MBE) was used in all these studies for the growth, between 400 and 750°C, of variable thickness $Si_{1-x}Ge_x$ layers (with a concentration of Ge between 15 and 70%). Their degree of pseudomorphicity with the underlying Si(001) substrate was determined by X-ray diffraction (XRD) and/or by Rutherford back scattering [BEA 84, HOU 91]. Researchers from the American Bell laboratories were then able to propose the well-known "People and Bean" relationship giving the critical thickness for plastic relaxation h_c at 550°C as a function of the lattice parameter mismatch f between SiGe and Si [PEO 85, PEO 86]:

$$h_c \cong \left(\frac{1.9x10^{-2}\,\AA}{f^2}\right)\ln\left(\frac{h_c}{4\AA}\right), \text{[PEO 86] with } f = \left(\frac{a_{SiGe} - a_{Si}}{a_{Si}}\right) \approx 0.042x \qquad [1.8]$$

Recent studies would suggest that h_c is higher than the values predicted by People and Bean [BUC 06, TOM 10].

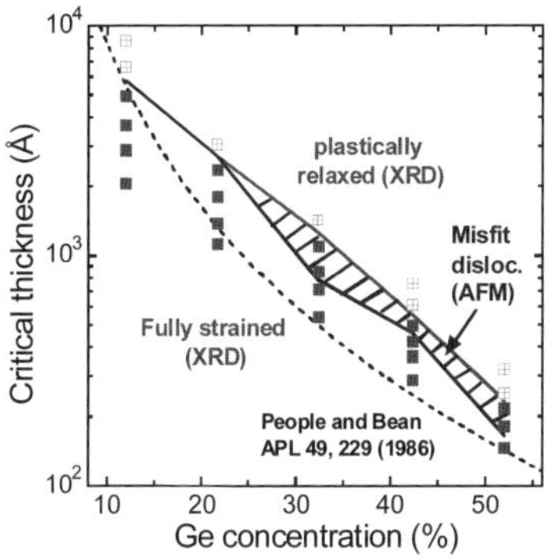

Figure 1.13. *The critical plastic relaxation thickness of $Si_{1-x}Ge_x$ epitaxial layers on Si (001) versus the Ge concentration x [PEO 86, HAR 11a]*

We have thus revisited the critical thickness for plastic relaxation of SiGe on Si(100). For this, 200 mm Si(001) substrates (of a higher crystalline quality than those used in the studies referenced in [HOU 91], that is with a lower pre-existing dislocation density) were used as templates for the RP-CVD of variable composition and thickness SiGe layers.

The growth temperature was reduced from 700 down to 550°C as the Ge concentration increased from 12 to 52% in order to reduce surface roughness. Because of the differences between SS-MBE and RP-CVD in terms of Si and Ge precursors (solid ingots of Si and Ge, which are sublimated by MBE ⇔ ultrapure gaseous precursors for CVD), from the way very low partial pressures of impurities such as oxygen or water are obtained (ultrahigh vacuum in MBE ⇔ large fluxes of ultrapure hydrogen in CVD), etc., we would expect differences in findings [BEA 84, HOU 91, HAR 11a].

Figure 1.13 shows the summary of the results [HAR 11a]. The dotted line shows those of Bean [BEA 84], the colored squares correspond to SiGe layers seen as fully compressively strained in XRD and the squares with a cross in their center represent partially relaxed SiGe layers. h_c is in fact two to three times higher than predicted by People and Bean. However, for high Ge concentrations, certain layers seen as pseudomorphic in XRD are actually characterized by the presence of a limited amount of misfit dislocations in their midst. The surface signature of their propagation ("plough" lines along the <110> crystallographic directions) was detected using atomic force microscopy (hatched zone shown in Figure 1.13).

1.3.3. *Surface preparation prior to epitaxy*

The substrate surface preparation that precedes any kind of epitaxy is of paramount importance. The goal is indeed to elongate the atomic columns of the substrate into the layer. The presence of amorphous silicon oxide layers (even very thin layers), polymers or etching residues (on masked substrates) on the surface has indeed a crippling effect on the quality of the epitaxy. Depending on the starting configuration, two strategies may be used to obtain Si surfaces free of oxide.

The starting substrate may be bulk Si or, to a lesser degree, an SOI-type substrate (i.e. a thin layer of Si/thermal buried layer of SiO_2/Si substrate stack, usually fabricated using the Smart Cut™ technique [BRU 95]). It is then possible to carry out high-temperature bakes, during which dozens of liters of ultrapure H_2 are injected into the reaction chamber (typically only a few ppb of impurities will be present because of dedicated gas purifiers), in order to remove the 0.8 to 1 nm layer of native or chemical SiO_2, which is initially present on the surface and smooth it [BEN 94]. The term chemical oxide is meant to design the oxide that is created by oxidizing wet cleanings such as RCA or dynamic diluted clean (DDC) [TAR 99, ROU 02], which remove surface particles, organic and metallic contaminants. The simplified chemical reaction for the removal of the oxide layer is:

$$SiO_2(s) + 2H_2(g) \rightarrow Si(s) + 2H_2O(g) \qquad [1.9]$$

An image taken by scanning tunneling microscopy (STM) of a Si(001) surface after DDC cleaning, followed by *in situ* H_2 bake at 1,100°C for 1 min, is shown in Figure 1.14. The surface is smooth at the atomic level (average roughness: 0.05 nm only). There are monoatomic steps, alternately smooth (S_A type) or rough (S_B type) [MO 91, VOI 97], limiting (1 × 2) or (2 × 1) reconstructed terraces slightly larger than 100 nm. The extended holes observed on the terraces are likely due to partial desorption of the silicon atoms during the H_2 bake.

Figure 1.14. *STM image (500 nm × 500 nm) of a Si surface (001) after DDC cleaning followed by H_2 bake at 1,100°C and 20 Torr for 1 min. The insert shows a (2 × 1) surface reconstruction*

The requirement for high thermal budgets (typically higher than 1,050°C for 1 min; 1,100°C for 2 min as the CEA-LETI standard) prohibits the use of this surface preparation on patterned wafers (i.e. with active Si zones surrounded by dielectrics) or ion-implanted substrates. The active zones could facet due to the strain generated by masking dielectrics, especially if they are in a Shallow Trench Isolation (STI) configuration. An exodiffusion of preimplanted atoms and a consequent autodoping of the epitaxial layer may also take place [CHA 85, JER 99]. In the case of SOI substrates, even blanket substrates, this type of bake is prohibited for thin Si films (< 20 nm). If these criteria are not followed, moat recess and islanding of the films will take place [ISH 02]. It is also wise to optimize the power delivered to the lamps used to heat up the SOI substrates so that the temperature at the surface is uniform; otherwise slip lines will form.

The surface preparation described here (where a native or chemical oxide present at the surface is removed by *in situ* H_2 bake in suitable conditions) is far from being universal given the high thermal budgets needed. In order to minimize the latter, the following sequence is then used:

1) The native SiO_2 is removed by an "HF-last" wet cleaning [ABB 04]. During the next to last step of such a cleaning, the wafer is dipped in hydrofluoric acid diluted in deionized and deoxygenated water

(the dilution is typically between 0.2 and 1%), to etch the surface SiO_2. The dissolution reaction is as follows:

$$SiO_2 + 4HF \rightarrow 2H^+ + SiF_6^{2-} + 2H_2O \text{ [BUH 97]} \quad [1.10]$$

The etch rate of thermal SiO_2 (the most resistant of all) is of the order of 1.2–1.4 nm/min for 0.2% HF. During the last step, the wafer is rinsed in deionized and deoxygenated water to remove all traces of HF. The wafer is then dried by isopropyl alcohol (IPA) vapors (at 190°C), before being charged as quickly as possible in the load-lock chambers of the epitaxy tool, in an inert atmosphere (purified N_2). After such a wet cleaning, approximately 85% of the Si dangling bonds are occupied by hydrogen atoms. The remaining 15% are mainly occupied by fluorine atoms as well as oxygen and carbon contaminants [MEY 90, TRU 90].

2) An H_2 *in situ* bake then takes place at temperatures close to 900°C to remove all O, F or C surface contaminants (if compatible with the technology it will be used in). A surface perfectly passivated by hydrogen atoms is then obtained, which is ideal for epitaxy.

An "HF-last" surface is only stable for a few dozens of minutes up to a couple of hours (before the regrowth of a native oxide a few Å thick can be detected by spectroscopic ellipsometry). Minimizing the time between an "HF-last" wet cleaning and the loading of wafers inside chambers purged with high-purity inert gases (such as N_2) is thus mandatory. A perfectly hydrogen-passivated Si surface obtained by a high-temperature H_2 bake is by contrast stable for several days [HER 01].

We will now focus on the impact of the H_2 bake temperature on the properties of the epitaxial layers and the interfacial contamination, for Si, SiGe and pure Ge surfaces of (100) and (110) crystallographic orientations.

The secondary ion mass spectrometry (SIMS) depth profiles of the C, O and F atoms present in 50 nm-thick Si layers grown at 650°C on Si (001) substrates after an "HF-Last" wet cleaning followed by a 2 min, 20 Torr H_2 bake, either at 750 or 775°C, are plotted in Figure 1.15 [ABB 04].

A temperature of 750°C is too low to get rid of F and O atomic contamination. Peaks are indeed present at the interface between the substrate and the epitaxial layer. No such peaks were observed at 775°C, however (the same as findings in [KOR 08]). These results are in full agreement with those of Brabant, who showed that 750°C H_2 bakes lasting as long as 10–20 min did not completely rid the Si(001) surface of O contamination [BRA 03].

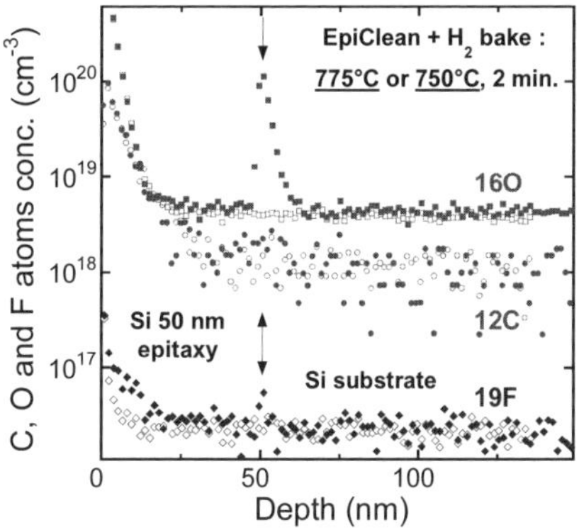

Figure 1.15. *SIMS depth profiles of the O, C and F atomic concentrations in epitaxial Si layers grown at 650°C on blanket Si(001) substrates, after a "HF-Last" wet cleaning and H_2 bake at 750°C (filled symbols) or 775°C (open symbols) for 2 min (P = 20 Torr)*

Figure 1.16. *STM images (500 nm × 500 nm) of Si(001) surfaces after a "HF-Last" wet cleaning (a), followed by H_2 bakes at 750°C (b) or 775°C (c), 20 Torr for 2 min*

500 nm × 500 nm STM images of the Si (001) surface after (1) "HF-last" wet cleaning followed by *in situ* H_2 bakes for 2 min at 750°C (2) or 775°C (3) are plotted in Figure 1.16. After an "HF-last" wet cleaning, the surface is

structured. After H_2 bake at 750°C, short terraces delimited by bi-atomic steps appear. After bake at 775°C, the surface is atomically smooth (average roughness: 0.09 nm), with, as in Figure 1.14, the appearance of monoatomic steps that are alternatively smooth (S_A type) and rough (S_B type) and that delimit several dozens of nm wide (2 × 1) or (1 × 2) reconstructed terraces [VOI 97].

It is interesting to note that the H_2 bake threshold temperature (between 750 and 775°C for Si (001)), above which the C, O and F surface contamination is eliminated, does not seem to be influenced by the crystallographic orientation of the Si surface. A low interfacial O peak was evidenced by Destefanis *et al.* [DES 08] after the H_2 bake at 775°C, for 2 min (after "HF-last" cleaning) of Si (110) substrates. This peak disappeared entirely at 800°C.

Similarly, changing, for (001) surfaces, from Si to SiGe or even pure Ge has little impact on the threshold temperature. O and F contamination peaks were evidenced by Abbadie *et al.* [ABB 04] after H_2 bakes at 775°C for 2 min (post "HF-last" cleaning) of SiGe 20 and 33% virtual substrates (VS) prior to their encapsulation with thin, tensily-strained Si layers [ABB 04]. These contamination peaks disappeared at 800°C. H_2 bakes at 750°C for 2 min enables us to get rid of interfacial contamination peaks in pure Ge (low temperature re-epitaxy of Ge on thick, cyclically annealed Ge layers, themselves on Si (001); [HAR 04a]); there is, however, a low O peak after H_2 bakes at 700°C.

The temptation here would be, given this information, to systematically adopt a high H_2 bake temperature (typically between 800°C and 900°C), for relatively long times (2 min), in order to be safe. This is not always possible.

Let us first deal with SiGe VS grown on Si (001), that is thick, linearly graded SiGe layers (used to confine misfit dislocations) capped with nearly fully relaxed constant Ge composition SiGe layers (several microns total thickness). The surface of such SiGe VS is usually mechanically and chemically polished prior to any re-epitaxy (of, for example, thin, tensily strained Si layers), in order to get rid of the <110> surface cross-hatch inherent to such stacks. A H_2 bake with a too high thermal budget leads to a resurgence of the surface roughness [HAR 08a]. When the H_2 bake temperature increases, the same type of surface roughening has clearly been shown for thick epitaxial layers of Ge grown directly on Si (001) and then polished [CLA 06]. The optimal compromises are, for virtual SiGe substrates, shorter H_2 bakes (15 s) at 850°C, and for thick layers of Ge on Si (001), H_2 bakes at 750°C for 2 min. For these two types of stacks, the "HF-Last" (where HF is hydrofluoric acid) wet cleaning used prior to H_2 bake and epitaxy must also be optimized. Indeed, (1) the etch rate in most of the solutions used in microelectronics to get rid of particular or metallic contamination increases exponentially with the Ge concentration [ABB 06] and (2) the HF concentration must be increased as the Ge concentration increases in order to obtain hydrophobic SiGe surfaces (i.e. without oxides) [ABB 08].

On ultrathin SOI substrates with mesa isolation, moat recess at the edges of the active Si zones and even some islanding of the Si film may occur if the thermal budget of the *in-situ* H_2 bake is too high. The H_2 bake is typically carried out at 650°C, 20 Torr for 2 min prior to the selective epitaxial growth of Si-raised sources and drains on ultrathin SOI (i.e. with a starting Si layer thickness of 3–4 nm) [JAH 05]. Too high a H_2 bake temperature may also lead, for MOSFET transistors, to a detrimental regrowth of the interfacial oxide layer present between the Si channel and the high permittivity gate dielectrics (such as HfO_2) or a detrimental diffusion of ions that had previously been implanted to dope the extensions.

What about the structural and electronic properties of the epitaxial layers on partially contaminated Si surfaces? What of the structural and electronic properties of epitaxial layers on partially contaminated Si surfaces? The impact of a H_2 bake temperature which is too low to completely remove O, C or F interfacial contamination (i.e. between 650°C and 750°C) is limited on (001) surfaces. The surface stays smooth after epitaxy [DES 08], Si/SiGe/Si stacks are of high crystalline quality in X-ray diffraction, the electrical performances of fully depleted transistors built on top of SOI or SiGe/Si dual channels [LE 11] with Si [AND 07] or SiGe:B [BAU 10] raised sources and drains are excellent, etc. However, H_2 bakes at too low a temperature (i.e. $T < 775$°C for 2 min at 20 Torr) have a catastrophic impact on the crystalline quality of epitaxial Si layers grown on Si(110) substrates [DES 08].

Whatever the surface epitaxy is carried out on, specific care must be taken during lithography and etch steps to remove all polymers, High K dielectrics (oxides from actual transistor gates), residues, etc. Failure to do so delays growth, prevents epitaxy from taking place if the H_2 bake temperature is too low, etc. [HAR 11b]. It may even lead to a significant deterioration of the layer crystalline quality [LIU 12].

As a conclusion for this section, let us note that it is possible to carry out the removal of the native or chemical SiO_2 not only *ex situ*, in automated wet benches (wafers processed either individually or in batches) [KOR 08], but also *in situ*. HF vapors in a dedicated chamber connected to the epitaxy cluster tool reactor can be used to get rid of surface oxide [KUI 92]. Dry plasma etching based on NH_3 and NF_3 gases in a "Siconi" chamber can also be used to that end [YAN 10].

1.3.4. *Low-temperature Si and SiGe growth: the comparison of three precursors (silane, disilane and dichlorosilane)*

Dichlorosilane (SiH_2Cl_2) is the silicon precursor of choice for the selective epitaxial growth of Si and SiGe in the Si windows of patterned Si substrates, the masking layers being dielectrics such as SiO_2 or Si_3N_4. This Si gaseous precursor is used for the growth of recessed or raised Si or SiGe:B sources and drains [HAR 04b, HAR 08b, BAU 10, HAR 11a], SiGe/Si epitaxial bilayers that constitute the core of localized silicon-on-insulator-type devices [MON 10] or the high-mobility channels of advanced p-type MOSFET transistors [HUT 10], etc. Dichlorosilane does, however, have relatively low growth rates at low temperatures [HAR 07], as we will see later on. Hydrogenated Si precursors such as silane (SiH_4) [HAR 02], disilane (Si_2H_6) [HAR 12] or trisilane (Si_3H_8) [BAU 07, GOU 09, TAK 10, VIN 10] yield much higher Si and SiGe growth rates at low temperatures. These precursors are not selective versus dielectrics, however. Monocrystalline Si and SiGe layers then grow in the Si windows and polycrystalline layers on the masked zones. This difficulty may be overcome with advanced cyclic deposition/etch (CDE) processes [LOU 12, HE 12]. During a CDE cycle, a few nanometer thick monocrystalline layer is then non-selectively deposited over the wafer surface. This layer is monocrystalline in the Si windows and polycrystalline on the masking dielectrics. The latter is removed selectively during the etch step that follows (with Cl_2 [BAU 12a] or a mixture of HCl + GeH_4 [BAU 12b]). By repeating the cycle several times, a perfect selectivity may be obtained for several tens of nanomolar thick layers. The use of hydrogenated precursors may be advantageous if a low thermal budget (and thus chemically abrupt interfaces) or high concentrations of substitutional carbon atoms are aimed for.

Trisilane, which yields very high growth rates at low temperatures, has, however, several flaws: (1) it is extremely expensive to produce with a quality high enough for nanoelectronics and (2) being liquid, it requires the use of a dedicated bubbler (in which high partial pressures of H_2 are used to obtain vapor phase Si_3H_8) prior to injection into the epitaxy reactor. It is also not always possible to take full advantage of its decomposition at very low temperatures; indeed, Si epitaxial layers grown with Si_3H_8 at temperatures below 500°C are of lesser crystalline and electronic quality [VIN 10]. Using Si_3H_8 fluxes (and thus growth rates) that are too high leads above 500°C to defective Si layers [GOU 09, SHI 12]. Disilane, on the contrary, is gaseous and yields higher growth rates than those obtained with silane (the Si-Si bond energy, 226 kJ mol^{-1}, is lower than the Si-H bond energy (318 kJ mol^{-1}); forming SiH_3 reaction by-products will therefore be easier with disilane than with silane). Disilane therefore seems like a good choice for the low temperature growth of Si and SiGe layers. In section 1.3.2, SiH_4, Si_2H_6 and SiH_2Cl_2 will be compared (in a 300 mm industrial RP-CVD reactor) for use in Si and SiGe growth. A more accurate description of their strengths and weaknesses will be given [HAR 12].

The growth rate of Si at 20 Torr is shown in Figure 1.17 (vs. the inverse of the absolute growth temperature). The flow of disilane is half that of silane or dichlorosilane, due to the fact that there are two Si atoms in a disilane molecule (i.e. $F(Si_2H_6)/F(H_2) = 0.006$ ⇔ $F(SiH_4$ or $SiH_2Cl_2)/F(H_2) = 0.012$). Similar growth rates, which have little dependence on the temperature, are obtained for the same mass flows of Si at higher temperatures (T > 950°C for SiH_2Cl_2 ⇔ T > 850°C for SiH_4 and Si_2H_6).

For SiH_4 and SiH_2Cl_2, as expected at low temperature, there is an exponential growth rate increase with temperature, the limiting factor being here the desorption of H and Cl atoms from the surface. Their activation energies are close to those predicted by the theory (E_a = 2.13 and 2.52 eV here ⇔ E_a = 2.17 and 2.30 eV in another 200 mm RP-CVD reactor, with 1 eV = 23.053 kcal·mol^{-1}). The situation with Si_2H_6 is more complicated, with a growth rate “plateau” between 575°C and 675°C (as with Si_3H_8 [GOU 09]). The growth rate increase with temperature is not the same below and above the plateau (E_a = 1.39 eV (T > 675°C) ⇔ E_a = 2.30 eV (T < 575°C)). For T < 575°C, growth rates are approximately 10 times higher with Si_2H_6 than with SiH_4, which are in turn approximately 10 times greater than the growth rates extrapolated for SiH_2Cl_2 (the practical Si growth rate limit of the order of 0.5 nm·min^{-1} is reached at ~450°C with Si_2H_6 ⇔ ~550°C with SiH_4 ⇔ ~650°C with SiH_2Cl_2).

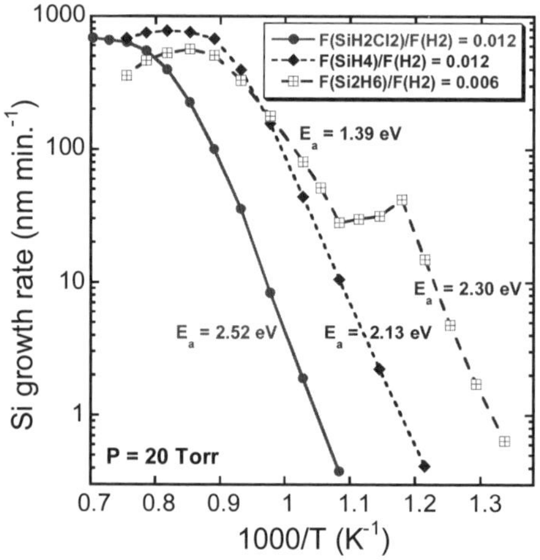

Figure 1.17. *The Si growth rate at 20 Torr versus the inverse of the absolute growth temperature. Same flows of Si ($F(SiH_2Cl_2)/F(H_2) = 0.012$ and $F(SiH_4)/F(H_2) = 0.012$ ⇔ $F(Si_2H_6)/F(H_2) = 0.006$)*

The three growth rate regimes observed for Si_2H_6 below 850°C may be explained this way: for T ≤ 575°C, the disilane molecules are adsorbed on the Si (001) surfaces according to $Si_2H_6(g) \rightarrow 2SiH_3(a)$. The adsorbed silyl, $SiH_3(a)$, decomposes into silylene, $SiH_2(a)$ plus an adsorbed H atom H(a) following $SiH_3(a) \rightarrow SiH_2(a) + H(a)$. SiH_2 is mobile on the underlying Si dimer rows and reacts to form Si monohydride following $2\ SiH_2(a) \rightarrow 2\ SiH(a) + H_2(g)$. The residual hydrogen atoms are desorbed following $2\ SiH(a) \rightarrow 2\ Si(a) + H_2(g)$; the two Si adatoms then contribute to growth [BRA 94]. For T ≥ 575°C (i.e. inside the "plateau"), the reaction $Si_2H_6(g) + 2\ SiH(a) \rightarrow 2\ SiH_4(g) + 2\ Si(s)$ suggested in [KUL 90] becomes predominant over $Si_2H_6(g) \rightarrow 2\ SiH_3(a)$. As the growth temperature increases, a gradual transition toward a behavior close to that of SiH_4 takes place. The growth rate of Si using Si_2H_6 is indeed close to that of SiH_4 for T ≥ 675°C (with the same Si mass flows).

The main features of the 20 Torr growth kinetics of SiGe using a gaseous mixture of dichlorosilane and germane (GeH_4) will be described in the following pages. The SiGe growth rate and Ge concentration functions of the $F(GeH_4)/F(SiH_2Cl_2)$ mass flow ratio (MFR) are provided in Figure 1.18, for growth temperatures between 550°C and 750°C. As shown previously, in another 200 mm epitaxial reactor [HAR 07], x, the concentration of Ge increases parabolically with the GeH_4 flux. Such behavior is accurately described by a relationship of the type $x^2/(1-x) = n \times F(GeH_4)/F(SiH_2Cl_2)$ [SUH 00], with n = 0.98 (700°C), 1.34 (650°C), 2.01 (600°C) and 2.52 (550°C).

The exponential increase of n as T (the absolute growth temperature) decreases may be modeled by $n = 5.2 \times 10^{-3} \exp(0.45\ eV/k_BT)$. The Suh and Lee model, which gives a physical description of the system and includes the $x^2/(1-x) = n \times F(GeH_4)/F(SiH_2Cl_2)$ formula, translates the fact that each surface Cl atom migrates toward a neighboring Ge atom, before being desorbed. The SiGe growth rate otherwise increases linearly with the mass flow of GeH_4. This is due to the catalyzed desorption of surface H and Cl atoms (freeing dangling bonds for the adsorption of Si and Ge atoms) due to the presence of Ge atoms (Ge-H and Ge-Cl bonds, respectively, 1.6 and 2.2 eV, are weaker than Si-H and Si-Cl bonds, 2.0 and 3.9 eV). However, the growth rate decreases and the Ge concentration increases when the growth temperature drops, making the 550°C epitaxy of SiGe layers rather complicated, at least with a $SiH_2Cl_2 + GeH_4$ mixture (prohibitively low growth rates for Ge contents below 35%).

The complexity of the situation increases when hydrochloric acid (HCl) is added to the gaseous mixture, which may be done, for example, to enable selective epitaxial growth versus Si_3N_4 [HAR 07]. At 650°C and 20 Torr, a growth rate drop of approximately 50% occurs with $F(HCl)/F(H_2) = 0.0015$ (Figure 1.19(a)). The SiGe growth rate increase with the germane flow, while being smaller, is also linear. Adding HCl leads to an increase in the concentration of Ge (Figure 1.19(b)).

In the gaseous phase, the added Cl atoms bond preferentially with Si, as the Si chlorides and chlorosilanes are more stable than the corresponding molecules formed with Ge (hence the Ge content increase). The increase of x with the $F(GeH_4)/F(SiH_2Cl_2)$ MFR is also parabolic, with $n = 2.06$. If adding B_2H_6 does lead to a slight increase in the growth rate (in this case, no more than 25%, with a 2×10^{20} cm^{-3} boron atomic concentration), it is still difficult to selectively grow SiGe:B recessed or raised sources and drains below 650°C (at least for Ge concentrations below 40% [HAR 08b, HAR 11b]).

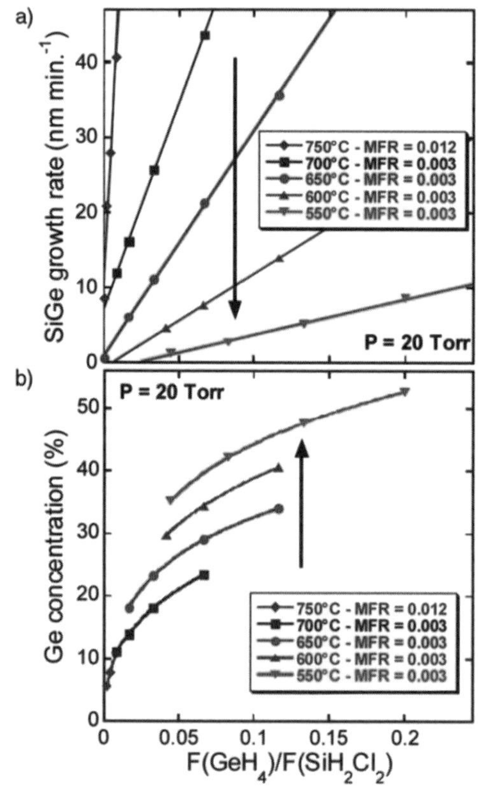

Figure 1.18. *a) SiGe growth rate and b) Ge concentration at 20 Torr versus the $F(GeH_4)/F(SiH_2Cl_2)$ mass flow ratio for different temperatures between 550°C and 750°C. The $F(SiH_2Cl_2)/F(H_2)$ MFR, equal to 0.003 for 550°C ≤ T ≤ 700°C, was multiplied by 4 at 750°C (i.e. 0.012) to obtain low Ge percentage*

Higher SiGe growth rates (and lower Ge concentrations) than those obtained with dichlorosilane may be achieved at 550°C with silane and especially disilane. The SiGe growth rate and the Ge concentration associated with SiGe layers grown at 550°C and 20 Torr with GeH_4 and either SiH_4, Si_2H_6 or SiH_2Cl_2 as the Si precursor are given in Figure 1.20 as a function of the $F(GeH_4)/F(\text{Si precursor})$ MFR.

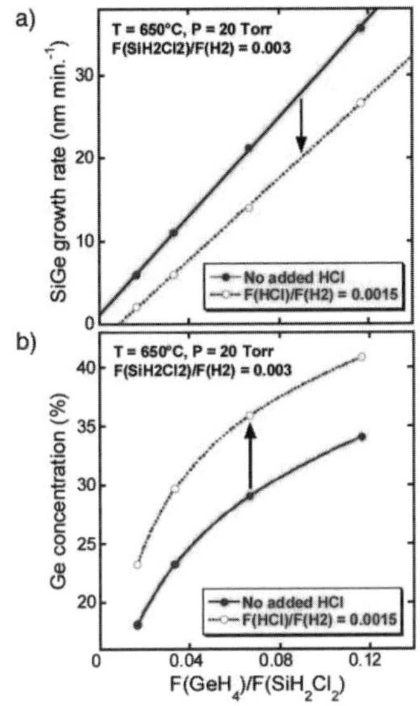

Figure 1.19. *a) SiGe growth rate and b) Ge concentration at 650°C and 20 Torr (vs. the F(GeH_4)/F(SiH_2Cl_2) Mass Flow Ratio) with or without HCl (F(HCl)/F(H_2) = 0.0015). The F(SiH_2Cl_2)/F(H_2) MFR is equal to 0.003*

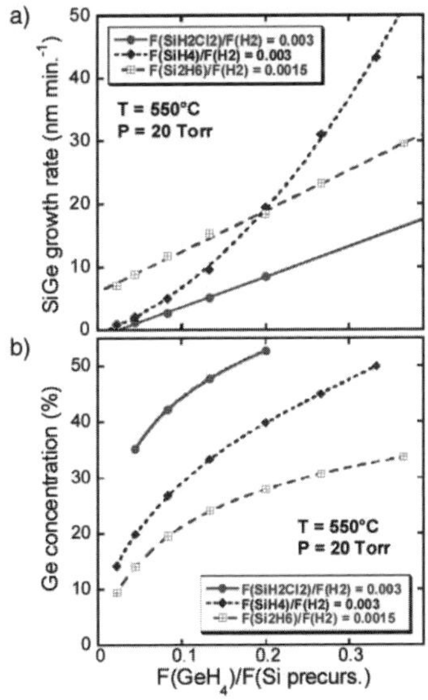

Figure 1.20. *a) SiGe growth rate and b) Ge concentration at 550°C and 20 Torr functions of the F(GeH_4)/F(Si precursor) MFR. The same Si flows were used ((F(SiH_2Cl_2)/F(H_2) and F(SiH_4)/F(H_2) = 0.003 ⇔ F(Si_2H_6)/F(H_2) = 0.0015)*

As expected with SiH_4, x increases linearly with the GeH_4 mass flux. Such a behavior is accurately described by $x/(1 - x) = m \times F(GeH_4)/F(SiH_4)$ relationships [ROB 91], with $m = 2.61$ at 550°C, 20 Torr (here) ⇔ $m = 2.7$ at 650°C, 20 Torr in another 200 mm epitaxy reactor [HAR 02]. It would seem that, for SiH_4 + GeH_4 gaseous mixtures, the growth temperature has only a small impact on the Ge concentration (at least between 550°C and 650°C). However, for reasons we still do not understand, the increase of x with the $F(GeH_4)/F$(Si precursor) MFR is parabolic for a Si_2H_6 + GeH_4 chemistry ($x^2/(1 - x) = 0.47 \times F(GeH_4)/2F(Si_2H_6)$ at 550°C, 20 Torr). For given mass flows, lower Ge concentrations are obtained with SiH_4 and especially Si_2H_6 than with SiH_2Cl_2. The SiGe growth rates increase linearly with the GeH_4 mass flow for SiH_2Cl_2 and Si_2H_6, and exponentially with SiH_4.

They are, for given Ge concentrations, much higher at 550°C and 20 Torr with SiH_4 and especially Si_2H_6 than with SiH_2Cl_2, as shown in Figure 1.21. Such results clearly illustrate the interest of disilane for the low temperature, non-selective epitaxial growth of high crystalline quality SiGe layers in RP-CVD reactors [HAR 12].

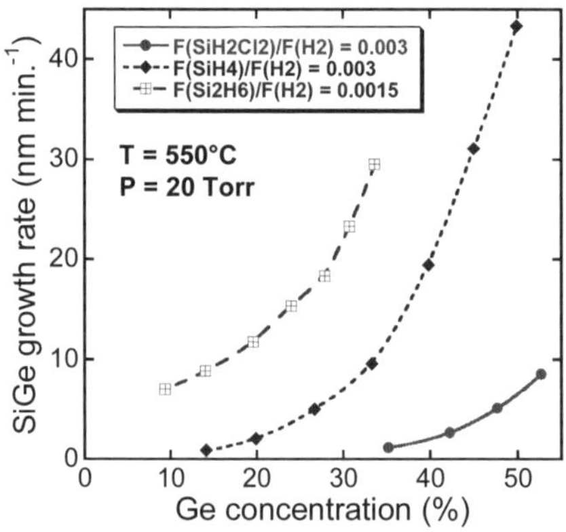

Figure 1.21. *SiGe growth rate (vs. the Ge concentration) for the three Si gaseous precursors investigated at 550°C and 20 Torr. The same Si flows are used (($F(SiH_2Cl_2)/F(H_2)$ and $F(SiH_4)/F(H_2) = 0.003$ ⇔ $F(Si_2H_6)/F(H_2) = 0.0015$)*

1.3.5. *Integration and conclusion*

The following paragraph will describe a successful use of low temperature H_2 bake and selective epitaxial growth in pMOSFET transistors with state-of-the-art electrical performances [LE 11]. To modulate the threshold voltage, V_T, and to have a high mobility channel for the holes, a {$Si_{0.8}Ge_{0.2}$ 3 nm/Si 2 nm} bilayer was first of all grown at 650°C and 20 Torr with SiH_2Cl_2 + GeH_4 (using the operating points from Figure 1.18) on a 300 mm ultrathin SOI substrate (the starting thickness of Si was around 3 nm). A "HF-Last" wet cleaning followed by an *in situ* H_2 bake at 650°C and 20 Torr for 2 min (in order to avoid any islanding of the very thin Si starting layer) was used to prepare the surface prior to growth. A second epitaxy was then carried out later on in the process flow (after gate stack deposition and patterning) in order to selectively grow highly *in situ* B doped $Si_{0.7}Ge_{0.3}$ raised sources and drains ([B] = 2×10^{20} cm^{-3}). This drastically reduces the access resistance of the transistors and to a certain extent, introduces a uniaxial compression in the SiGe channel, increasing the hole mobility. As with the channel, a temperature of 650°C was used for both the *in situ* H_2 bake and the SiGe:B epitaxy (with SiH_2Cl_2 + GeH_4 + HCl + B_2H_6 chemistry; see Figure 1.19). Cross-sectional transmission electron microscopy (TEM) images of a pMOSFET transistor (gate length: 30 nm) and of the channel beneath its gate are shown in Figure 1.22. The {$Si_{0.8}Ge_{0.2}$/Si} bilayer has a perfect crystallinity (no extended defects; abrupt interfaces). The deposition process used for the $Si_{0.7}Ge_{0.3}$:B raised sources and drains is otherwise perfectly selective versus the Si_3N_4 spacers. {111} facets are, however, present at the boundaries between the source and drain regions and the gate.

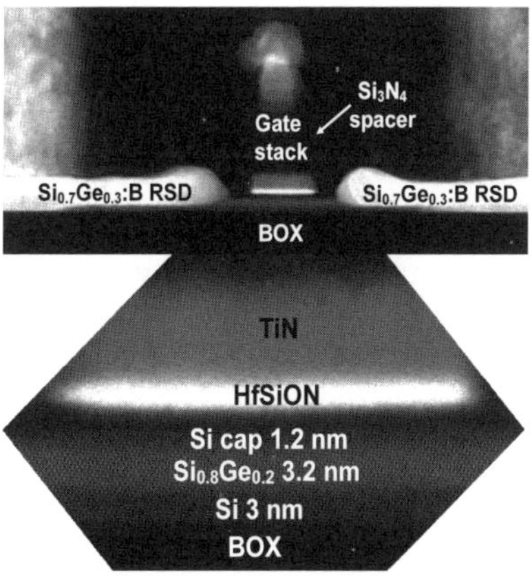

Figure 1.22. *Cross-sectional high-resolution TEM images of a pMOSFET transistor (of 30 nm gate length) with a SiGe/Si bilayer as channel for the holes and silicided, 18 nm thick $Si_{0.7}Ge_{0.3}$:B raised sources and drains*

1.4. Contact areas between the gate and the "source" and "drain"

1.4.1. *Generalities*

1.4.1.1. *A history of silicides in microelectronics*

The first metallic silicides were initially introduced to microelectronics at the end of the 1970s to improve the contact of MOS devices on the polysilicon gate [IWA 02]. The polycide/polysilicon gate was directly patterned using photolithography and etching of a deposited polycide/polysilicon stack layer according to a quite simple integration scheme. $MoSi_2$ was the first polycide to be used and was replaced by WSi_2 from the 1980s onwards, due to its lower resistivity (about 70 µΩ·cm vs. approximately 100 µΩ·cm for $MoSi_2$). However, the device miniaturization was accompanied by an increase in series resistance in the source and drain region, so a metallic silicide layer with low resistivity at both the gate and the source and drain zones was introduced. Thus, the self-aligned silicidation process (SALICIDE process) was introduced during the 1990s for sub-1 µm technological nodes. This ingenious process allows the simultaneous manufacturing of silicide for the gate, as well as the source and drain, by reaction between the metal deposited and the silicon without a photolithographic stage [LAU 82]. The silicide is formed all over the source and drain surfaces, increasing the contact area on the silicon compared with an approach without silicide (see Figure 1.23): contact resistance is thus considerably improved. Furthermore, since the silicide has metallic conduction properties, the doped junction layer's contribution (R_{HDD}) to the resistance is then reduced as, in this case, the metallic layer is brought as close as possible to the channel.

Titanium silicide ($TiSi_2$), was the first silicide to be integrated by the self-aligned process as it presents low resistivity in the C54 crystalline phase (10–15 µΩ·cm). The formation of $TiSi_2$-C54 is done via a first more resistive crystalline phase, called $TiSi_2$-C49. The polymorphic transformation of the C49 phase into C54 is controlled by the nucleation of the grain boundaries and so becomes difficult when the dimensions of the silicide zones are reduced [GAM 98]. For this reason, from 0.25 µm nodes onwards, $TiSi_2$ was replaced by cobalt silicide ($CoSi_2$) that has a similar resistivity but with fewer nucleation problems. However, limitations in the use of this type of silicide also appeared for devices smaller than 0.1 µm: the large amounts of silicon are required to form $CoSi_2$ (1 nm of Co consumes approximately

3.6 nm of Si to form ~3.5 nm of $CoSi_2$ [MAE 95]) as well as its high formation temperature (above 800°C) make it incompatible with the presence of ultra-shallow junctions.

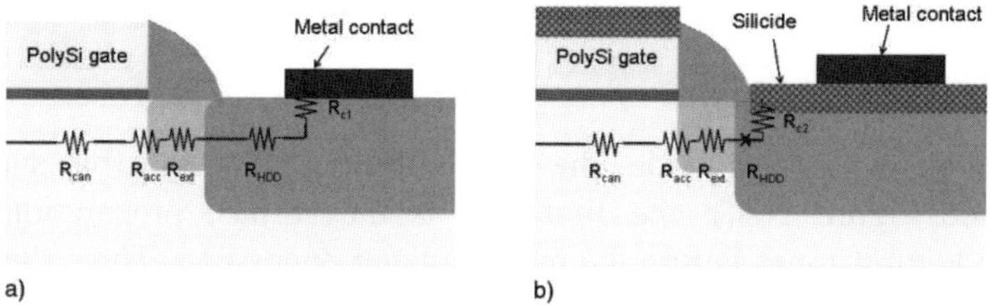

Figure 1.23. *Schematic diagrams of half a MOS transistor showing the different contributors to the series resistance in the source and drain region a) a transistor with direct metallic contact on silicon; b) a transistor with silicided zones. R_{can}: resistance of the channel's layer; R_{acc}: accumulation resistance; R_{ext}: resistance of the lightly doped silicon extensions' layer; R_{HDD}: resistance of the highly doped silicon drain; R_{c1}: resistance of the metal contact on silicon; R_{c2}: resistance of the silicide on silicon; as a first approach, the contact resistance of metal on silicide may be neglected*

1.4.1.2. *Nickel silicide*

Nickel silicide (NiSi) had been known to science for a long time [TU 75] and has become the silicide of reference for the 65 nm node for CMOS on bulk silicon substrates [FRO 04], such as CMOS on SOI substrates [STR 07]. Compared with its predecessors $TiSi_2$ and $CoSi_2$, NiSi has many advantages: its formation does not consume much silicon (1 nm of Ni consumes ~1.8 nm of Si to form ~2.2 nm of NiSi [MAE 95]) and it is formed at relatively low temperatures (~300–400°C). Unlike $TiSi_2$ and $CoSi_2$, its growth is not controlled by nucleation but by the diffusion of the metal, which leads to smoother layers. Its resistance is similar to that of $CoSi_2$ (~15–20 $\mu\Omega\cdot$cm) and is compatible with the formation of contacts on active zones made up of SiGe that, as explained later on in this chapter, is crucial. Since its integration into 65 nm node, numerous adaptations of the manufacturing protocol have taken place in order to adapt to the ever changing specifications of the semiconductor industry. NiSi is, to this day, still the silicide of choice for the most advanced technologies [ITR 11].

This section will describe the chemical treatments used during the nickel silicide (NiSi) formation process, the SALICIDE process. The chemical mechanisms will be succinctly described, particularly the adaptations of the SALICIDE process that were developed in order to overcome the constraints of NiSi integration. The impact of the recent technological orientation and the miniaturization of technology will also be discussed.

1.4.2. *Introduction to the conventional NiSi process for sub-90 nm nodes*

1.4.2.1. *General description of the self-aligned silicidation process*

Applied to nickel silicide, the SALICIDE process has five successive steps (see Figure 1.24): the first step is the surface preparation of the semiconducting zones before the metallic layer deposition. Then, the deposit of a thin metal layer (less than 20 nm thick) is followed by that of a capping layer of titanium nitride (TiN). A first low-temperature annealing leads to the formation of a nickel-rich silicide. The TiN is then wet-etched and the excess nickel (especially the non-transformed nickel present on the insulating zones) removed via a second wet-etching process, selective toward the silicide. Finally, a second thermal treatment is performed to transform all the Ni- rich silicide into the monosilicide NiSi.

This chapter is dedicated to steps that use chemical treatments: step 1 (surface preparation) and 4 (wet etching). Steps 2 (deposition), 3 (first annealing) and 5 (second annealing) will only be described when they influence or are influenced by the chemical treatment steps.

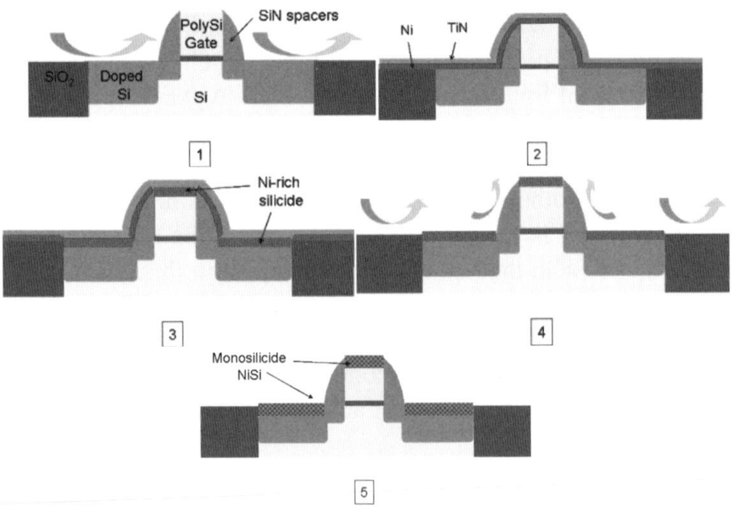

Figure 1.24. *Schematic diagrams of successive self-aligned silicidation process steps (SALICIDE) applied to a MOS transistor*

1.4.2.2. *The Ni/Si system and the formation of silicide by solid state reaction*

1.4.2.2.1. Growth mechanisms

The annealing steps 3 and 5 lead to solid state reactions that constitute the core of the silicidation process. This part will first give some general information about Ni/Si systems and silicide formation. For more details in this area, the reader should refer [DHE 86], to which the following paragraphs refer.

The phase diagram of the Ni/Si system at thermal equilibrium is relatively complex as it is made up of 11 phases of which only six are stable at standard temperature [NAS 87]. These phases are likely to grow simultaneously or competitively when the system undergoes a thermal treatment but only the less resistive NiSi phase is desired at the end of the SALICIDE process. The nickel silicide growth during thermal treatments occurs typically in three steps:

– silicide nucleation.

– lateral growth of the nuclei into a uniform layer.

– increase in the layer's thickness by reactive diffusion (reaction at the interface and diffusion through the formed layer) following Deal and Grove's model [DEA 65] developed for silicon oxides.

Thankfully, thin film growth only leads to a limited number of phases. Indeed, some phases that are present on the binary diagram are sometimes unable to form because a too high energy or a too low diffusion coefficient is required for their nucleation [DHE 86]. Furthermore, the thin film growth of the silicide is generally sequential rather than simultaneous [GOE 82, DHE 86, GAS 93] due to the competition between reaction kinetics at the interface and the diffusion of metal in the growing phase. From the reactive diffusion equations, it can be shown that for an M/Si system leading to two phases I and II, phase I must reach a minimum critical thickness before phase II starts growing. This minimum thickness is of the same order as the initial thickness of the deposited metal. Consequently, phase II will only appear when the metal has been entirely consumed by the reaction [GAS 93]. There are never more than two phases present simultaneously.

Taking into account the specifics of thin film growth, the usual sequence of nickel silicides growth follows reaction [1.11] [OLO 76]:

$$Ni \xrightarrow{250°C} Ni_2Si \xrightarrow{350°C} NiSi \xrightarrow{800°C} NiSi_2 \qquad [1.11]$$

The transformation curve of the Ni/Si system (which represents the resistance of the system's layer vs. the annealing temperature) shows the apparition of the three various phases (see Figure 1.25). The NiSi phase with the lowest resistivity corresponds to intermediate temperatures. For the smallest initial nickel thickness, the progressive increase in resistance for temperatures above 500°C corresponds to the agglomeration of the monosilicide NiSi. This phenomenon is the main cause of NiSi degradation and must be avoided [KIT 03, MUR 03]. The agglomeration of the silicide is influenced by many parameters such as the metal's thickness, the doping of the silicon, the impurities content in the silicon and/or in the layer of nickel, whether it is controlled or not.

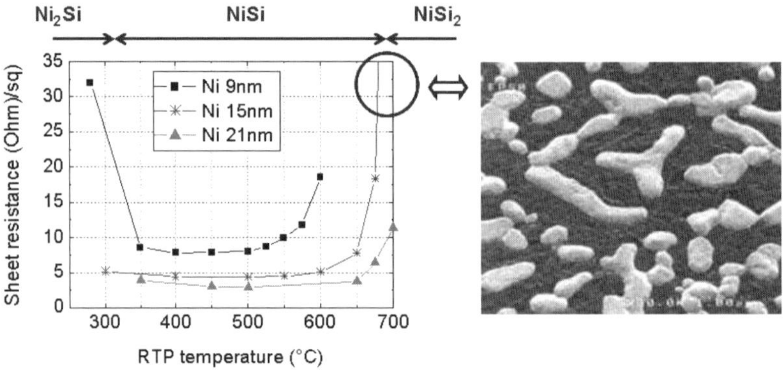

Figure 1.25. *a) Evolution of the sheet resistance of nickel silicide versus the silicidation temperature (60 s in argon) for three initial thicknesses of deposited nickel: 9 nm, 15 nm, 21 nm; b) Scanning electron microscopy (SEM) of a 15 nm thick Ni sample surface annealed at 700°C. RTP = rapid thermal process*

While the growth of Ni_2Si and NiSi is controlled by the metal diffusion in the silicide, the growth of $NiSi_2$ is controlled by a nucleation step. In general, the nucleation takes place above 800°C (see Figure 1.25), however a localized or extended decrease in nucleation activation energy (due to the presence of grain boundaries, dislocations, amorphous zones, etc.) may favor the apparition of $NiSi_2$ at lower temperatures (<500°C), thus changing the usual process sequence [DHE 88]. Under certain conditions, other phases with nucleation controlled growth have been transiently observed during the thin film formation of nickel silicides: $Ni_{31}Si_{12}$ [KIT 07], Ni_3Si_2 [LAV 03] as well as the θ-Ni_2Si phase (non-stoichiometric, actually Ni_2Si_{1+x}), sometimes mistaken for the Ni_3Si_2 phase [MAN 09].

1.4.2.2.2. Adapting to the constraints of integration

Management of the nickel diffusion by two-step annealing. Referring only to the transformation curves obtained on a Si(100) substrate and with no device (see Figure 1.25), the formation of the NiSi monosilicide could be limited to a surface preparation, a nickel deposition and a thermal treatment in the 400–500°C range. However, these annealing conditions applied on a masked substrate favor a strong inhomogeneity in the silicide's thickness both on the gate and active zones on the same chip. This phenomenon is linked to the nickel diffusion that is the main diffusing species during the Ni_2Si and NiSi growth [IWA 02]. On a masked substrate, the nickel present on the insulating zones such as Shallow Trench Isolation (STI) or nitride spacers diffuse toward the silicon, leading to an excessive silicidation at the pattern edges (see Figure 1.26) [IWA 02, FRO 04]. As line width decreases, the contribution of the edges becomes more and more dominant until a thicker silicide forms as the edges meet. Consequently, industrial SALICIDE process with two annealing steps has been used from the 65 nm node onwards [FRO 04]. The sequence of Ni_2Si and NiSi phases makes such a choice even more relevant: the first annealing step is carried out at low temperature (below 350°C) and leads to the formation of a nickel-rich phase, generally Ni_2Si. At low temperature, the lateral diffusion of nickel atoms is much slower as diffusion phenomena are thermally activated. The second annealing step, which aims to transform Ni_2Si into NiSi, is carried out in the 350–500°C range after the selective removal of the metal. Afterwards, there is no more metal on the insulators to diffuse toward the silicide. By allowing a better control of the nickel diffusion, the SALICIDE process with two annealing steps reduces the junction leakages in the MOS transistors.

NiSi stabilization by platinum incorporation. The main issue with nickel monosilicide is its low morphological (i.e. agglomeration) and thermodynamic (transforms into $NiSi_2$) stability (Figure 1.25). Indeed, during the manufacturing process of chips, the silicide formed must be able to withstand the highly cumulated thermal budgets, especially during contacts fabrication and interconnection levels. Many studies have been carried out on the effect of incorporating a third element to the NiSi structure (W, Ti, Ta, Co, Pd, Pt, etc.) on its thermal and morphological stability [DET 06, DED 07]. Platinum was chosen by most of the suppliers [STR 07]. Adding platinum into NiSi was initially studied to delay the nucleation of the disilicide, $NiSi_2$ [MAN 99]. Because NiSi and PtSi phases have the same crystallographic structure, they can form a solid solution $Ni_{1-x}Pt_xSi$ where some nickel atoms are substituted by platinum atoms from the crystal lattice. On the contrary, Pt is insoluble in $NiSi_2$, and so delays the phase formation, which only appears on a large scale above 900°C. More recent studies show that the use of platinum improves the morphological stability of monosilicide and delays its agglomeration at higher temperatures [DET 04]. As such, platinum shows a double advantage and the majority of the industry now uses between 5 and 10 atomic percent platinum in the SALICIDE process.

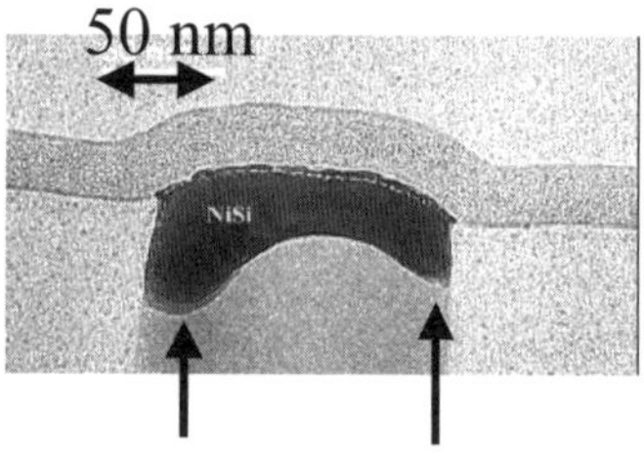

Figure 1.26. *Scanning electron microscopy (SEM) cross-section image that illustrates the edge effects induced by the lateral diffusion of Ni: the silicide formed at the edge is thicker due to a greater nickel injection from the insulating zones. This figure is extracted from [FRO 04]*

1.4.2.3. *Steps of the SALICIDE process: techniques and processes*

1.4.2.3.1. Surface preparation

The surface preparation before deposition is carried out aiming to remove the native silicon oxide and to entirely clean the silicon surface before the metallic layer deposition. The presence of oxygen at the Si/Ni interface may change the phase sequence and even stop the silicide growth if the amount is too large [CHA 04]. Moreover, oxygen at low content may also lead to a rougher silicide surface.

The situation is similar to surface preparation before epitaxy (see Chapter 3) except for the fact that thermal cleaning in the presence of H_2

fluxes is never carried out prior to metal deposition due to the presence of ultra-shallow junctions. Three types of surface preparations may be used:

– *Wet chemical cleaning.* Wet chemical cleaning is a complex sequence of successive chemical treatments aiming to prepare the silicon surface for silicidation while preserving the patterns (such as spacers, the gate and insulators). The last step is a chemical deoxidization using a HF solution) diluted in deionized and deoxygenated water (the dilution is in the range of 0.2–1%), which will be described in Chapter 3 of this book. The overall dissolution reaction may be summed up in reaction [1.12]:

$$SiO_2 + 4HF \rightarrow 2H^+ + SiF_6^{2-} + 2H_2O \qquad [1.12]$$

Chemical etching is preceded by rinsing and cleaning steps that will not be described here. They are carried out aiming to remove any possible traces of organic or particular contamination from the silicon surface. The last active step is the HF-based treatment, leading to the "HF-Last" name of this type of process. The wafer is then rinsed with deionized and deoxygenated water and dried by anhydrous vapor fluxes (such as isopropylalcohol (IPA)). After this treatment, the silicon atoms at the surface are exposed and the dangling bonds are passivated by hydrogen atoms[1]. Such a surface is stable for dozens of minutes that leaves enough time to load the wafers into the deposition tool where they are protected from oxidation by an inert atmosphere (N_2 or Ar generally). Any exposure to air before loading must be minimized.

– *Physical cleaning with argon plasma.* Often performed in addition to chemical cleaning, plasma cleaning is based on mechanical abrasion (by Ar^+ ions bombardment) of the silicon's surface [FRO 04, IMB 07, BON 07, YAN 10]. The aim is to remove the final oxygen and fluorine residues on the silicon surface that may still remain after the HF cleaning, as well as any possible traces of carbon. The plasma chamber is connected to the deposition equipment that allows the sequence of cleaning and metallic deposition steps. Direct plasma cleaning may, however, lead to some problems: due to the bombardment, defects may appear on the Si surface. In addition, injection of charges in the gate oxide may alter the electric behavior of the components. Plasma cleaning may also lead, under certain plasma generation conditions, to the creation of a layer presenting a high defects density near

1 For more details about the mechanisms of SiO_2 etching with hydrofluoric acid, the reader may refer to Chapter 3 of this book, which is dedicated to the chemistry of surface preparations.

the silicon surface [IMB 07, BON 07]. This layer contains implanted argon atoms as well as contaminants such as oxygen, fluoride or carbon. It will induce kinetic modifications to the nickel diffusion and during the first silicidation steps will favor the apparition of Ni_3Si_2 rather than Ni_2Si. On a blanket substrate, this change in the sequence has little influence *in fine* on the texture and resistivity of the monosilicide [IMB 07]. However, it negatively impacts the yield of the electrical devices [BON 07].

– *Surface preparation by reactive plasma.* Surface preparation by reactive plasma (or remote plasma) was recently introduced to overcome the drawbacks of argon direct plasma cleaning. As with argon plasma, the cleaning chamber is connected to the deposition equipment and so allows *in situ* substrates surface cleaning without any exposure to air before the metal deposition. The cleaning is no longer physical but chemical as the active species are separately generated in the reactor and directed toward the substrate. The most common cleaning based on reactive plasma used for the silicidation is called Siconi™ that is based on the reaction between a fluoride component (NH_4F) and silicon oxide [YAN 10]. The "SICONI" process consists of three main steps that are shown in reactions [1.13], [1.14] and [1.15]:

– the generation of active species in the plasma chamber

$$NF_3 + NH_3 \rightarrow NH_4F + NH_4F.HF \qquad [1.13]$$

– the chemical etching of the silicon oxide at approximately 30°C

$$NH_4F + SiO_2 \rightarrow (NH_4)_2SiF_6 + H_2O \qquad [1.14]$$

– sublimation of the solid species $(NH_4)_2SiF_6$ at about 100°C

$$(NH_4)_2SiF_6 \xrightarrow{100°C} SiF_4 + NH_3 \qquad [1.15]$$

Reactive plasma cleaning has improved the electrical performances and the yields of the devices to which it is applied [BON 07, YAN 10]. Several explanations have been proposed: for one, SICONI does not damage the silicon's surface, unlike argon plasma. The lack of defect on the surface may limit the random diffusion of the nickel atoms, which reduces the junction leakage. Moreover, it has been proven that the "SICONI" process leaves a non-negligible quantity of fluorine atoms at the surface [IMB 07]. This contamination could act as a barrier to the nickel's diffusion, further reducing junction leakage.

1.4.2.3.2. Deposition of the metallic layer and the silicidation annealing

Metallic layers are generally deposited under vacuum at room temperature by physical vapor deposition techniques such as magnetron sputtering. The metallic target is pulverized by an Ar+ ions beam and atoms of metal that are ripped from the surface are then deposited on the silicon substrate. The metallic thickness required for sub-65 nm technology is less than or equal to a dozen nanometers. The nickel (or Ni(Pt) alloy with a few percentage of platinum atoms) sputtering is usually followed by the deposition (in a different chamber of the same equipment) of a titanium nitride capping layer (TiN). The TiN is supposed to protect the nickel layer from oxidization during the exposure to air that precedes the charging in the annealing equipment.

The silicidation annealing is done by rapid thermal annealing (RTA) technique under inert conditions (i.e. in the presence of Ar or N_2). It lasts less than 120 s, which is more than enough to fully transform the metal layer into a silicide layer. The temperature is chosen depending on the phase needed: between 250°C and 350°C for a nickel-rich phase and between 350°C and 550°C for the monosilicide NiSi, according to the transformation curves of the Ni/Si system (see Figure 1.25).

1.4.2.3.3. Selective wet etching of the metallic layer[2]

After the first silicide is formed, the TiN cap as well as the non-transformed metal that is present on the insulating zones (SiO_2 insulators and Si_3N_4 spacers) must be removed. The aim of this step is to chemically etch the TiN and metal with very high selectivity with regard to the silicide and the other constituting elements of the device (spacers, insulators). The process used must meet these specifications. The TiN layer is generally etched by a mix of aqueous ammonium hydroxide NH_4OH and hydrogen peroxide H_2O_2, commonly known in microelectronics as SC1 (standard clean 1) [KER 70]. It reacts poorly with nickel and mainly contributes to passivate the Ni surface via the formation of a hydroxide layer. Several processes may then be used to remove the Ni or Ni(Pt).

Sulfuric acid/hydrogen peroxide mixture (SPM). The chemical solution that is most commonly used to etch nickel selectively versus the silicide is a

2 The chemistry described in this chapter is detailed in Chapter 3.

mixture of concentrated sulfuric acid (H_2SO_4 96%) and hydrogen peroxide (H_2O_2) diluted at 30% with deionized water, commonly named SPM (sulfuric peroxide mixture). The mixing of the H_2SO_4 and H_2O_2/H_2O is highly exothermic and leads to the formation of Caro's acid H_2SO_5 by oxidation of the sulfuric acid following reaction [1.16]:

$$H_2SO_4 + H_2O_2 \Leftrightarrow H_2SO_5 + H_2O \qquad [1.16]$$

Caro's acid is a highly oxidizing compound: the standard potential of the H_2SO_5/H_2SO_4 couple is 1.81 V (relative to the standard hydrogen electrode (SHE) against 1.776 V for H_2O_2/H_2O [POU 63]). For the selective etching step, the SPM is used in varying proportions of H_2SO_4 96%/H_2O_2 30%, ranging from 8:1 to 2:1 with the mixture sometimes being diluted in deionized water. Without dilution, the temperature can reach 130°C in the minutes following mixing. The decomposition of the active species (especially H_2O_2) is fast, meaning the chemical etching of nickel must be carried out straightaway after mixing in the reactor. The selectivity of the etching is based on the formation at the silicides surface of a passivating layer of SiO_2 under the oxidizing action of H_2O_2 and/or H_2SO_5 (depending on the dilution) [RAN 74, SHE 98]. On the insulating zones, the nickel is rapidly oxidized and then dissolved into the solution (see reaction [1.17]), while a competition between the oxidization/dissolution of the nickel and oxidization of silicon happens on the surface of the silicide. The oxidation of silicon, which is thermodynamically more favorable, (see reaction [1.18]) is instantaneous and the oxide formed protects the underlying silicide from etching. Figure 1.30 illustrates the NiSi passivation surface mechanism. The selectivity, which may be defined as the ratio of the nickel etching rate to that of silicide, is quasi-infinite [FRO 03].

$$Ni + H_2O_2 + 2H^+ \Leftrightarrow Ni^{2+} + 2H_2O \qquad [1.17]$$

$$2Si + H_2O_2 + 2H_2O \Leftrightarrow 2SiO_2 + 6H^+ \qquad [1.18]$$

Hydrochloric acid (HCl)-based mixtures: aqua regia and HPM. The incorporation of platinum stabilizes the nickel silicide but also creates additional constraints for the use of SPM for the selective etching step. Platinum, as a noble metal is thus difficult to oxidize and so is not etched when using this solution. Platinum atoms that are contained in the nickel layer, sputtered from a Ni(Pt) alloy target, are then etched by the "lift-off" phenomenon during the selective SPM removal: as they are in a small enough amount, they are removed during the oxidation/dissolution reaction of the surrounding nickel atoms. However, platinum residues sometimes remain at the surface of the insulating zones after the SPM etching. These residues are thus potential sources of short circuits between the gate and the active zones of the transistor. Their chemical etching requires the presence of platinum complexing agents in the selective etching solution that may lower the apparent oxidation potential of the metal. Using chloride-based chemistries is a possibility since platinum at oxidation states +2 and +4 forms soluble complexes with chlorides, respectively, tetra- and hexachloroplatinates ($PtCl_4^{2-}$ and $PtCl_6^{2-}$).

Experimental selective etchings based on aqua regia (a well-known platinum etcher based on a mixture of nitric acid (HNO_3) and hydrochloric acid (HCl)) were carried out successfully [GOH 07]. However, the treatment of the effluent streams is somewhat delicate and aqua regia is seldom used in microelectronics to this day. The mixture of hydrochloric acid and hydrogen peroxide (also called SC2 or hydrochloric peroxide mixture (HPM)) may equally be a possible alternative. In this case, H_2O_2 oxidizes the platinum atoms that are then dissolved in the form of hexachloroplatinic acid (H_2PtCl_6). This chemistry is not, however, without its own problems: under certain conditions of silicon manufacturing, it leads to an important oxidization of the silicon at the surface of the silicide [IMB 08]. This phenomenon appears mainly at very low silicidation temperatures, when the silicide formed is nickel-rich. The silicon oxide's thickness can measure up to several dozens of nanometers. It highlights a decomposition of the underlying silicide: the nickel and the silicon of the silicide are both oxidized, respectively, in Ni^{2+}, which is then dissolved, and SiO_x. The remaining platinum in the SiO_x layer could catalyze the decomposition of the silicide into SiO_x by galvanic corrosion.

As such, SPM remains despite everything the favored solution nowadays to perform the selective etching step for the latest technological nodes. Actual studies are leading toward the use of high temperature SPM mixtures (>150°C) that are useful for removing the final platinum residue in cases in which Ni(Pt) layers have a platinum content of 10% [IMB 09]. The mechanisms of platinum dissolution are still not well understood.

1.4.2.4. *Abnormal migration of nickel under the gate: how chemistry rescued the latest dimensions*

The integration of silicides in tridimensional devices such as MOS transistors induced a whole series of modifications in the formation conditions of silicide. As described previously, the phenomena of the masked substrates (such as the lateral diffusion of nickel from the insulating zones) have lead the industry to adapt the SALICIDE process. At the very smallest dimensions, a new adverse phenomenon of transistor yield appeared (see Figure 1.27): the abnormal nickel migration under the gate and/or the spacers of the transistor [STR 07, SEG 05]. This phenomenon is called "encroachment" and its frequency of apparition increases considerably with use of smaller dimensions and high structures density. It is very difficult to detect as it may only involve one transistor in 100,000 and its effect is mainly seen in yield measurements (percent of functional devices). The nickel encroachment is due to numerous parameters: the substrates orientation (i.e. (110) or (100)), the dopants type (boron, arsenic, phosphorus, etc.), the presence of species such as oxygen, fluorine or hydrogen in the silicon or the silicide, the thickness of the deposited metal layer, the platinum content in the metal layer and the annealing conditions of the silicidation process (temperature, time). Several "encroachment" morphologies have been observed: in some cases, the phase formed was analyzed and found to be $NiSi_2$ that has certain privileged orientations, notably [110] [KUD 08]. Other times, the spherical defects corresponding to the NiSi phase were observed near contact pads [IMB 10]. Without going into too much detail, it appears that the suppression of this diffusive phenomenon requires the preferential formation of Ni_2Si during the first anneal. It is crucial to avoid the formation of other nickel-rich species. Therefore, argon plasma surface preparation is to be avoided as it favors, as previously explained, the apparition of compounds such as Ni_3Si alongside Ni_2Si. The replacement of argon plasma by SICONI cleaning (after initial cleaning with a hydrofluoric acid) favors the formation of Ni_2Si exclusively [IMB 07] and contributes to reduce the Ni encroachment [BON 07]. Better yields are then obtained [BON 07, YAN 10]. The encroachment may be reduced by increasing the platinum content, using even thinner Ni(Pt) layers and by reducing the temperature of the first annealing [IMB 10].

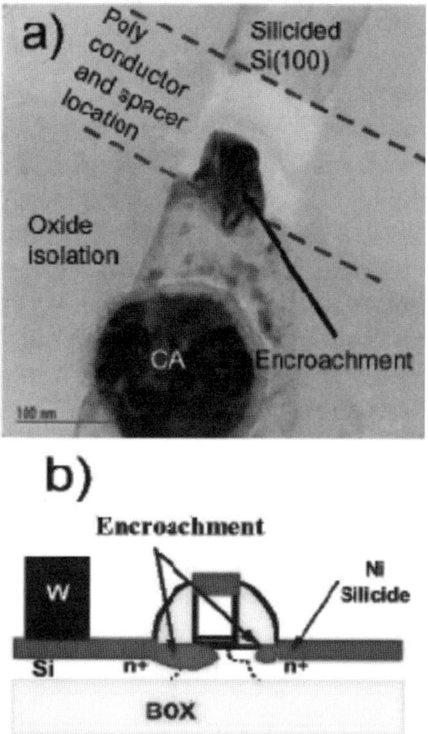

Figure 1.27. *a) A transmission electron microscopy (TEM) top-view image of a device after the removal of the gate. The dotted line represents the edges of the spacers. The encroachment phenomenon is clearly visible (it corresponds to an abnormal migration of the silicide in the channel). b) Schematic diagram of the cross-section of a MOS device showing the encroachment. This figure is extracted from [STR 09]*

1.4.3. *Implications for the SALICIDE process of the recent technology evolutions*

1.4.3.1. *Introduction of germanium into MOS devices*

From the mid 2000s and onwards, an increasing interest was taken in germanium-based electronic devices [THO 04, ZHA 05, LI 06, ZHU 05, DE 07]. The manufacturing of a MOSFET device with a Ge channel has become very worthwhile in terms of electrical performance due to the higher mobility of the electrons and holes in Ge compared with Si [LI 06, ZHU 05]. As well as pure germanium, an $Si_{1-x}Ge_x$ alloy also has very interesting channel, source and drain properties, especially when made by epitaxial growth on Si(001) [ZHA 05, LI 06]. One of the most common applications for SiGe is the filling of the etched cavities by heteroepitaxy in the silicon's active zones [THO 04]. This technique consists of generating lateral and uniaxial strain on the silicon channel in order to increase the mobility of the holes in a pMOS[3] device. The main obstacle to the integration of germanium is the chemical instability of its GeO_2 oxide. Indeed, it is soluble in water and requires the implementation of adapted surface cleanings that are specific to germanium and germanium-rich SiGe alloys (Ge $\geq$ 30%) [BRU 08a]. Aqueous and oxidizing chemistries (especially those with a high H_2O_2 content) are forbidden as they lead to a high degradation of the Ge layer via the dissolution of the GeO_2 formed at the surface following reaction [1.19]:

$$GeO_2 + 2H_2O \rightarrow H_2GeO_3 \qquad [1.19]$$

As with silicon, Ge and SiGe devices require the formation of low resistance contacts in the source and drain regions. Many metals have been evaluated by Gaudet for the formation of the contacts on germanium (called germanide or germanosilicide) [GAU 06]. Nickel is, once again, considered to be the most useful due to the resistance and thermal stability of the germanide obtained, subject to a few adaptations to be done in the self-aligned process. The cleaning of germanium before the metallic deposition is relatively simple due to the fact that the native germanium oxide is soluble in water. Diluted HF solutions [CAR 06] or, more simply, water-based cleanings [BRU 08b] are enough to remove the native oxide. Unlike the silicon process, the residual presence of oxygen does not decrease the formation rate of nickel germanide [NEM 08]. The transformation curve of the Ni/Ge system is shifted toward lower temperatures with respect to that of Ni/Si (see Figure 1.28). Below 250°C, Ni_5Ge_3 and NiGe phases coexist and grow simultaneously [BRU 08b, NEM 06]. Above 250°C, NiGe is the only phase present [BRU 08b]. The agglomeration of the germanide layer takes place at temperatures 100°C cooler than for the silicide and leads to a rapid increase in the resistance of the layer. The ideal temperature range to form a robust nickel germinide layer is approximately 50–100°C less than that of NiSi. In the case of germanosilicide, the situation is intermediate [ZHA 04, CAR 06]. The higher the germanium content in SiGe, the more the transformation curve looks like that of pure Ge. However, for the same amount of germanium, the strain and the relaxation level of the SiGe layer

3 SiGe heteroepitaxy is covered in section 1.3 of this chapter.

greatly affect the morphological stability (or the agglomeration) of the germanosilicide: relaxation increases the morphological stability [ZHA 04].

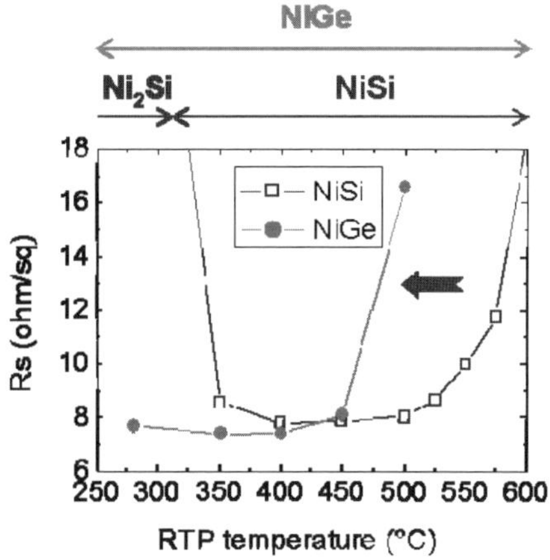

Figure 1.28. *Evolution of the nickel silicide NiSi and of nickel germanide NiGe sheet resistance versus the annealing temperature (60 s in Ar). Ni initial thickness is 9 nm*

As previously explained, the selectivity of the nickel chemical etching performed by SPM mixtures toward the silicide is due to the formation of SiO_2 at the silicide's surface that protects it from etching. In the presence of germanium, the highly oxidizing power of hydrogen peroxide (H_2O_2) leads to the formation of GeO_2; as this compound is very unstable unlike SiO_2, the NiGe is rapidly etched (see Figure 1.30). A germanium content in $Si_{1-x}Ge_x$ above which SPM is incompatible may be defined. Figure 1.29 illustrates this phenomenon is the case of diluted SPM [CAR 06]. For a Ge content of 20% in $Si_{1-x}Ge_x$ (x = 0.2), SPM treatment applied to germanosilicide only leads to a small increase in the layers resistance, mainly at low annealing temperatures, corresponding to the formation of the Ni_2SiGe phase. So the germanosilicide is mainly left intact. For a Ge content of 30%, the increase in resistance after SPM is high, but still acceptable. At 50% Ge, the high resistance leads to a complete etching of not only the germanosilicide (when formed at low temperature (<400°C)), but also of the underlying SiGe. In the temperature range corresponding to mono-NiSiGe formation (T>350°C), the increase in resistivity is less but still leads to a strong deterioration of the phase. Beyond 500°C, the redistribution of the germanium between the substrate and thc germanosilicide leads to a reduced content of Ge in the NiSiGe [JAR 02]. It is thus less sensitive to SPM treatment, meaning the variation in the resistivity of the layer is smaller. However, high annealing temperatures are not compatible with the fabrication of a MOS device. As a result, germanosilicide must be formed at less than 450°C and this germanosiliciure must resist the selective etching step.

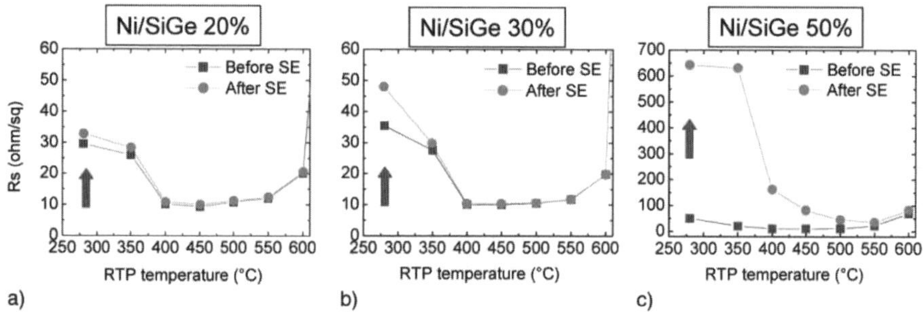

Figure 1.29. *a) Evolution (vs. the annealing temperature) of the sheet resistance of a 9 nm layer of nickel deposited on a relaxed SiGe layer with an atomic concentration of 20% Ge; b) 30% Ge; c) 50% Ge. In squares: after annealing (and before SPM etching); in circles: after SPM etching. The treatment is a diluted SPM at 65°C. For the ratio H_2SO_4 96%/ H_2O_2 30%/EDI = 4:1:60 annealing lasts 60 s in argon. This figure is extracted from [CAR 06]*

Several approaches for selective etching in the presence of germanium may be found in the literature. A simple strategy is to use non-selective chemical reactions and to base the process on precise control of the etching time. "Selectivity" in these cases is based on the difference in the layer thickness of the obtained NiGe and the deposited Ni layers: NiGe is 2.4 times thicker than Ni so it is possible to completely etch the nickel layer while leaving enough NiGe. Aqueous and acid mixtures (HNO_3/H_2O) [LI 06] or alkaline ($NH_4OH/H_2O_2/H_2O$) [ZHU 05] may be used. However, the process control is delicate as for the same process time a slight variation of temperature or mixture composition may strongly modify the final thickness of the remaining NiGe.

Another approach is to use acid aqueous chemical reactions that are poorly oxidizing, such as a mixture of HCl/H_2O with [BRU 08b] or without few percents of HF [NEM 08]. In this case, nickel etching is favored by the presence of Cl^- ions that are complexing agents of Ni^{2+}. The oxidation potential of Ni is thus lowered and Ni is oxidized by the H_3O^+ ions, whereas in the case of Ge (or rather, of its oxide) the dissolution is very slow.

Finally, a third approach is to use anhydrous and non-oxidizing mixtures. Concentrated sulfuric acid at 96% is a good example [CAR 06] as it slowly etches the nickel while it leads to the formation of the insoluble GeO_2 oxide on the germanide's surface. The germanide is then protected from etching during the treatment. In this case, the main NiGe consumption comes from the rinsing step performed in deionized water after the sulfuric acid treatment. During this step, the surface GeO_2 is dissolved in water following reaction [1.19]. It is, however, mandatory to carry out the rinsing step in order to remove the sulfide residues left by the decomposition of the sulfuric acid. It leads to a consumption of less than 10% of the initial NiGe layer. Figure 1.30 summarizes schematically the different surface evolutions of NiSi and NiGe under SPM or 96% H_2SO_4 treatment followed by rinsing with deionized water.

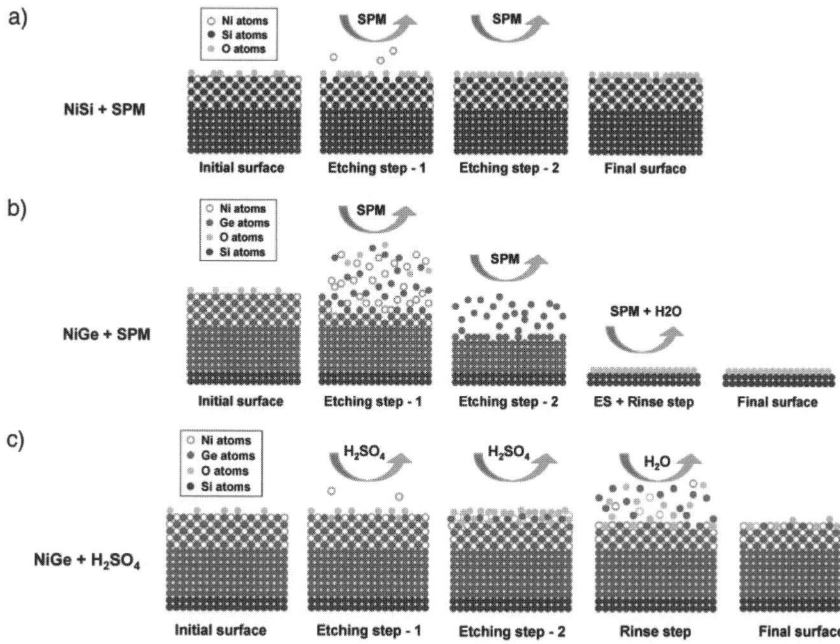

Figure 1.30. *Schematic representation of the surface evolution of the NiSi layer a) and the NiGe layer b) and c) during the nickel selective etching process. Etching performed using diluted SPM a) and b) and with 96% H_2SO_4 c) followed by a rinsing step with deionized water. This figure is extracted from [CAR 06]*

1.4.3.2. *Planar and 3D co-integration pMOS/nMOS*

As previously explained, with a few adaptations to the self-aligned NiSi process, an equivalent NiGe process would be possible and the literature shows numerous examples of such a MOS device on Ge with NiGe metallization [WEB 05, BRU 08]. The use of strained SiGe layers has become widespread, including at industrial level [AND 05, LE 11]. However, germanium is not interesting for nMOS transistors, meaning the requirement to co-integrate on the same substrate a Ge-based pMOS and a Si-based nMOS has emerged. Examples can be found of planar and three-dimensional (3D) co-integration [LE 08] (see Figure 1.31). The aim of 3D integration is to increase the density of the devices on a chip. Contact metallurgy for pMOS and nMOS transistors are carried out separately. As nMOS is buried, it must be able to resist the thermal budget of the pMOS manufacturing. With this approach, very good results in terms of electrical performance have been obtained for superior pMOS on a GeOI (germanium on insulator) substrate as well as SOI substrate [BAT 09]. In the case of the planar approach, a simultaneous silicidation of the pMOS and nMOS may be performed. The self-aligned metallization process must meet the specifications of silicon as well as germanium or SiGe alloy. However, it is possible to find an acceptable compromise on the formation temperature of the silicide and on the selective etching mixture. Here, as well, examples may be found of successful planar nMOS on SOI and pMOS on GeOI co-integration with nickel metallization that show the incredible flexibility of the self-aligned process |LER 10].

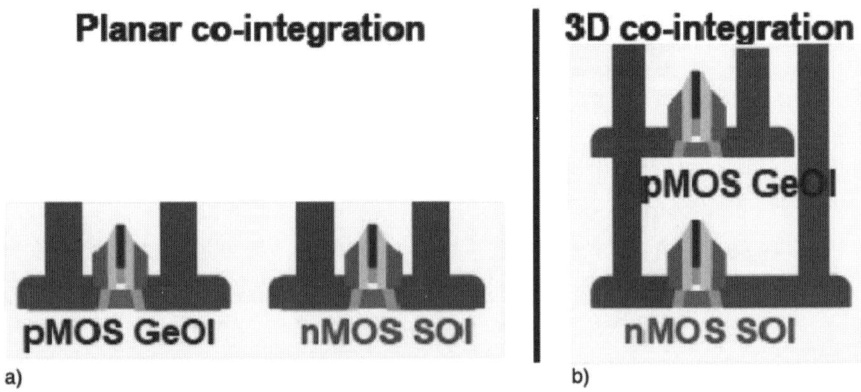

Figure 1.31. *Schematic representation of two pMOS Ge/nMOS Si co-integration strategies; co-integration is a) planar or b) 3D [LER 08]*

1.4.4. *Conclusion*

To meet the ever increasing specifications such as miniaturization and performance of devices, the Ni SALICIDE process, which was introduced initially for the 65 nm node, has since been routinely adapted. For every technological node and each new integration approach, the steps of the process had to be redesigned. The previous sections have shown a few examples of the evolution of the various chemical treatments steps (for the surface preparation and the selective etching of the metal) used in the process. These steps are very sensitive to the integration of new materials (platinum to improve the morphological stability of NiSi, germanium to increase the carrier mobility in the channel). For future generations, one of the main challenges is the ever decreasing size of the contact surface between the metal and the junction [ITR 11], the effect of which is increased in the case of devices on SOI where the contact may be lateral. In these conditions, the reduction of the silicide/doped silicon contact resistance is a dominant parameter and several minimization approaches are available: engineering of the silicide/silicon interface via the incorporation of new species, double silicidation (one for the pMOS and the other for the nMOS with a low Schottky barrier height for, respectively, the holes and the electrons). Here, as well, the processes will need to adapt to the integration of new materials. Finally, another challenge for future generations is the control of the junction consumption by the silicidation process. As the junction is no bigger than a dozen nanometers, the thickness of the metal must be less than 5 nm and its uniformity controlled, a difficult feat to achieve with sputtering deposition techniques. These techniques may have to give way to chemical deposition techniques that allow the selective deposition of the metal, such as MOCVD or electrochemical deposition. To integrate such techniques into general production would certainly mean the end of the SALICIDE process as we know it. This prospect is still distant, however, and the SALICIDE process still has a few years ahead of it before being dropped from use.

1.5. General conclusion

This chapter has described the first stages in the manufacturing of a transistor. The "front end of the line" steps are essential as they define the final transistor performance and could be described as the precise dimensioning of the heart of the transistor in order to give it what will be its main characteristic: either high performance (with a strong passing current and an acceptable leakage current) or low consumption (with an acceptable passing current and a very low leakage current). The gate stack is a key element of the FEOL, as are the growth of the materials by epitaxy and the manufacturing of efficient contacts, two other crucial sequential modules that have been described in this chapter.

From there, the latest transistors designed with 20 nm or 14 nm gate length show already some significant improvements over the previous generation. New "3D" shape transistors (called FinFETs) have been recently developed and presented [YAM 11, AUT 12, JAN 12] when standard planar CMOS technology also improved in parallel through a fully depleted SOI technology by keeping the historical planar integration scheme and extending the high performance and the low power consumption targets through innovative pathways [LIU 11, CHE 12, GRE 12, LE 12]. Finally, going down the road of the technology roadmap, new transistors are also on their way [KUH 12], "nanowires" transistors being currently a more widely investigated option in research [ERN 08, SAI 10, BAR 12, RIE 12].

1.6. List of Abbreviations

MOS:	Metal Oxide Semiconductor
MOSFET:	Metal Oxide Semiconductor Field Effect Transistor
FinFET:	Fin (shape) Field Effect Transistor
CMOS:	Complementary Metal Oxide Semiconductor (N-type+P-type)
NMOS:	N-tpye Metal Oxide Semiconductor
PMOS:	P-type Metal Oxide Semiconductor
EOT:	Equivalent Oxide Thickness
CVD:	Chemical Vapour Deposition
PVD:	Physical Vapour Deposition
MOCVD:	Metal Organic Chemical Vapour Deposition
ALD:	Atomic Layer Deposition
PEALD:	Plasma Enhanced Atomic Layer Deposition
ALCVD:	Atomic Layer Chemical Vapor Deposition
RPCVD:	Reduced Pressure Chemical Vapor Deposition
BDMAS:	Bis_DiMethylAmino Silane
TDMAS:	Tris(DiMethylAmino) Silane

TDMAT:	Tetrakis(DiMethylAmino) Titanium
TEMAT:	Tetrakis(EthylMethylAmino) Titanium
TDEAT:	Tetrakis(DiEthylAmino) Titanium
GPC:	Growth Per Cycle
MFR	Mass Flow Ratio
RSD:	Raised Source Drain
SSMBE:	Solid Source Molecular Beam Epitaxy
XRD:	X-Ray Diffraction
DDC:	Diluted Dynamic Clean
STM:	Scanning Tunelling Microscopy
IPA:	IsoPropyl Alcohol
HF:	HydroFluoric acid
SIMS:	Secondary Ion Mass Spectrometry
VS:	Virtual Substrate
CDE:	Cyclic Deposition Etch
TEM:	Transmission Electron Microscopy
STI:	Shallow Trench Isolation
RTA:	Rapid Thermal Annealing

1.7. Bibliography

[AAR 99] AARIK J., AIDLA A., KIISLER A.A., *et al.*, "Influence of substrate temperature on atomic layer growth and properties of HfO_2 thin films", *Thin Solid Films*, vol. 340, nos. 1–2, pp. 110–116, 1999.

[ABB 04] ABBADIE A., HARTMANN J.-M., HOLLIGER P., *et al.*, "Low thermal budget surface preparation of Si and SiGe", *Applied Surface Science*, vol. 225, nos. 1–4, pp. 256–266, 2004.

[ABB 06] ABBADIE A., HARTMANN J.-M., DI NARDO C., *et al.*, "An efficient wet-cleaning of SiGe virtual substrates and of thick, pure Ge layers on Si (001) after a chemical mechanical planarization step", *Microelectronic Engineering*, vol. 83, pp. 1986–1993, 2006.

[ABB 08] ABBADIE A., HARTMANN J.-M., BESSON P., *et al.*, "Wet cleaning and surface characterization of $Si_{1-x}Ge_x$ virtual substrates after a CMP step", *Applied Surface Science*, vol. 254, no. 21, pp. 6793–6798, 2008.

[AKI 07] Akiyama K., "Vfb roll off in HfO_2 gate stack after high temperature annealing process – a crucial role of out diffused oxygen from HfO_2 to Si", *Proceedings of the 2007 VLSI Conference*, pp. 72–73, 2007.

[AND 05] Andrieu F., Ernst T., Lime F., *et al.*, "Experimental and comparative investigation of low and high field transport in substrate- and process-induced strained nanoscaled MOSFETs", *Proceedings of the 2005 VLSI Conference*, pp. 176–177, 2005.

[AND 07] Andrieu F., Faynot O., Rochette F., *et al.*, "Impact of mobility boosters (XsSOI, CESL, TiN gate) on the performance of <100> or <110> oriented FDSOI cMOSFETs for the 32 nm node", *Proceedings of the 2007 VLSI Conference*, pp. 50–51, 2007.

[AUT 12] Auth C., Allen C., Blattner A., *et al.*, "A 22 nm high performance and low power CMOS technology featuring fully depleted tri-gate transistors, self aligned contacts and high density MIM capacitors", *Proceedings of the 2012 VLSI Conference*, pp. 131–132, 2012.

[BAR 04] Barrett N., Renault O., Damlencourt J.-F., *et al.*, "Thermal stability of the HfO_2/SiO_2 interface for sub 0.1 μm CMOS gate oxide stack: a valence band and quantitative core-level study by soft X ray photoelectron spectroscopy", *Journal of Applied Physics*, vol. 96, no. 11, pp. 6362–6369, 2004.

[BAR 06] Barrett N., Renault O., Besson P., *et al.*, "Band offsets of nitrided ultrathin hafnium silicate films", *Applied Physics Letters*, vol. 88, no. 16, pp. 162906–162906_03, 2006.

[BAR 12] Bartsch S.T., Dupre C., Ollier E., *et al.*, "Resonant-body silicon nanowire field effect transistor without junctions", *Proceedings of the 2012 IEDM Conference*, abstract 15.2, 2012.

[BAT 09] Batude P., Vinet M., Pouydebasque A., *et al.*, "GeOI and SOI 3D monolithic cell integrations for high density applications", *Proceedings of the 2009 VLSI Conference*, pp. 166–167, 2009.

[BAU 07] Bauer M., Machkaoustan V., Arena C., "Highly tensile strained silicon–carbon alloys epitaxially grown into recessed source drain areas of NMOS devices", *Semiconductor Science and Technology*, vol. 22, no. 1, pp. S183–S187, 2007.

[BAU 10] BAUDOT S., ANDRIEU F., WEBER O., *et al.*, "Fully depleted strained silicon-on-insulator p-MOSFETs with recessed and embedded silicon–germanium source/drain", *Electron Device Letters*, vol. 31, no. 10, pp. 1074–1076, 2010.

[BAU 11] BAUDRANT A., *Silicon Technologies Ion Implantation and Thermal Treatment*, ISTE Ltd, UK/John Wiley & Sons, Inc, 2011.

[BAU 12a] BAUER M., THOMAS S.G., "Low temperature Si:C co-flow and hybrid process using Si_3H_8/Cl_2", *Thin Solid Films*, vol. 520, no. 8, pp. 3133–3138, 2012.

[BAU 12b] BAUER M., THOMAS S.G., "Low temperature catalyst enhanced etch process with high etch selectivity for amorphous silicon based alloys over single-crystalline based alloys", *Thin Solid Films*, vol. 520, no. 8, pp. 3139–3143, 2012.

[BEA 84] BEAN J.C., FELDMAN L.C., FIORY A.T., *et al.*, "Ge_xSi_{1-x}/Si strained layer superlattice grown by molecular beam epitaxy", *Journal of Vacuum Science and Technology A*, vol. 2, no. 2, pp. 436–440, 1984.

[BEN 94] BENDER H., VERHAVERBEKE S., CAYMAX M., *et al.*, "Surface reconstruction of hydrogen annealed (100) silicon", *Journal of Applied Physics*, vol. 75, no. 2, pp. 1207–1209, 1994.

[BER 10] BERSUKER G., PARK C.S., WEN H.C., *et al.*, "Origin of the flatband-voltage roll-off phenomenon in metal / high k gate stacks", *Transactions on Electronic Devices*, vol. 57, no. 9, pp. 2047–2056, 2010.

[BLI 03] BLIN D., ROCHAT N., ROLLAND G., *et al.*, "Analytical characterization of process parameter influence on the initial growth and cristallinity of atomic layer deposition of HfO_2 thin films", *Electrochemical Society Proceedings*, vol. 2003, no. 3, ALTECH, Paris, pp. 233–242, 2003.

[BON 07] BONNETIER S., IMBERT B., HOPSTAKEN M., *et al.*, "Silicide preclean effects on NiPtSi thermal stability for 65 nm technologies and beyond", *Microelectronic Engineering*, vol. 84, no. 11, pp. 2528–2532, 2007.

[BRA 94] BRAMBLETT T.R., LU Q., KARASAWA T., *et al.*, "Si (001) 2×1 gas source molecular beam epitaxy from Si_2H_6: growth kinetics and boron doping", *Journal of Applied Physics*, vol. 76, no. 3, pp. 1884–1888, 1994.

[BRA 03] BRABANT P., WEN J., ITALIANO J., *et al.*, "Achieving a SiGe HBT epitaxial emitter with novel low thermal budget technique", *Applied Surface Science*, vol. 224, nos. 1–4, pp. 347–349, 2003.

[BRU 95] BRUEL M., "Silicon on insulator material technology", *Electronics Letters*, vol. 31, no. 14, pp. 1201–1202, 1995.

[BRU 08a] BRUNCO D.P., DE JAEGER B., ENEMAN G., *et al.*, "Germanium MOSFET devices: advances in materials understanding, process development, and electrical performance", *Journal of the Electrochemical Society*, vol. 155, no. 7, pp. H552–H561, 2008.

[BRU 08b] BRUNCO D.P., OPSOMER K., DE JAEGER B., *et al.*, "Observation and suppression of nickel germanide overgrowth on germanium substrates with patterned SiO_2 structures", *Electrochemical and Solid State Letters*, vol. 11, no. 2, pp. H39–H41, 2008.

[BUC 06] BUCA D., FESTE S.F., HOLLANDER B., *et al.*, "Growth of strained Si on He ion implanted Si/SiGe heterostructures", *Solid State Electronics*, vol. 50, no. 1, pp. 32–37, 2006.

[BUH 97] BÜHLER J., STEINER F.-P., BALTES H., "Silicon dioxide sacrificial layer etching in surface micromachining", *Journal of Micromechanics and Microengineering*, vol. 7, no. 1, p. R1, 1997.

[CAR 06] CARRON V., RIBEIRO M., BESSON P., *et al.*, "Nickel selective etching studies for self-aligned silicide process in Ge and SiGe-based devices", *ECS Transactions*, vol. 3, no. 7, pp. 643–654, 2006.

[CAR 07] CARRON V., BESSON P., PIERRE F., "Wet etching step evolutions for selective removal on silicide or germanide applications", *ECS Transactions*, vol. 11, no. 2, pp. 309–320, 2007.

[CAU 08] CAUBET P., BLOMBERG T., BENABOUD R., *et al.*, "Low temperature, low resistivity PEALD TiN using TDMAT under hydrogen reducing ambient", *Journal of the Electrochemical Society*, vol. 155, no. 8, pp. H625–H632, 2008.

[CHA 85] CHANG H.-R., "Autodoping in silicon epitaxy", *Journal of the Electrochemical Society*, vol. 132, no. 1, pp. 219–224, 1985.

[CHA 04] CHAMIRIAN O., LAUWERS A., KITTL J.A., *et al.*, "Ni silicide morphology on small features", *Materials Research Society*, vol. 810, pp. 147–152, 2004.

[CHE 12] CHENG K., KHAKIFIROOZ A., LOUBET N., *et al.*, "High performance extremely thin SOI (ETSOI) hybrid CMOS with Si channel NFET and strained SiGe channel PFET", *Proceedings of the 2012 IEDM Conference*, abstract 18.1, 2012.

[CHO 11] CHOI J.H., MAO Y., CHANG J.-P., "Development of hafnium based high k materials: a review", *Materials Science and Engineering Review*, vol. 72, no. 6, pp. 97–136, 2011.

[CLA 06] CLAVELIER L., DEGUET C., LE ROYER C., *et al.*, "Review of some critical aspects of Ge and GeOI substrates", *ECS Transactions*, vol. 3, no. 7, pp. 789–805, 2006.

[COS 04] COSNIER V., DABERTRAND K., BLONKOWSKI S., *et al.*, "Atomic Vapour Deposition (AVD™) process for high performance HfO_2 Dielectric Layers", *MRS Proceedings,* vol. 811, D9.8, 2004.

[CUB 04] CUBAYNES F., Ultra-thin plasma nitrided oxide gate dielectrics for advanced MOS transistors, Phd Thesis, Catholic University of Louvain, Belgium, 2004.

[DEA 65] DEAL B.E., GROVE A., "General relationship for the thermal oxidation of silicon", *Journal of Applied Physics*, vol. 36, no. 12, pp. 3770–3778, 1965.

[DE 07] DE JAEGER B., KACZER B., ZIMMERMAN P., *et al.*, "Ge deep sub-micron HiK/MG pFETs with superior drive compared to Si HiK/MG state-of-the-art reference", *Semiconductor Science and Technology*, vol. 22, no. 1, pp. S221–S226, 2007.

[DED 07] DEDUYTSCHE D., DETAVERNIER C., VAN MEIRHAEGHE R.L., *et al.*, "Formation and morphological stability of NiSi in the presence of W, Ti, and Ta alloying elements", *Journal of Applied Physics*, vol. 101, no. 4, pp. 044508-1–044508-8, 2007.

[DEL 06] DELABIE A., CAYMAX M., BRIJS B., *et al.*, "Scaling to sub-1 nm equivalent oxide thickness with hafnium oxide deposited by atomic layer deposition", *Journal of the Electrochemical Society*, vol. 153, no. 8, pp. F180–F187, 2006.

[DES 08] DESTEFANIS V., HARTMANN J.-M., HOPSTAKEN M., *et al.*, "Low-thermal surface preparation, HCl etch and Si/SiGe selective epitaxy on (110) silicon surfaces", *Semiconductor Science and Technology*, vol. 23, no. 10, pp. 105018.1–105018.9, 2008.

[DET 04] DETAVERNIER C., LAVOIE C., "Influence of Pt addition on the texture of NiSi on Si (001)", *Applied Physics Letters*, vol. 84, no. 18, pp. 3549–3551, 2004.

[DET 06] DETAVERNIER C., DEDUYTSCHE D., JORDAN-SWEET J., *et al.*, "Influence of alloying elements on the formation and stability of NiSi", *ECS Transactions*, vol. 3, no. 2, pp. 131–138, 2006.

[DHE 86] D'HEURLE F.M., GAS P., "Kinetics of formation of silicides: a review", *Journal of Materials Research*, vol. 1, pp. 205–221, 1986.

[DHE 88] D'HEURLE F.M., "Nucleation of a new phase from the interaction of two adjacent phases: some silicides", *Journal of Materials Research*, vol. 3, pp. 167–195, 1988.

[ELA 03] ELAMA J.W., SCHUISKYA M., FERGUSONA J.D., *et al.*, "Surface chemistry and film growth during TiN atomic layer deposition using TDMAT and NH_3", *Thin Solid Films*, vol. 436, pp. 145–156, 2003.

[ERN 08] ERNST T., DURAFFOURG L., DUPRE C., *et al.*, "Novel Si-based nanowire devices: will they serve ultimate MOSFETs scaling or ultimate hybrid integration?", *Proceedings of the 2008 IEDM Conference*, 2008.

[FED 07] FEDORENKO Y., SWERTS J., MACS J.W., *et al.*, "Atomic layer deposition of hafnium silicate from $HfCl_4$, $SiCl_4$, and H_2O", *Electrochemical and Solid-State Letters*, vol. 10, no. 5, H149–H152, 2007.

[FER 02] FERRARI S., SCAREL G., WIEMER C., *et al.*, "Chlorine mobility during annealing in N_2 in ZrO_2 and HfO_2 films grown by atomic layer deposition", *Journal of Applied Physics*, vol. 92, pp. 7675–7677, 2002.

[FER 06] FERRARI S., FANCIULLI M., "Diffusion reaction of oxygen in $HfO_2/SiO_2/Si$ stacks", *Journal of Physical Chemistry B*, vol. 110, pp. 14905–14910, 2006.

[FIL 05] FILLOT F., MOREL T., MINORET S., *et al.*, "Investigations of titanium nitride as metal gate material, elaborated by metal organic atomic layer deposition using TDMAT and NH_3", *Microelectronics Engineering*, vol. 82, nos. 3–4, pp. 248–253, 2005.

[FRO 03] FROMENT B., MULLER M., BRUT H., *et al.*, "Nickel vs. cobalt silicide integration for sub-50 nm CMOS", *33rd European Solid-State Device Research Conference*, pp. 215–218, 2003.

[FRO 04] FROMENT B., CARRON V., MORAND Y., *et al.*, "Optimized nickel silicide process formation for high performance sub-65 nm CMOS nodes", *Proceedings of the Electrochemical Society*, vol. 1, pp. 191–201, 2004.

[GAM 98] GAMBINO J.P., COLGAN E.G., "Silicides and ohmic contacts", *Materials Chemistry and Physics*, vol. 52, no. 2, pp. 99–146, 1998.

[GAN 04] GANEM J.-J., TRIMAILLE I., VICKRIDGE I.C., *et al.*, "Study of thin hafnium oxides deposited by atomic layer deposition", *Nuclear Instruments and Methods in Physics Research Section B*, vols. 219–220, pp. 856–861, 2004.

[GAR 05] GARROS X., REIMBOLD G., LOUVEAU O., *et al.*, "Process damages in HfO_2/TiN stacks: the key role of H_0 and H_2 anneals", *Proceedings of 2005 IEDM Conference*, pp. 1–4, 2005.

[GAR 08] GARROS X., CASSE M., REIMBOLD G., *et al.*, "Guidelines to improve mobility performances and BTI reliability of advanced high-k/metal gate stacks", *Proceedings of the 2008 VLSI Conference*, pp. 68–69, 2008.

[GAS 93] GAS P., D'HEURLE F.M., "Formation of silicide thin films by solid state reaction", *Applied Surface Science*, vol. 73, no. 2, pp. 153–161, 1993.

[GAU 06] GAUDET S., DETAVERNIER C., KELLOCK A.J., *et al.*, "Thin film reaction of transition metals with germanium", *Journal of Vacuum Science and Technology A*, vol. 24, no. 3, pp. 474–485, 2006.

[GIR 07] GIRARD J.-M., BLASCO N., DUSSARAT C., *et al.*, "CVD and ALD precursor chemistry for advanced gate stack and memory capacitors", *EuroAsia Semiconductor*, p. 25, 2007/2008.

[GOE 82] GOESELE U., TU K.N., "Growth kinetics of planar binary diffusion couples", *Journal of Applied Physics*, vol. 53, no. 4, pp. 3252–3260, 1982.

[GOH 07] GOH F., SIH V.K.T., TAN W.L., *et al.*, "Challenges in nickel platinum silicide wet etching for sub-65 nm device technology", *ECS Transactions*, vol. 11, no. 2, pp. 321–326, 2007.

[GOU 09] GOUYE A., KERMARREC O., HALIMAOUI A., *et al.*, "Low-temperature RPCVD of Si, SiGe alloy, and $Si_{1-y}C_y$ films on Si substrates using trisilane (Silcore®)", *Journal of Crystal Growth*, vol. 311, no. 13, pp. 3522–3527, 2009.

[GRE 12] GRENOUILLET L., VINET M., GIMBERT J., *et al.*, "UTBB FDSOI transistors with dual STI for a multi-V_t strategy at 20 nm node and below", *Proceedings of the 2012 IEDM Conference*, abstract 3.6, 2012.

[GRO 12] GROSJEAN M., BOUVERT P., GRAOUI H., *et al.*, "Modelization of hafnium silicate vapor deposition using tetrakis-diethyl-amino-hafnium and tetrakis-dimethyl-amino-silane", *Thin Solid Films*, vol. 520, pp. 2594–2599, 2012.

[HAL 98] HALIMAOUI A., HENRISEY E., HERNANDEZ C., *et al.*, "Tailoring of the Nitrogen profile in thin gate oxides using substrate nitridation by nitric oxide", *MRS Proceedings*, vol. 532, pp. 159–164, 1998.

[HAR 02] HARTMANN J.-M., LOUP V., ROLLAND G., *et al.*, "SiGe growth kinetics and doping in reduced pressure-chemical vapor deposition", *Journal of Crystal Growth*, vol. 236, nos. 1–3, pp. 10–20, 2002.

[HAR 04a] HARTMANN J.-M., ABBADIE A., PAPON A.M., *et al.*, "Reduced pressure chemical vapor deposition of Ge thick layers on Si(001) for 1.3–1.55 µm photo-detection", *Journal of Applied Physics*, vol. 95, no. 10, pp. 5905–5913, 2004.

[HAR 04b] HARTMANN J.-M., CLAVELIER L., JAHAN C., *et al.*, "Selective epitaxial growth of boron and phosphorous-doped Si and SiGe for raised sources and drains", *Journal of Crystal Growth*, vol. 264, p. 36, 2004.

[HAR 07] HARTMANN J.-M., "Low temperature growth kinetics of high Ge content SiGe in reduced pressure – chemical vapor deposition", *Journal of Crystal Growth*, vol. 305, no. 1, pp. 113–121, 2007.

[HAR 08a] HARTMANN J.-M., BOGUMILOWICZ Y., ABBADIE A., *et al.*, "Impact of the H_2 bake temperature on the structural properties of tensily strained Si layers on SiGe", *Journal of Crystal Growth*, vol. 310, no. 10, pp. 2493–2502, 2008.

[HAR 08b] HARTMANN J.-M., GONZATTI F., FILLOT F., *et al.*, "Growth kinetics and boron-doping of very high Ge content SiGe for source/drain engineering", *Journal of Crystal Growth*, vol. 310, no. 1, pp. 62–70, 2008.

[HAR 11a] HARTMANN J.-M., ABBADIE A., FAVIER S., "Critical thickness for plastic relaxation of SiGe on Si(001) revisited", *Journal of Applied Physics*, vol. 110, no. 8, pp. 083529–083529_8, 2011.

[HAR 11b] HARTMANN J.-M., PY M., BARNES J.P., *et al.*, "Low temperature boron and phosphorous doped SiGe for recessed and raised sources and drains", *Journal of Crystal Growth*, vol. 327, no. 1, pp. 68–77, 2011.

[HAR 12] HARTMANN J.-M., BENEVENT V., DAMLENCOURT J.F., *et al.*, "A benchmarking of silane, disilane and dichlorosilane for the low temperature growth of group IV layers", *Thin Solid Films*, vol. 520, no. 8, pp. 3185–3189, 2012.

[HE 12] HE H., BRABANT P., CHUNG K., *et al.*, "High strain embedded-SiGe via low temperature reduced pressure chemical vapor deposition", *Thin Solid Films*, vol. 520, no. 8, pp. 3175–3178, 2012.

[HEG 97] HEGDE R., MAITI B., TOBIN P.J., "Growth and film characteristics of N_2O and NO oxynitride gate and tunnel dielectrics", *Journal of the Electrochemical Society*, vol. 144, no. 3, pp. 1081–1086, 1997.

[HER 01] HERSAM M.C., GUISINGER N.P., LYDING J.W., *et al.*, "Atomic-level study of the robustness of the Si (100)-2×1:H surface following exposure to ambient conditions", *Applied Physics Letters*, vol. 78, no. 7, pp. 886–888, 2001.

[HOU 91] HOUGHTON D.C., "Strain relaxation kinetics in $Si_{1-x}Ge_x$/Si heterostructures", *Journal of Applied Physics*, vol. 70, no. 4, pp. 2136–2151, 1991.

[HUT 10] HUTIN L., LE ROYER C., ANDRIEU F., *et al.*, "Dual strained channel co-integration into CMOS, RO and SRAM cells on FDSOI down to 17 nm gate length", *Proceedings of 2010 IEDM Conference*, pp. 253–256, 2010.

[IMB 07] IMBERT B., GUICHET C., BONNETIER S., *et al.*, "Impact of surface preparation on nickel–platinum alloy silicide phase formation", *Microelectronic Engineering*, vol. 84, no. 11, pp. 2523–2527, 2007.

[IMB 08] IMBERT B., ZOLL S., GARNIER P., *et al.*, "Self-aligned nickel–platinum silicide oxidation", *Materials Science and Engineering B*, vols. 154–155, pp. 155–158, 2008.

[IMB 09] IMBERT B., ZOLL S., MOROTE C., *et al.*, "High temperature control of platinum", Issue III, pp. 17–21, 2009, available at www.euroasiasemiconductor.com.

[IMB 10] IMBERT B., PANTEL R., ZOLL S., *et al.*, "Nickel silicide encroachment formation and characterization", *Microelectronic Engineering*, vol. 87, no. 3, pp. 245–248, 2010.

[INU 05] INUMIYA S., AKASAKA Y., MATSUKI T., *et al.*, "A thermally stable sub-0.9 nm $TaSi_x$/HfSiON Gate Stack with high electron mobility, suitable for gate-first fabrication of hp45 LOP devices", *Proceedings of 2005 IEDM Conference*, pp. 23–26, 2005.

[ISH 02] ISHIKAWA Y., KUMEZAWA M., NURYADI R., *et al.*, "Effect of patterning on thermal agglomeration of ultrathin silicon-on-insulator layer", *Applied Surface Science*, vol. 90, nos. 1–4, pp. 11–15, 2002.

[ITR 11] International Technology Roadmap for Semiconductors, http://www.itrs.net/Links/2011ITRS/2011Chapters/2011FEP.pdf

[ITR 12] International Technology Roadmap for Semiconductors, available at www.itrs.net.

[IWA 02] IWAI I., OHGURO T., OMHI S.I., "NiSi salicide technology for scaled CMOS", *Microelectronic Engineering*, vol. 60, nos. 1–2, pp. 157–169, 2002.

[JAH 05] JAHAN C., FAYNOT O., TOSTI L., *et al.*, "Agglomeration control during the selective epitaxial growth of Si raised sources and drains on ultra-thin silicon-on-insulator substrates", *Journal of Crystal Growth*, vol. 280, nos. 3–4, pp. 530–538, 2005.

[JAN 12] JAN C.-H., BHATTACHARYA U., BRAIN R., *et al.*, "A 22 nm SoC platform technology featuring 3-D Tri-gate and high k / metal gate, optimized for ultra low power, high performance and high density SoC applications", *Proceedings of the 2012 IEDM Conference*, p. 3.1, 2012.

[JAR 02] JARMAR T., SEGER J., ERICSON F., *et al.*, "Morphological and phase stability of nickel–germanosilicide on $Si_{1-x}Ge_x$ under thermal stress", *Journal of Applied Physics*, vol. 92, no. 12, pp. 7193–7199, 2002.

[JER 99] JERIER P., DUTARTRE D., "Boron autodoping in single-wafer epitaxy of silicon at reduced pressure", *Journal of the Electrochemical Society*, vol. 146, no. 1, pp. 331–335, 1999.

[JUN 05] JUNG H.-S., LEE J.-H., HAN S.K., *et al.*, "A highly manufacturable MIPS (metal inserted poly-Si stack) technology with novel threshold voltage control", *Proceedings of the 2005 VLSI Conference*, pp. 232–233, 2005.

[KAM 08a] KAMADA H., TANIMURA T., TOYODA S., *et al.*, "Control of oxidation and reduction reactions at HfSiO/Si interfaces through N exposure or incorporation", *Applied Physics Letters*, vol. 93, no. 21, p. 212903, 2008.

[KAM 08b] KAMIYAMA S., ISHIKAWA D., KUROSAWA E., *et al.*, "Systematic study of V_{th} controllability using ALD Y_2O_3, La_2O_3, MgO layers with HfSiON / metal gate first n-MOSFETs for hp 32 nm bulk devices", *Proceedings of 2008 IEDM Conference*, pp. 1–4, 2008.

[KAT 08] KATAMREDDY R., FEIST B., TAKOUDIS C., "Bis (diethylamino) silane as the silicon precursor in the atomic layer deposition of $HfSiO_x$", *Journal of the Electrochemical Society*, vol. 155, no. 8, pp. G163–G167, 2008.

[KER 70] KERN W., POUTINEN D.A., "Cleaning solution based on hydrogen peroxide for use in silicon semiconductor technology", *RCA Review*, vol. 31, pp. 187–205, 1970.

[KIM 02] KIM H., CABRAL C., LAVOIE C., *et al.*, "Diffusion barrier properties of transition metal thin films grown by plasma-enhanced atomic-layer deposition", *Journal of Vacuum Science Technology B*, vol. 20, no. 4, pp. 1321–1326, 2002.

[KIM 03] KIM J.-Y., CHOI G.-H., KIM Y.-D., *et al.*, "Comparison of TiN films deposited using Tetrakisdimethylaminotitanium and Tetrakisdiethylaminotitanium by the atomic layer deposition method", *Japanese Journal of Applied Physics*, vol. 42, pp. 4245–4248, 2003.

[KIM 04a] KIM H., MARSCHALL A., MC INTYRE P., "Crystallisation kinetics and microstructure dependent leakage current behavior of ultrathin HfO_2 dielectrics: *in-situ* annealing", *Applied Physics Letters*, vol. 84, no. 12, pp. 2064–2066, 2004.

[KIM 04b] KIM H., MC INTYRE P., CHUI C.O., *et al.*, "Engineering chemically abrupt high k metal oxide/silicon interfaces using an oxygen-gettering metal overlayer", *Journal of Applied Physics*, vol. 96, no. 6, pp. 3467–3472, 2004.

[KIR 06a] KIRSCH P.D., QUEVEDO-LOPEZ M.A., KRISHNAN S.A., *et al.*, "Band edge n-MOSFET's with high K / metal gate stacks scaled to EOT = 0.9 nm with excellent carrier mobility and high temperature stability", *Proceedings of 2006 IEDM Conference*, pp. 439–442, 2006.

[KIR 06b] KIRSCH P.D., QUEVEDO-LOPEZ M.A., LI H.-J., *et al.*, "Nucleation and growth study of atomic layer deposited HfO_2 gate dielectrics resulting in improved scaling and electron mobility", *Journal of Applied Physics*, vol. 99, no. 2, p. 023508, 2006.

[KIR 08] KIRSCH P.D., SIVASUBRAMANI P., HUANG J., *et al.*, "Dipole model explaining high k/metal gate field effect transistor threshold voltage tuning", *Applied Physics Letters*, vol. 92, no. 9, pp. 09201–09201_3, 2008.

[KIT 03] KITTL J.A., LAUWERS A., CHAMIRIAN O., *et al.*, "Ni and Co-based silicides for advanced CMOS applications", *Microelectronic Engineering*, vol. 70, nos. 2–4, pp. 158–165, 2003.

[KIT 07] KITTL J.A., LAUWERS A., DEMEURISSE C., *et al.*, "Work function of Ni_3Si_2 on $HfSi_xO_y$ and SiO_2 and its implication for Ni fully silicided gate applications", *Applied Physics Letters*, vol. 90, no. 3, p. 32103, 2007.

[KOR 08] KORMANN T., GARNIER P., CHABANNE G., *et al.*, "*Ex-situ* wet clean and *in-situ* hydrogen clean for Si and SiGe epitaxy", *Thin Solid Films*, vol. 517, no. 1, pp. 269–271, 2008.

[KRI 11] KRISHNAN S., KWON U., MOUMEN N., *et al.*, "A manufacturable dual channel (Si and SiGe) high k metal gate CMOS technology with multiple oxides for high performance and low power applications", *Proceedings of 2011 IEDM Conference*, pp. 634–637, 2011.

[KUD 08] KUDO S., HIROSE Y., HASHIKAWA N., *et al.*, "Analysis of Ni silicide abnormal growth mechanism using advanced TEM techniques", *46th IEEE International Reliability Physics Symposium (IRPS)*, pp. 580–583, 2008.

[KUH 12] KUHN K., AVCI U., CAPPELLANI A., *et al.*, "The ultimate CMOS device and beyond", *Proceedings of the 2012 IEDM Conference*, abstract 8.1, 2012.

[KUI 92] KUIPER A.E.T., LATHOUWERS E.G.C., "Room-temperature HF vapor-phase cleaning for low-pressure chemical vapor deposition of epitaxial Si and SiGe layers", *Journal of the Electrochemical Society*, vol. 139, no. 9, pp. 2594–2599, 1992.

[KUL 90] KULKARNI S.K., GATES S.M., GREENLIEF C.M., *et al.*, "Kinetics and mechanics of Si_2H_6 surface decomposition on Si", *Journal of Vacuum Science and Technology A*, vol. 8, pp. 2956–2959, 1990.

[LAU 82] LAU C.K., SEE Y.C., SCOTT D.B., *et al.*, "Titanium disilicide self-aligned source/drain + gate technology", *Proceedings of the 1982 IEDM Conference*, vol. 28, pp. 714–717, 1982.

[LAV 03] LAVOIE C., D'HEURLE F.M., DETAVERNIER C., *et al.*, "Towards implementation of a nickel silicide process for CMOS technologies", *Microelectronic Engineering*, vol. 70, nos. 2–4, pp. 144–157, 2003.

[LE 08] LE ROYER C., VINET M., POUYDEBASQUE A., *et al.*, "Germanium-on-insulator MOS transistor integration: challenges and opportunities for advanced technologies", *Proceeding of the European Material Research Society (E-MRS) Conference, Spring Meeting, Strasbourg, France*, 2008.

[LE 11] LE ROYER C., VILLALON A., CASSE M., *et al.*, "First demonstration of ultrathin body c-SiGe channel FDSOI pMOSFETs combined with SiGe(:B) RSD: drastic improvement of electrostatics (V_{th}, p tuning, DIBL) and transport (μ_0, I_{sat}) properties down to 23 nm gate length", *Proceedings of 2011 IEDM Conference*, pp. 16.5.1–16.5.4, 2011.

[LE 12] LE TIEC Y., VINET M., GRENOUILLET L., *et al.*, "FDSOI technology for sub-22 nm nodes: a surface preparation view", *Proceedings of the 12th Surface Preparation and Cleaning Conference*, Austin, TX, 2012.

[LI 06] LI R., YAO H.B., LEE S.J., *et al.*, "Metal-germanide Schottky source/drain transistor on germanium substrate for future CMOS technology", *Thin Solid Films*, vol. 504, nos. 1–2, pp. 28–31, 2006.

[LIU 11] LIU Q., MONSIEUR F., KUMAR A., *et al.*, "Impact of back bias on ultra-thin-body and box (UTBB) devices", *Proceedings of the 2011 VLSI Conference*, pp. 160–161 2011.

[LIU 12] LIU J.F., CAMACHO-AGUILERA R., BESSETTE J.T., *et al.*, "Ge-on-Si optoelectronics", *Thin Solid Films*, vol. 520, no. 8, pp. 3354–3360, 2012.

[LOC 06] LOCQUET J.-P., MARCHIORI C., SOUSA M., *et al.*, "High k dielectric for the gate stack", *Journal of Applied Physics*, vol. 100, no. 5, pp. 05610–056610_14, 2006.

[LOU 03] LOUP V., HARTMANN J.-M., ROLLAND G., *et al.*, "Reduced pressure – chemical vapor deposition of Si / SiGeC heterostructures using a chlorinated chemistry", *Semiconductor Science and Technology*, vol. 18, no. 4, pp. 352–360, 2003.

[LOU 12] LOUBET N., ADAM T., RAYMOND M., *et al.*, "Ultra-low resistivity *in-situ* phosphorous doped Si and SiC epitaxy for source/drain formation in advanced 20 nm n-type field effect transistor devices, *Thin Solid Films*, vol. 520, no. 8, pp. 3149–3154, 2012.

[MAE 95] MAEX K., VON ROSSUM M., *Properties of Metal Silicides (EMIS Data Reviews)*, vol. 14, p. 335, INSPEC, London, 1995.

[MAN 99] MANGELINCK D., DAI J.Y., PAN J.S., *et al.*, "Enhancement of thermal stability of NiSi films on (100)Si and (111)Si by Pt addition", *Applied Physics Letters*, vol. 75, no. 12, pp. 1736–1738, 1999.

[MAN 09] MANGELINCK D., HOUMMADA K., BLUM I., "Kinetics of a transient silicide during the reaction of Ni thin film with (100)Si", *Applied Physics Letters*, vol. 95, no. 18, pp. 181902–181904, 2009.

[MEY 90] MEYERSON B.S., HIMPSEL F.J., URAM K.J., "Bistable conditions for low-temperature silicon epitaxy", *Applied Physics Letters*, vol. 57, no. 10, pp. 1034–1036, 1990.

[MIG 08] MIGITA S., WATANABE Y., OTA H., *et al.*, "Design and demonstration of very high k (k = 50) HfO_2 for ultra-scaled Si CMOS", *Proceedings of the 2008 VLSI Conference*, pp. 152–153, 2008.

[MO 91] MO Y.W., LAGALLY M.G., "Anisotropy in surface migration of Si and Ge on Si (001)", *Surface Science*, vol. 248, no. 3, pp. 313–320, 1991.

[MON 10] MONFRAY S., FENOUILLEt-BERRANGER C., BIDAL G., *et al.*, "Thin-film devices for low power applications", *Solid State Electronics*, vol. 54, no. 2, pp. 90–96, 2010.

[MUR 03] MURAKAMI T., CARRON V., YOO W.S., "Advanced devices using low-temperature NiSi formation", *Solid State Technology*, vol. 46, no. 9, pp. 32, 2003.

[MUS 09] MUSSCHOOT J., XIE Q., DEDUYTSCHE D., *et al.*, "Atomic layer deposition of titanium nitride from TDMAT precursor", *Microelectronic Engineering*, vol. 86, no. 1, pp. 72–77, 2009.

[NAR 04] NARAYANAN V., CALLEGARI A., MC FEELY F.R., *et al.*, "Dual work function metal gate CMOS using CVD metal electrodes", *Proceedings of the 2004 VLSI Conference*, pp. 192–193, 2004.

[NAS 87] NASH P., NASH A., "The Ni-Si (nickel-silicon) system", *Bulletin of Alloy Phase*, vol. 8, no. 1, pp. 6–14, 1987.

[NEM 06] NEMOUCHI F., MANGELINCK D., LABAR J., *et al.*, "A comparative study of nickel silicides and nickel germanides: phase formation and kinetics", *Microelectronic Engineering*, vol. 83, pp. 2101–2106, 2006.

[NEM 08] NEMOUCHI F., CARRON V., MORAND Y., *et al.*, "Nickel metallization process for low resistance contact formation on planar cointegration nMOS-Si and pMOS-Ge on insulator", *MRS Spring Meeting*, San Francisco, CA, Abstract E2.4, 2008.

[NIE 01] NIEMINEN M., PUTKONEN M., NINISTO L., "Formation and stability of lanthanum oxide thin films deposited from beta-diketonate precursor", *Applied Surface Science*, vol. 174, no. 2, pp. 155–165, 2001.

[NIN 06] NINISTO J., PUTKONEN M., NINISTO L., *et al.*, "HfO_2 films grown by ALD using cyclopentadienyl precursors and H_2O or O_3 as oxygen source", *Journal of the Electrochemical Society*, vol. 153, no. 3, pp. F39–F45, 2006.

[OLO 76] OLOWOLAFE J.O., NICOLET M.A., MAYER J.W., "Influence of the nature of the Si substrate on nickel silicide formed from thin Ni films", *Thin Solid Films*, vol. 38, no. 2, pp. 143–150, 1976.

[PAC 09] PACKAN P., AKBAR S., ARMSTRONG M., *et al.*, "High performance 32 nm logic technology featuring 2nd generation high k + metal gate transistors", *Proceedings of 2009 IEDM Conference*, pp. 659–662, 2009.

[PAR 02] PARK J.-S., PARK H.-S., KANE S.-W., "Plasma enhanced atomic layer deposition of Ta-N Films", *Journal of the Electrochemical Society*, vol. 149, no. 1, pp. C28–C32, 2002.

[PAR 03] PARK S.-G., LEE Y.-K., KANG S.-B., *et al.*, "Performance improvement of MOSFET with HfO_2-Al_2O_3 laminate gate dielectric and CVD-TaN metal gate deposited by TAIMATA", *Proceedings of 2003 IEDM Conference*, p. 13.6, 2003.

[PAR 08] PARK T.J., KIM J.-H., JANG J.-H., *et al.*, "Effective work function and interfacial reactions with underlying HfO_2 layer of plasma-enhanced atomic layer deposited TaC_xN_y films", *Applied Physics Letters*, vol. 92, no. 20, pp. 202902–202902_3, 2008.

[PAS 63] PASCAL P., ALBERT P., *Nouveau traité de chimie minérale - Tome IX, Titane, Zirconium, Hafnium, Thorium*, Masson Editeur, Paris, 1963.

[PEO 85] PEOPLE R., BEAN J.C., "Calculation of critical layer thickness versus lattice mismatch for GexSi1-x/Si strained layer heterostructures", *Applied Physics Letters*, vol. 47, no. 3, pp. 322–324, 1985.

[PEO 86] PEOPLE R., BEAN J.C., "Erratum: Calculation of critical layer thickness versus lattice mismatch for GexSi1-x/Si strained-layer heterostructures", *Applied Physics Letters*, vol. 49, no. 4, p. 229, 1986.

[POU 63] POURBAIX M., *Atlas d'équilibres électrochimiques*, Gauthier-Villars & Cie, Paris, 1963.

[PUU 04] PUURUNEN R., "Analysis of hydroxyl group controlled atomic layer deposition of hafnium dioxide from hafnium tetrachloride and water", *Journal of Applied Physics*, vol. 95, no. 9, pp. 4777–4786, 2004.

[RAG 05] RAGNARSSON L., SEVERI S., TROJMAN L., *et al.*, "High performing 8Å EOT HfO_2 / TaN low thermal-budget n-channel FETs with solid-phase epitaxially regrown (SPER) junctions", *Proceedings of the 2005 VLSI Conference*, p. 234, 2005.

[RAN 74] RAND M.J., ROBERTS J.F., "Observations on formation and etching of platinum silicide", *Applied Physics Letters*, vol. 24, no. 2, pp. 49–51, 1974.

[REN 02] RENAULT O., SAMOUR D., DAMLENCOURT J.-F., *et al.*, "HfO_2/SiO_2 interface chemistry studied by synchrotron radiation x-ray photoelectron spectroscopy", *Applied Physics Letters*, vol. 81, no. 19, pp. 3627–3629, 2002.

[RIE 12] RIEL H., MOSELUND K.E., BESSIRE C., *et al.*, "InAs-Si heterojunction nanowire tunnel diodes and tunnel FETs", *Proceedings of the 2012 IEDM Conference*, abstract 16.6, 2012.

[RIT 99] RITALA M., LESKELA M., RAUHALA E., *et al.*, "Atomic layer epitaxy growth of TiN thin films", *Journal of the Electrochemical Society*, vol. 142, no. 8, pp. 2731–2737, 1995.

[ROB 91] ROBBINS D.J., GLASPER J.L., GULLIS A.G., *et al.*, "A model for heterogeneous growth of $Si_{1-x}Ge_x$ films from hydrides", *Journal of Applied Physics*, vol. 69, no. 6, pp. 3729–3732, 1991.

[ROB 00] ROBERTSON J., "Band offsets in wide band gap oxides and implications for future electronic devices", *Journal of Vacuum Science Technology B*, vol. 18, no. 3, pp. 1785–1791, 2000.

[ROU 02] ROUCHON D., ROCHAT N., GUSTAVO F., *et al.*, "Study of ultrathin silicon oxide films by FTIR-ATR and ARXPS after wet chemical cleaning processes", *Surface and Interface Analysis*, vol. 34, no. 1, pp. 445–450, 2002.

[SAG 96] SAGNES I., LAVIALE D., REGACHE M., *et al.*, "Study of nitrogen incorporation in gate oxides using the resistance to oxidation method", *MRS Proceedings*, vol. 429, p. 251, 1996.

[SAI 10] SAITOH M., NAKABAYASHI Y., ITOKAWA H., *et al.*, "Short channel performance and mobility analysis of <110> and <100> oriented tri-gate nanowire MOSFETs with raised source/drain extensions", *Proceedings of the 2010 VLSI Conference*, pp. 169–170, 2010.

[SEG 05] SEGER J., HELLSTROM P.-E., "Lateral encroachment of Ni-silicides in the source/drain regions on ultrathin silicon-on-insulator", *Applied Physics Letters*, vol. 86, no. 25, pp. 253507–253507_3, 2005.

[SHE 98] SHEIN A.B., "Electrochemical behavior of nickel monosilicide in the electrolyte containing sulfuric acid and fluoride ions", *Russian Journal of Electrochemistry*, vol. 34, no. 8, pp. 807–810, 1998.

[SHI 12] SHINRIKI M., CHUNG K., HASAKA S., *et al.*, "Gas phase particle formation and elimination on Si(100) in low temperature reduced pressure chemical vapor deposition silicon-based epitaxial layers", *Thin Solid Films*, vol. 520, no. 8, pp. 3190–3194, 2012.

[SRE 06] SREENIVASAN R., MC INTYRE P.C., KIM H., *et al.*, "Effect of impurities in the fixed charge of nanoscale HfO_2 films grown by atomic layer deposition", *Applied Physics Letters*, vol. 89, no. 11, pp. 112903–112905, 2006.

[STO 06] STOUT P.J., ADAMS V., VENTZEK P.L.G., "Modelling HfO_2 atomic layer chemical vapor deposition on blanket wafer, via, and trench structures using $HfCl_4/H_2O$", *Journal of Vacuum Science and Technology B*, vol. 24, no. 5, pp. 2372–2380, 2006.

[STR 07] STRANE J., BROWN D., LAVOIE C., *et al.*, "Implementation of robust nickel alloy salicide process for high-performance 65 nm SOI CMOS manufacturing", *Proceedings of the 2007 VLSI Conference*, 2007.

[SUH 00] SUH K.Y., LEE H.H., "Ge composition in $Si_{1-x}Ge_x$ films grown from SiH_2Cl_2/GeH_4 precursors", *Journal of Applied Physics*, vol. 88, no. 7, pp. 4044–4047, 2000.

[SWE 07] SWERTS J., FEDORENKO Y., MAES J., *et al.*, "ALD La-based oxides for Vt-tuning in high-k/metal gate stacks", *ECS Transactions on Atomic Layer Deposition Applications*, vol. 11, no. 7, pp. 201–211, 2007.

[TAK 10] TAKEUCHI S., NGUYEN N.D., GOOSENS J., *et al.*, "$Si_{1-x}Ge_x$ growth using Si_3H_8 by low temperature chemical vapor deposition", *Thin Solid Films*, vol. 518, no. 6, pp. S18–S22, 2010.

[TAL 08] TALLARIDA M., KARAVAEV K., SCHMEISSER D., "The initial atomic layer deposition of HfO_2/Si(001) as followed *in situ* by synchrotron radiation photoelectron spectroscopy", *Journal of Applied Physics*, vol. 104, no. 6, pp. 064116–064116_8, 2008.

[TAR 99] TARDIF F., LARDIN T., ABOLAFIA Y., *et al.*, "New aspects of the diluted dynamic clean process", *Solid State Phenomena*, vol. 65–66, pp. 19–22, 1999.

[THO 04] THOMPSON S.E., ARMSTRONG M., AUTH C., *et al.*, "A logic nanotechnology featuring strained-silicon", *Electron Device Letters*, vol. 25, no. 4, pp. 191–193, 2004.

[TOM 10] TOMASINI P., MACHKAOUTSAN V., THOMAS S.G., *et al.*, "Stability of silicon germanium stressors", *Thin Solid Films*, vol. 518, no. 6, pp. S133–S135, 2010.

[TRI 05] TRIYOSO D.H., HEGDE R.I., WHITE B.E., *et al.*, "Physical and electrical characteristics of atomic layer deposited hafnium dioxide formed using hafnium tetrachloride and tetrakis (ethylmethylaminohafnium)", *Journal of Applied Physics*, vol. 97, no. 12, pp. 124107–124107_9, 2005.

[TRI 06] TRIYOSO D.H., TOBIN P.J., WHITE B.E., *et al.*, "Impact of film properties of atomic layer deposited HfO_2 resulting from annealing with a TiN capping layer", *Applied Physics Letters*, vol. 89, no. 13, p. 132903, 2006.

[TRI 08] TRIYOSO D.H., GREGORY R., PARK M., *et al.*, "Physical and electrical properties of Atomic-layer-deposited $Hf_xZr_{1-x}O_2$ with TEMAHf, TEMAZr and ozone", *Journal of the Electrochemical Society*, vol. 55, no. 1, pp. H43–H46, 2008.

[TRU 90] TRUCKS G.W., RAGHAVACHARI K., HIGASHI G.S., *et al.*, "Mechanism of HF etching of silicon surfaces: a theoretical understanding of hydrogen passivation", *Physical Review Letters*, vol. 65, no. 4, pp. 504–507, 1990.

[TU 75] TU K.N., CHU W.K., MAYER J.W., "Structure and growth kinetics of Ni_2Si on silicon", *Thin Solid Films*, vol. 25, no. 2, pp. 403–413, 1975.

[UED 06] UEDONO A., NAITO T., OTSUKA T., *et al.*, "Introduction of defects into HfO_2 gate dielectrics by metal-gate deposition studied using x-ray photoelectron spectroscopy and positron annihilation", *Journal of Applied Physics*, vol. 100, no. 6, p. 064501, 2006.

[VAN 75] VAN DER MERWE J.H., BALL C.A.B., in MATTHEWS J. (eds.) *Epitaxial Growth*, Part B, Academic, New York, NY, 1975.

[VAN 04a] VAN ELSHOCHT S., BAKLANOV M.R., BRIJS B., *et al.*, "Bulk properties of MOCVD-deposited HfO_2 layers for high k dielectric applications", *Journal of the Electrochemical Society*, vol. 151, no. 10, pp. F228–F234, 2004.

[VAN 04b] VAN DER LEE A., DURAND J., BLIN D., *et al.*, "Crystallisation kinetics and density profiles in ultra thin hafnia", *The European Physical Journal B*, vol. 39, no. 2, pp. 273–277, 2004.

[VEL 11] VELOSO A., RAGNARSSON L.A., CHO M.J., *et al.*, "Gate-last vs. gate-first technology for aggressively scaled EOT logic/RF CMOS", *Proceedings of the 2011 VLSI Conference*, pp. 34–35, 2011.

[VIN 10] VINCENT B., LOO R., VANDERVORST W., *et al.*, "Low temperature Si homo-epitaxy by reduced pressure chemical vapor deposition using dichlorosilane, silane and trisilane", *Journal of Crystal Growth*, vol. 312, no. 19, pp. 2671–2676, 2010.

[VOI 97] VOIGTLANDER B., WEBER T., SMILAUER P., *et al.*, "Transition from island growth to step-flow growth for Si/Si(100) epitaxy", *Physical Review Letters*, vol. 78, no. 11, pp. 2164–2167, 1997.

[WEB 05] WEBER O., BOGUMILOWICZ Y., ERNST Y., *et al.*, "Strained Si and Ge MOSFETs with high k / metal gate stack for high mobility dual channel CMOS", *Proceedings of 2005 IEDM Conference*, pp. 137–140, 2005.

[WEB 10] WEBER O., ANDRIEU F., MAZURIER J., *et al.*, "Work-function engineering in gate first technology for multi-Vt dual-gate FDSOI CMOS on UTBOX", *Proceedings of 2010 IEDM Conference*, pp. 3.4.1–3.4.4, 2010.

[WIL 01] WILK G.D., WALLACE R.M., ANTHONY J.M., "High K gate dielectric: current status and materials properties considerations", *Journal of Applied Physics*, vol. 89, no. 10, pp. 5243–5275, 2001.

[WU 10] WU C.C., LIN D.W., KESHAVARZI A., *et al.*, "High performance 22/20 nm FINFET CMOS devices with advanced high K / metal gate scheme", *Proceedings of 2010 IEDM Conference*, pp. 27.1.1–27.1.4, 2010.

[YAM 11] YAMASHITA T., BASKER V.S., STANDAERT T., *et al.*, "Opportunities and challenges of FinFET as a device structure candidate for 14 nm node CMOS technology", *Electrochemical Chemical Society Transactions*, vol. 34, no. 1, pp. 81–86, 2011.

[YAN 10] YANG R., SU N., BONFANTI P., *et al.*, "Advanced *in situ* pre-Ni silicide (Siconi) cleaning at 65 nm to resolve defects in $NiSi_x$ modules", *Journal of Vacuum Science and Technology B*, vol. 28, no. 1, pp. 56–61, 2010.

[ZHA 04] ZHAO Q.T., BUCA D., LENK S., *et al.*, "Formation of ternary Ni-silicide on relaxed and strained SiGe layers", *Microelectronic Engineering*, vol. 76, nos. 1–4, pp. 285–289, 2004.

[ZHA 05] ZHANG S.L., *Advanced Gate Stack, Source/Drain and Channel Engineering for Si-Based CMOS: New Materials, Processes and Equipment*, GUSEV E.P. *et al.* (Eds), The Electrochemical Society Proceedings Series, Pennington, NJ, p. 597, 2005.

[ZHA 06] ZHANG Z., SONG S.C., QUEVEDO-LOPEZ M.A., *et al.*, "Co-optimization of the metal gate / high k stack to achieve high-field mobility >90% of SiO_2 universal mobility with an EOT = ~1 nm", *Electron Device Letters*, vol. 27, no. 3, pp. 185–187, 2006.

[ZHU 05] ZHU S., LI R., LEE S.J., *et al.*, "Germanium pMOSFETs with Schottky-barrier germanide S/D, high K gate dielectric and metal gate", *Electron Device Letters*, vol. 26, no. 2, pp. 81–83, 2005.

Chapter 2

Chemistry in Interconnects

2.1. Introduction

While we often associate integrated circuits with transistors, we should not ignore another aspect of circuits, known as interconnects, or "back end of the line" (BEOL), based on their position in the production chain. Indeed, modern integrated circuits can be separated into two groups: transistors, a schematic of which is shown in Figure 2.1, and interconnects, represented in Figure 2.2. Transistors enable the construction of logical functions. They act largely as a faucet of information, authorizing, or otherwise, a signal's passage. The effort of technologists has long been centered on the production of the transistors themselves, however, the vast increase in their number in an increasingly reduced space makes connecting these transistors more and more complex. In order to operate, the signal must be able to circulate between transistors, which is the role of interconnects (see Figure 2.2). They comprise a conductive element encased in an insulator. Horizontal lines of conductors are superposed over each other, linked by conductive pads or vias. A line and via set is commonly known as a "metal level".

By 2016, according to ITRS predictions [ITR 00], the 22 nm technological node should be developed, whereby a hundred billion transistors will be integrated onto a chip with 13 metal levels.

Chapter written by Vincent JOUSSEAUME, Paul-Henri HAUMESSER, Carole PERNEL, Jeffery BUTTERBAUGH, Sylvain MAÎTREJEAN and Didier LOUIS.

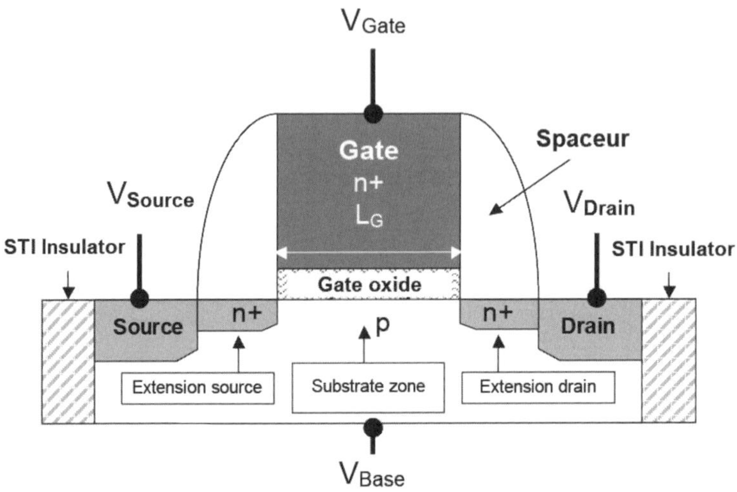

Figure 2.1. *Diagram of a conventional nMOSFET transistor*

Figure 2.2. *Diagram of an integrated circuit with interconnects [CAR 06]*

In the first section, we will review some generalities and the backgrounds concerning interconnects. At this point, constraints applicable to their development will be raised.

In the following sections, we will tackle the heart of the subject: realization processes and the associated materials, particularly insulating film deposition, the development of conductors and the cleaning stages[1].

2.2. Interconnects: generalities and background

2.2.1. *What conditions are required for an interconnect?*

2.2.1.1. *Introduction*

As we have briefly discussed in the introduction, an interconnect comprises of a grid of metallic lines, separated by an insulator. The connecting lines between two levels are linked by vias. Such a stack is illustrated in Figure 2.3 and is exposed on the circuit breaker (see Figure 2.4). Two principal sets of characteristics are required of interconnects: good signal transfer performance and a good level of electrical and mechanical reliability. These characteristics and their implication in material terms are discussed in sections 2.2.1.2 and 2.2.1.3.

Figure 2.3. *Section of a MOS circuit at an interconnect*

1 To impart is to choose. By choice, and in order to enable a complete description of the relevant topics, we have ignored two significant aspects of integration processes: dry or plasma etching and chemical mechanical polishing (CMP). For more information, we refer the reacter to [LI 07] and [BAK 12].

Figure 2.4. *Transmission electron microscopy section of a 65 nm technology (Joint collaboration between the CEA-Leti and STMicroelectronics Crolles [ARN 07]) interconnect grid. The commonly accepted nomenclature of the various interconnect levels is shown*

2.2.1.2. *Keeping the signal intact*

Ideally, the signal should be transferred without delay, distortion or energy consumption. Focusing on the delay, an interconnect grid can be compared with an RLC circuit. Using the numerical resolution of Maxwell's equations and by modeling the input and output transistor charges, Farcy [FAR 07] proposes a simplified circuit model, which is shown in Figure 2.5. The channel access resistance is modeled using R_{acc}, the contribution of the vertical interconnects is taken into account through the metallization contact resistance R_{ct} and the resistance of vias R_{via}, R_{int}, L_{int} and C_{int} are the resistance, inductance and capacitance of the line, respectively. C_{inv} is a capacitance representing the charge in the channel. Calculations have been carried out using 32 nm node technology dimensions, being an interline space and line width of the order of 32 nm for the line and via heights of 100 and 80 nm, respectively. The relative influence of the various parameters on the signal's propagation time is translated into the figure for different line lengths. We first note the significance of interconnect line characteristics compared with device characteristics. The effect of the interline capacitance (C_{int}) is immediately visible, whatever the interconnect lengths. When the line resistance increases with their lengths, the product $C_{int} \times R_{int}$ becomes a good approximation of the response time. A response time command passes through an interconnect capacitance and resistance control.

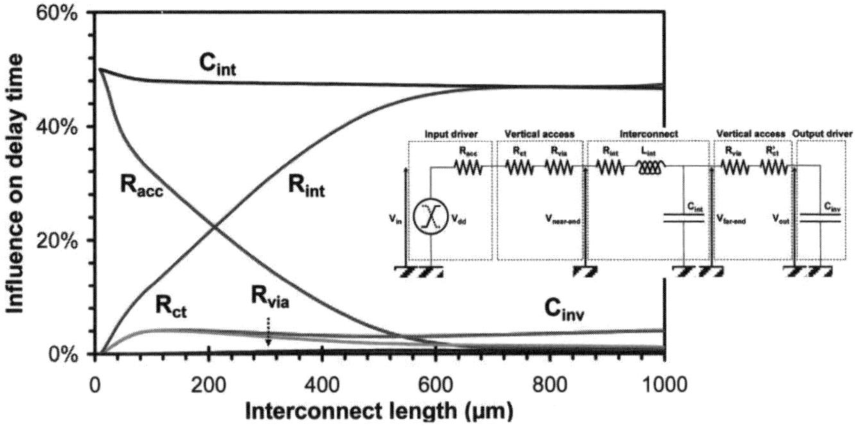

Figure 2.5. *Influence of circuit parameters on signal propagation delay [FAR 07]*

If two lines are relatively close, some crosstalk effects may also occur: a parasitic signal resulting from line 1 will appear in line 2. This signal may have two repercussions: causing an erroneous change of state of a device or inducing a supplementary lag in signal propagation in line 2. The electromagnetic simulations conducted by Blampey on 32 nm node dimension devices clearly demonstrate the significance of interline and interlevel capacitance control on the crosstalk value [BLA 06]: a reduction in capacitances enables a reduction in crosstalk.

Another parameter needs to be considered: using Ohm's law, the drop in voltage of a 100 μm long (using 45 nm node dimensions) line can easily be calculated. This is around 0.2 V and is of course proportional to the line resistance. The recommended power supply voltage is between 0.7 and 1 V. Such a drop becomes critical for transistor control.

Finally, it is necessary to be aware of the dissipated power in a circuit. As a first approximation, it comprises two terms: dynamic and static power. The second is essentially the leakage current within the transistors, the first, P, can perhaps be viewed as being proportional to CV^2f, where C is a capacitance of either the transistor or the line of interconnects, V is the working voltage and f is the clock speed.

Using a traditional circuit design in 2002, Chandra established an evaluation of the distribution of power consumption and its values in 180 and 50 nm technology circuits [CHA 02], predicting a fourfold increase in the dissipated power. The contribution of the interconnects (the distribution of the clock signal and calculation signal propagation) represents 70% for the 180 nm node and 32% for the 50 nm node (see Figure 2.6). While their projection shows that a major part of the power consumed is found in the active parts of the circuit, it is however essential to reduce the consumption of each part of the circuit. An interline capacitance command is necessary for this. In order to support this fact, we can use the title of Prof. Bill Dailly's (Willard R. and Inez Kerr Bell chair at Stanford and Scientific Officer for the NVidia Corporation) address to the 2012 IITC Conference: "It's about power: an architecture point of view of interconnects".

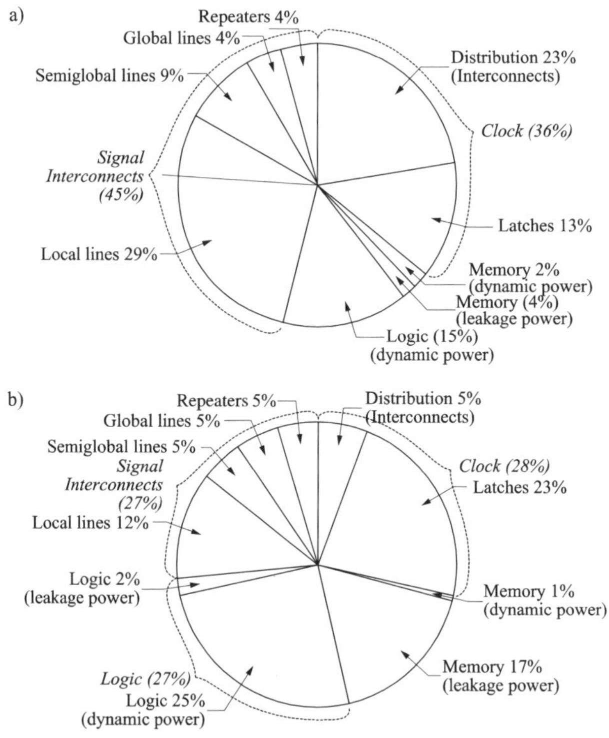

Figure 2.6. *2002 projection of the distribution of power consumption in an integrated circuit for a) 180 nm and b) 50 nm nodes [CHA 02]*

From this section, we deduce that interconnect performance is directly linked to the values of line resistance and interline capacitance. Line resistance and interline capacitance will need to be the lowest possible values. In terms of intrinsic physical material properties, low conductor resistivity and low polarizability or dielectric constant (denoted κ or κ in microelectronics) of the insulator are therefore necessary.

2.2.1.3. *Keeping the structure intact*

Whatever the material or device, if it is continuously subjected to a "stress" (thermal, electrical, chemical, mechanical, radiative, etc.), it will be weakened over time so as to finally no longer assume its initial function. Understanding, quantifying, minimizing and anticipating this behavior is the key to dealing with reliability. Producing reliable interconnects has always been a challenge. When discussing the reliability of an integrated circuit, we try to distinguish two principal conditions to which the circuit is exposed. There is, of course, an electrical stress: under the effect of an electrical field, the dielectric can lose its insulating properties. Under the effect of a current, with regard to electromigration behavior, the metal can be caused to diffuse, thus creating cavities (leading to line rupture) or extrusions (leading to short circuit). The second type of stress is mechanical. On the one hand, we can observe the creation of cavities in the metal conductors from the internal constraints, or "stress voiding". On the other hand, fissure propagation in the insulating material, or at an interface, has become a critical point in modern interconnects.

We must, however, note that the mechanical–electrical distinction is a crude one. Thus, we have quickly established that the global stress state of the system plays a part in the electromigration behavior of the metal [AIN 72]. There is also a strong relationship between metallic film adhesion and electromigration activation energy [LAN 03]. Similarly, a variation in stress state at an interface may modify the failure mode of the insulator [GUE 07]. In the following sections, we will focus on the electromigration and mechanical reliability of interconnects.

Electromigration behavior was first identified as a particular issue relating to Al interconnect reliability in the 1960s [BLE 66]. Some 40 years later, its control and command remain essential aspects of integrated circuit reliability. The main reason for this is the continual reduction in device sizes, accompanied by a significant increase in current densities required to operate circuits: during operation, at a temperature of 105°C, between 1966 and 2010, line current densities changed from 1×10^5 to some 1×10^6 A/cm^2. Figure 2.7 shows us the continual evolution in current densities required for current and future generations of integrated circuits.

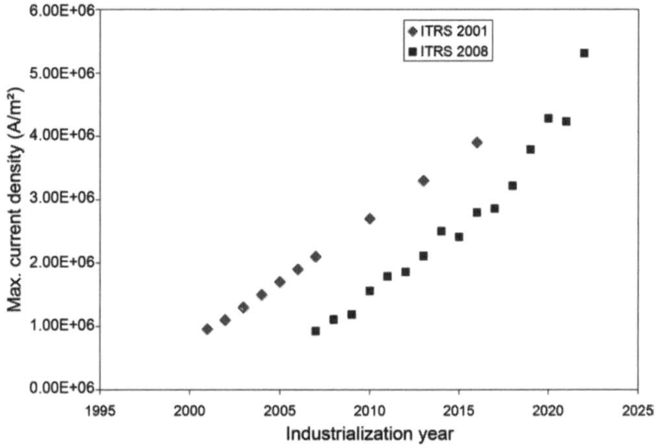

Figure 2.7. *Evolution of maximum current densities expected in interconnect lines, based on ITRS 2001 and 2008 values [ITR 00]*

The physical principle of metal electromigration relies on the displacement of ionized metal atoms, due to Coulomb interactions with the conduction electron flow. Using the Nernst–Einstein equation, this flow of matter J_{ion} can be expressed in the form:

$$J_{ion} = n\frac{D_{eff}}{kT}Z^{*}eE$$

where n is the atomic density, κ is the Boltzmann constant, T is the temperature, D_{eff} is the effective atomic diffusion coefficient, e is the absolute value of the electronic charge, E is the applied electrical field and Z^* is the apparent effective charge number. Z^* denotes the effective ion valence, which is a behavioral parameter indicating the significance of interactions between the charge carriers and the ions.

As we can see from the equation, the behavior is thermally activated through the ability of species to diffuse. Diffusion is a critical point.

Effective diffusion, D_{eff}, which was initially defined by Hu *et al.* [HU 95], is calculated by estimating the flows for each diffusion path. In an interconnect, there are three diffusion paths: the material itself, the interface between the line of copper and the encapsulation barrier and the grain boundaries. Table 2.1 shows the activation and self-diffusion energies in Cu and Al, based on diffusion path. It seems clear that sensitivity to electromigration is highly dependent on the diffusion path followed.

Diffusion path	Ea (eV) for pure aluminum	Ea (eV) for copper
Volume	1.4	2.3
Grain boundaries	0.5 (Al–Cu: 0.65– 0.85)	0.85, 0.92 and 1.2
Surface	0.28 (inhibited by Al_2O_3)	Highly variable
Interfaces	0.84 at the Al/Al_2O_3 interface	0.8
Dislocations	>0.62	1.53

Table 2.1. *Comparison of diffusion activation energy values for aluminum and copper, based on the diffusion path. Based on [BER 01] with an actualization of diffusion activation energy at the interfaces*

The production and use of an interconnect circuit inevitably leads to the generation of mechanical stresses, for example the deposit of a refractory metal such as W creates a GPa constraint [FLO 02, NIX 09]. The materials used in an interconnect have (widely) different thermal expansion coefficients. The change from a high deposit temperature to a lower storage or utilization temperature therefore has a corresponding deformation differential [ORA 04]. During electromigration, there is a voltage state where the matter disappears, whereas a compression state is present where it is accumulated [AIN 72, BLE 76a, BLE 76b]. During production, a certain number of stages are accompanied by a strong mechanical interaction between the mesoscopic and macroscopic worlds of the device: the layers can be removed by mechanochemical abrasion [HOM 03, YOO 04, LED 05], it is necessary to electrically connect the circuit to the outside world during the packaging stages [MER 03]. During operation, the circuit temperature varies cyclically by more than 50°, which then leads to the emergence of fatigue behaviors.

The continuous reduction in dimensions may once more be damaging in more than one respect:

1) From the structure form factor, the area to volume ratio (i.e. the interface "quantity") increases linearly as the critical dimensions reduce. We thus see the number of potential defect areas and mechanical singularity points increase.

2) For very small dimensions, in the case of encapsulated metals, the plastic relaxation of constraints in metallic lines is virtually null, due to a triaxial stress state [BAL 04], or, as observed for thin films [NIX 09, NIX 89], due to a limitation of relaxations by surface diffusion. As a result, stress and/or overall deformation levels may increase.

3) Finally, and we will return to this in the following chapters, maintaining electrical performance necessitates the use of materials that unfortunately have very poor mechanical properties.

The circuit is indeed subject to instant (such as a mechanochemical polishing step), cyclical (such as a variation in operating temperature) and continuous (such as stress voiding) mechanical stresses. These occur in an environment that may be chemically aggressive due to the gasses and solutions used during the production stages. From this, the physical integrity of the circuit is threatened if the structure is not sufficiently mechanically robust.

From these considerations, we can identify a number of requirements that need to be satisfied in order to obtain an integrated interconnect; on the one hand, the conductor must have a low sensitivity to electromigration. The metallic material, in its integrated form, should therefore limit as far as possible the matter diffusion paths. On the other hand, the structure is subjected to numerous mechanical stresses. It should therefore be capable of accommodating and/or limiting plastic and elastic deformations, while resisting fissuration and delamination.

2.2.2. *The main technological advancements*

2.2.2.1. *From polysilicon to aluminum*

Traditionally, the active elements of a circuit were connected by polycrystalline silicon lines, commonly known as "polysilicon", which were strongly doped in order to ensure electrical conduction. In order to reduce line resistance, polysilicon was then alloyed with a metal to form a silicide; a reduction in resistivity was thus obtained. At the end of the 1970s, the first aluminum (Al) or A1 alloy interconnect grids were finally introduced. Here, the Al or Al alloy would be in direct contact with the gate silicide or the transistor sources and drains. Conductors were insulated using silicon oxide. A particular advantage of aluminum alloys is their ability to limit the silicon oxide through formation of a layer of passivating alumina. If we accept that the complexification of integrated circuits, which is connected to the device reduction, is a driving force behind the semiconductor industry, this leads to new stresses on the interconnect grid; these stresses, which are still valid, are linked to:

1) a lack of space, within a two-dimensional area, for the signal to be spread between unitary devices or predefined cells;

2) an increase in conductor line resistance;

3) an increase in interline capacitance;

4) an increase in current density.

While the last three above-mentioned constraints are linked to circuit performance and reliability, the first, lack of space, has a direct impact on the degree of integration; on the one hand, it is no longer possible to use the whole surface of the Si by integrating the maximum number of devices; on the other hand, complex circuit design becomes problematic. In order to resolve this problem, the concept of multilevel metallization was proposed, developed and integrated. The first conference on this subject was held in 1984: the IEEE VLSI Multilevel Interconnect Conference, followed in 1998 by the IEEE International Interconnect Technology Conference (IITC).

By the middle of the 1980s, the number of metal levels was at that point limited to two and the critical dimensions were greater than 1 μm.

At the start of the 1990s, the use of three or four interconnect levels had become commonplace. The critical dimensions were then of the order of 0.5 μm. Aluminum alloys were the materials of choice and their barriers (titanium and tantalum) were used for metallic lines, tungsten for vias and silicon oxide (SiO_2) as insulators. Other layers with antireflective (for the photolithographic stage for circuit image production) properties, passivating properties, or which enabled etching control, were added. The production of interconnects thus necessitated a wide variety of processes: (1) metal depositing and etching, (2) dielectric depositing and etching, (3) surface preparation or cleaning before or after deposition and (4) annealing, enabling an improvement of the intrinsic material and stack properties (such as microstructure, resistivity, reduction of impurities and defects, and adhesion increase) or stabilization of the structure. Each family of processes continues to evolve and there is a wide choice available to technologists for construction of multilevel grids. Among the major advances, the introduction of so called "planarization" techniques has been an essential technological step in the production of multiple interconnect levels. Indeed, the simple superposition of film on a structure after etching leads to a significant topography, which may have two serious consequences:

1) Lithographic definition of structures will become difficult, or even impossible. This necessitates a virtually flat, long distance surface.

2) The creation of cavities in films due to non-perfect step covering, and from this, the failure of particular lines.

Several planarization techniques have been developed and used, for example:

1) Planarization through reflow of the deposited layer, often called "thermal reflow". The film temperature is raised sufficiently to enable atom movement and topography smoothing. This technique is used for Al as well as for SiO_2 (with P and B dopant often added in order to reduce the reflow temperature).

2) Planarization through etching after deposition, known as "dep-etch dep" and/or "etch back". The surplus material is etched. A tip here is to use a sacrificial material and to artfully adjust etching selectivity between the different layers.

3) Finally, chemical mechanical polishing (CMP), which is in its early stages of development. The plate surface is set into rotation and in contact with a polyurethane "pad". This operation takes place in a chemical solution, called the slurry", loaded with colloidal, often silicon-based (silicon oxide) particles. Choosing the best chemical solution and processing conditions enables a virtually perfectly flat surface to be obtained over not only a short (local roughness < 0.5 nm) but also a long distance (see Figure 2.8). Broadly speaking, we can say that there are two stages: a first stage during which the solution chemically modifies the surface in order to modify its mechanical properties and a second stage during which particles draw out the chemically modified section. This technique will be

increasingly used in the technologies which follow, ultimately becoming one of the main production center costs.

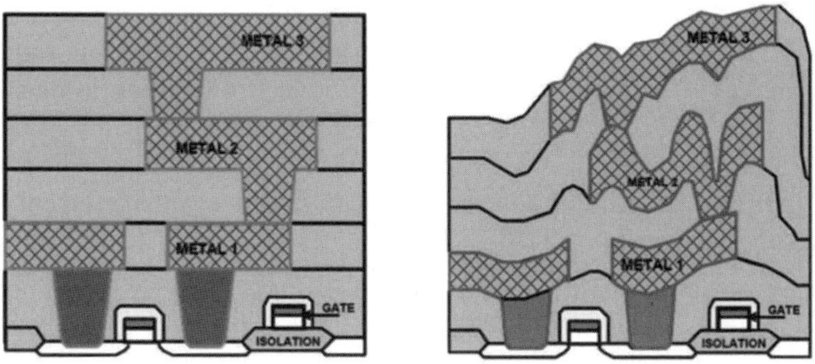

Figure 2.8. *Microelectronic structure with (left) and without (right) CMP*

The result of the past 30 years of research is a complex yet comprehensive multilevel Al-based interconnect technology. The technical and scientific reasons justifying (partly) the numerous technological choices available can be found in Wilson's work [WIL 93], or in various reference works from the same period [MUR 97], where an idea of the processes used in Figure 2.9 can be obtained. This describes a traditional production sequence for producing an aluminum metal level. The principal stages are as follows:

1) the interline dielectric (ILD) is deposited onto the preceding level (a line of aluminum). A deposit of SiO_2 by chemical vapor deposition (CVD), using plasma, is often used.

2) the dielectric is planarized by CMP, or "dep-etch-dep".

3) the via is defined using photolithography and then anisotropic plasma etching.

4) the excess photoresist is stripped out. A plasma and/or chemical cleaning solution process may be used.

5) a layer of TiN and of W are deposited. With regard to TiN, a reactive magnetron sputtering process is often used. With regard to W, in order to obtain good via refilling, a CVD process is preferred. TiN has a dual role as

a W adhesion layer and as a barrier to the diffusion of possible fluoride species resulting from the deposition of W.

6) the excess of W and TiN are removed, either by etching or by CMP.

7) a Ti/Al(Cu)/Ti/TiN stack is deposited using magnetron sputtering. Using an Al(Cu) alloy enables sensitivity to electromigration to be strongly reduced. The Ti and TiN films can be used as electrical paths in cases of cavity creation as a result of electromigration or stress voiding. The lower Ti layer improves the Al(Cu) adhesion and can create an interdiffusion barrier through the formation of $TiAl_3$. The upper Ti layer prevents Al nitridation during the deposition of TiN. The TiN acts as a barrier layer, an antireflective layer for lithography and a layer preventing etching of the via; the line is defined by photolithography, then by anisotropic plasma etching of the Ti/Al(Cu)/Ti/TiN stack, and we can then repeat this technological process for the next level.

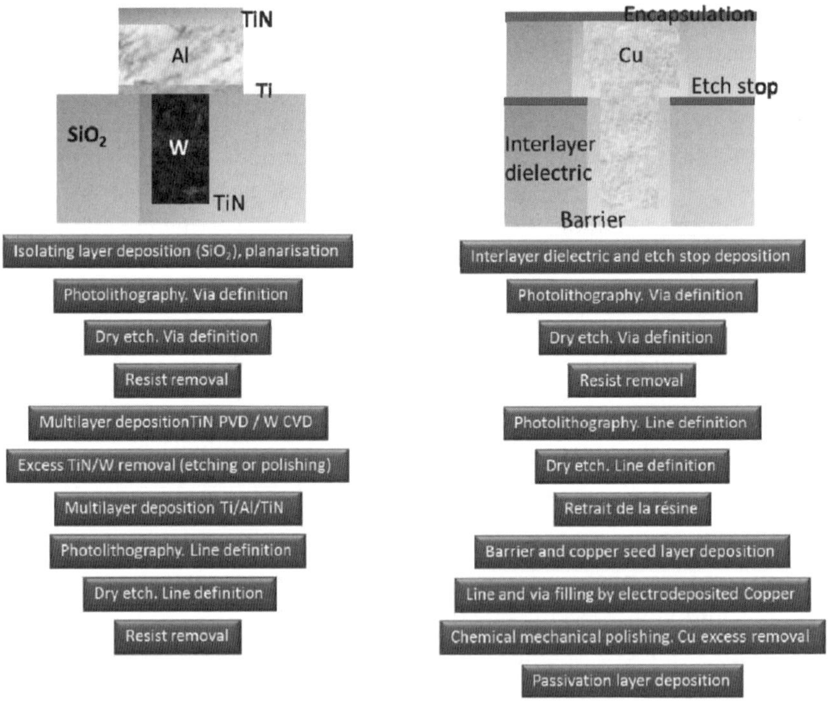

Figure 2.9. *Production sequence for an aluminum and copper metal level*

2.2.2.2. *From aluminum to copper*

During the 1990s, multilevel Al/Si interconnects enabled the design and integration of complex functions onto a microchip. They offered a high degree of freedom in circuit design and it was possible to integrate more transistors per unit of area and to connect them. The counterbalance to this accrued density of integration was a reduction in performance. Figure 2.10 is from a study undertaken in 1994 [JEN 94] and shows the rapid degradation of the RC time constant based on the reduction in dimensions. It seemed clear that a major new technological evolution in interconnect performance maintenance was necessary.

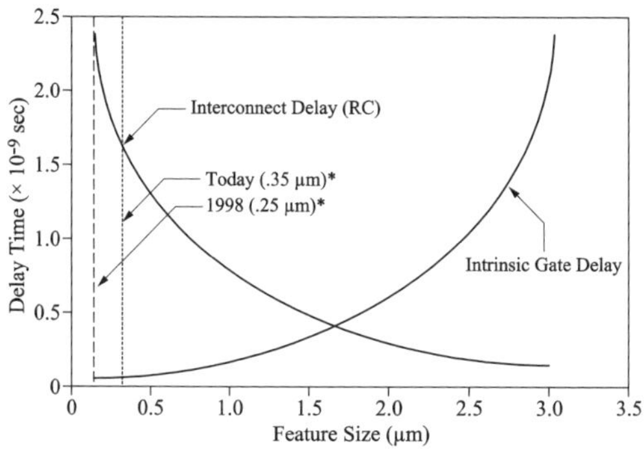

Figure 2.10. *Degradation of the RC time constant as a function of dimension reduction [JEN 94]*

Since the end of the 1980s, the Si community has been researching the technological solutions that would enable the Al/SiO_2 metallization performance deficits to be. Two views are naturally emphasized and studied in parallel: the choice of metal which is the least resistive and most resistant to electromigration, and the choice of insulator with the lowest dielectric constant.

Part of the effort is being directed toward the replacement of aluminum. Copper is being considered owing to its low resistivity (1.72 µΩ·cm compared with 2.73 µΩ·cm for solid aluminum) and its good resistance in solid form to electromigration. A large number of technical difficulties have made its integration a challenge in the following ways:

1) there are no halogen-containing Cu species that are sufficiently volatile to enable etching following simple processes at low temperature;

2) unlike aluminum, it does not create any passivating oxide and remains sensitive to corrosion.

Without natural passivation layer, it diffuses easily into SiO_2 and is a very harmful contaminant to transistor operation.

Despite these difficulties, after approximately 10 years of development, IBM presented its first integrated circuit using multilevel Cu metallization in 1997, during the IEDM conference [EDE 97]. This 0.22 µm node CMOS logic circuit comprised six metal levels. Edelstein wrote, "The Cu wiring has advantages of significantly lower resistance, higher allowed current density, and increased scalability, relative to comparable Ti/Al(Cu) wiring". This was verified by the results presented (see Figure 2.11): the line resistance was lower than that obtained using Al and the electromigration line performance is superior for Cu lines.

The first circuits were made available in 1998. Motorola [VEN 97], on the one hand, and the joint CEA-Leti and STMicroelectronics research team, on the other hand, presented similar results [MOR 00].

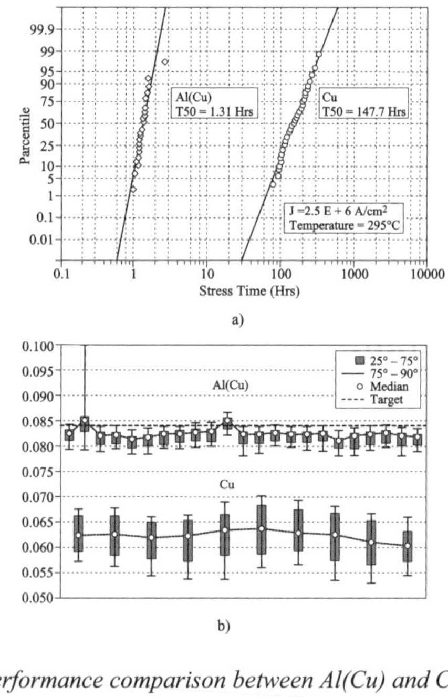

Figure 2.11. *Performance comparison between Al(Cu) and Cu metallization: a) Normalized resistance of lines of width 0.32 µm; b) Electromigration datafor 0.3 µm lines. The distribution of time to failure is reported*

Aside from the technological performance, what is remarkable about the IBM paper is the fact that the process and integration choices presented within it are essentially those still in effect today for integrated circuit manufacturers: the technology presented was based on a so-called dual-Damascene approach. The Cu was deposited through an electrochemical channel, as opposed to the physical vapor deposition (PVD) and CVD options followed by Motorola and the CEA-Leti/STMicroelectronics alliance, respectively. CMP was used on the copper.

The Damascene approach was inspired by the Damasquinage technique, which has been used for jewelry for more than 2,000 years. The design or line is etched into the surface, in our case the insulator. The metal thus fills the cavity and the surplus metal is removed by polishing. Let us now set out a simplified version of the Damascene technology, such as that illustrated in Figure 2.9. More details on the integration diagram will be given in Chapter 4:

1) We are starting from a metallization that is perfectly planarized and encapsulated by a layer that is insulating and a barrier to Cu diffusion.

2) The ILD is deposited onto the preceding level. An insulating layer that will be able to be used as an etching or polishing control may be added.

3) Through photolithography and etching, the via is designed and partially etched. The resin is removed.

4) A new lithographic and etching stage enables line definition and complete etching of the via. This etching occurs twice: first, when the ILD is removed, and then when the Cu line is opened by removal of the encapsulation layer.

5) A wet cleaning stage is often necessary in order to remove the remaining etching residues.

6) Within a vacuum environment, three stages are linked: first, a plasma process removes the residual copper oxide from the front of the via. Then, a conductive layer with copper diffusion barrier properties is deposited. A reactive magnetron sputtering technology is often used to form a TaN/Ta dual layer. The TaN is used to ensure the Cu diffusion barrier and the Ta is used to maintain good adhesion with the Cu. A thin Cu layer, called a "seed layer" is finally deposited. This is necessary for the next stage.

7) The final filling of the cut is performed by electrochemical deposition of Cu. The Cu "seed layer" acts as a cathode. Sensible choice of additives,

which we will return to shortly, promotes Cu increase at the front of the shape and thus the absence of cavities in vias and lines.

8) The wafer is then subjected to annealing at between 150°C and 400°C. Its principal aim is to obtain Cu grain increase in order to reduce conductor resistivity.

9) The surplus Cu is removed using CMP. There are generally three stages to polishing: removal of the "largest" pieces, landing on the barrier and, after changing the polishing solution, removal of the barrier. The wafer is then chemically treated in order to clean the residues and to limit corrosion (see section 2.5 for precise details).

10) Finally, after the plasma deoxidization treatment, an insulating and diffusion barrier layer is deposited on the Cu line.

2.2.2.3. *From SiO_2 to low dielectric constant films*

The Cu having been integrated, it is now necessary to reduce the dielectric constants of the insulators in order to maintain the interconnect performance, that is to reduce or maintain the propagation delay, crosstalk and dissipated power.

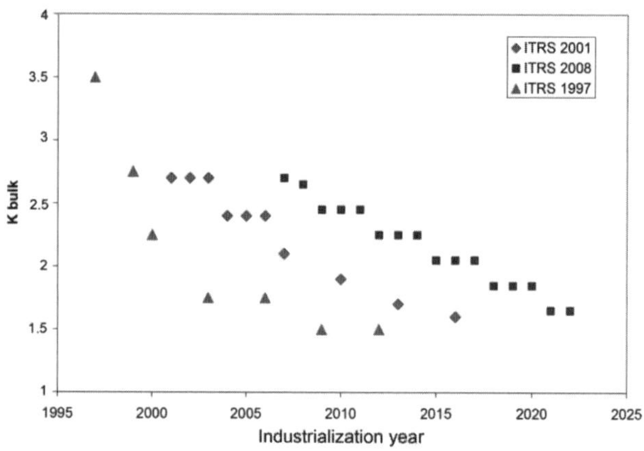

Figure 2.12. *Value of the interline insulator dielectric constant required for interconnect lines, based on ITRS 1997, 2001 and 2008 values*

Until the beginning of the 21st Century, the insulator of choice for the Si industry was Si oxide. It had the advantage of being chemically inert, thermally stable and very well known. It was a very good insulator with a low leakage current (1×10^{-11} A/cm² at 1 MV/cm), a strong breakdown field (6 MV/cm) and good mechanical properties (the Young's modulus being equal to 70 GPa). There was a great technical know-how around its deposition through plasma-assisted CVD, etching and polishing. However, its dielectric constant was around 4.2, which was relatively high. Predictions from various versions of the ITRS shown in Figure 2.12 provided us with an idea of the dielectric constant that was then estimated to be necessary and attainable. The solutions imagined and implemented are presented in section 2.3.

2.2.3. *Conclusion*

In order to address the performance and reliability needs of interconnects, a large number of technological solutions have been proposed over more than four decades. Today, the most advanced interconnects rely on a Damascene integration for which the conductor is Cu and the insulator is a low dielectric constant material.

We should also note that these technologies are nowadays reused and reactualized for external interconnect structure production, which enable the linking of multiple devices, such as the crossing vias within the so-called 3D integration for three-dimensional circuit stacking approach [LED 08].

The following sections will illustrate how the control and use of chemical behaviors and mechanisms is accompanying the development of interconnect technologies.

2.3. Dielectric deposits

2.3.1. *Dielectric generalities*

2.3.1.1. *Introduction*

As discussed in the first section, with the reduction in the critical dimensions of integrated circuits in the most advanced technologies, the choice of suitable candidates to act as interconnect dielectrics is a crucial stage, particularly in minimizing the delay, crosstalk and consumption between two metal lines. Since the advent of 130 nm generation technology, the dielectric properties of Si oxide have no longer been sufficient to satisfy interconnect insulator specifications (dielectric constant $\kappa = 4.2$ for SiO_2 deposited by plasma-enhanced CVD (PECVD)) and new low dielectric constant (known as "low κ") materials are nowadays used. As we will see in this section, this problem has been resolved by introducing chemical bonds that have low polarizability and by strongly reducing material density through the introduction of porosity. Various families of materials (in particular organic–inorganic hybrid materials) have thus been developed and PECVD has been confirmed as the industrial technique for interconnect dielectric deposition.

As we have previously mentioned, there are numerous stages involved in the manufacture of integrated circuits. Structures contain certain materials in order to deal with specific operations linked to integration. This is so for the dielectric barrier, which acts as both copper diffusion barrier and an etching and CMP stopping layer. Silicon nitride is traditionally used for this, but, again, its dielectric properties (dielectric constant $\kappa > 6$) no longer enable specifications to be satisfied for the more advanced technologies and new dielectric barriers with low dielectric constants are now used.

2.3.1.2. *Origins of the dielectric constant*

Before we discuss the various dielectric materials in interconnect technology, we need to better understand the physical origins of the dielectric constant. The dielectric constant (or relative permittivity ε_r) of a dielectric material is the measurement that describes the modification in the distribution of charges when a dielectric is placed into an electrical field. In a dielectric material, the charge carriers (electrons or ions) are principally linked charges. Under the effect of an applied electrical field, modifications in the distribution of these linked charges are produced, which are due to deformation or orientation of the various groupings (atoms, molecules or ions). With the barycenter (center of mass) of the positive and negative charges no longer being coincident, electric dipoles appear and the dielectric body is polarized. This mechanism leads to the concept of polarizability (expressed in $Å^3$ in the CGS system and m^3 in the MKSA system), which indicates the aptitude of a structure to being deformed under the influence of a field and to create a dipole.

In the case of a diluted medium, where the interaction between different polarizable species may be ignored, relative permittivity ε_r is expressed as:

$$\varepsilon_r = 1 + N\mu/\varepsilon_0, \qquad [2.1]$$

where N and μ are the density and polarizability, respectively, of the polarizable species. Permittivity of a material is thus directly linked to its polarizability and generally comprises three contributions: electronic (μ_e), ionic (μ_{ion}) and dipolar (μ_{dip}). Electronic polarizability is linked to deformation of the electronic cloud under the electrical field. For an isolated atom, the positive charge is concentrated into the quasi-punctual nucleus and the negative charge is distributed round the nucleus. The barycenters of the positive and negative charges coincide and the dipole moment is null. When an atom is subjected to an electrical field E, the field induces opposing forces on the nucleus and the electrons. The electronic cloud is deformed so that its barycenter no longer coincides with the nucleus (see Figure 2.13(a)) and the atom acquires a dipole moment.

Ionic polarizability is linked to ionic bond vibration under the effect of the electrical field. Under the effect of an external field, the relative positions of atoms can be modified and thus bring about a new contribution to the dipole moment (see Figure 2.13(b)). Dipolar or orientation polarizability is the result of molecules having a permanent electrical dipole moment, which may change orientation in an applied electrical field. Under the effect of an electrical field, each dipole is subjected to coupling that orients it in the direction of the field (see Figure 2.13(c)). This orientation process is permanently counterbalanced by thermal agitation, which, conversely, tends to randomly distribute the dipole directions.

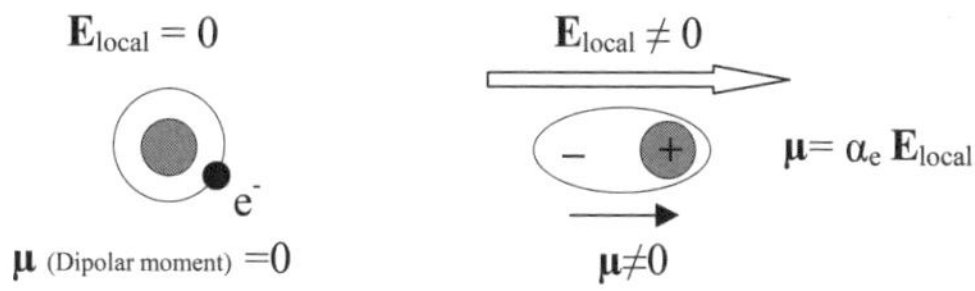

Figure 2.13a. *Electronic polarization (electronic cloud deformation under the electrical field)*

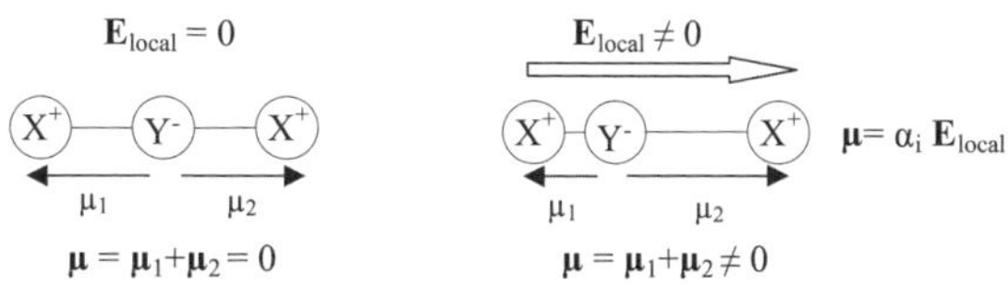

Figure 2.13b. *Ionic polarization (ionic bond elongation under the effect of the field)*

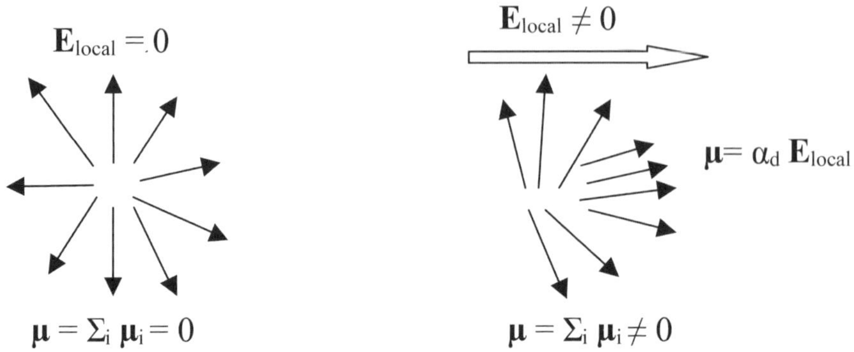

Figure 2.13c. *Dipolar polarization (alignment of dipoles in the direction of the field)*

In the low-density case of polarizable species, relative permittivity is thus the sum of three contributions equation [2.2]:

$$\kappa = \varepsilon_r = 1 + \varepsilon_e + \varepsilon_{ion} + \varepsilon_{dip} \qquad [2.2]$$

where ε_e, ε_{ion} and ε_{dip} are the electronic, ionic and dipolar contributions to material permittivity, respectively. The dielectric constant of a material is therefore linked to the types of atoms that comprise it and to the nature of their bonds. From a practical point of view, it is therefore necessary to find atoms and chemical bonds that will present low polarizability. Moreover, κ is now linked to the number of bonds per unit of volume (n in equation [2.2]). The lower the density of a material, the lower its dielectric constant. Table 2.2 shows that the C–C and C–F bonds present low electronic polarizability, which increases dramatically when dual or triple bonds are present, or with electronegativity (in particular for the O–H and Si–O bonds). With regard to dipolar polarizability, a compromise needs to be found between the volume occupied by the chemical groups that enable the density of the dielectric constant to be reduced (–F, $-CH_3$, for example), even if their dipole moment is significant. Materials that present hydrocarbon (fluoride or otherwise) groups are therefore in theory good potential "low κ" materials. O–H bonds should be avoided since their polar contribution is significant. Moreover, the dielectric constant of a hydrophilic material will strongly increase over time due to water absorption.

Chemical bond	Dipole moment (D)	Electronic polarizability (Å^3)	Bond energy (Kcal·mol^{-1})
C–C	0	0.531	83
C–F	1.41	0.555	116
C–O	0.74	0.584	84
C–H	1.46	0.652	99
C=O	2.3	1.02	176
C=C	0	1.64	146
C≡C	0	2.04	200
C≡N	3.5	2.24	213
Si–F	4.14		128
Si–O	3.1		88
Si–H	0.12		70
O–H	1.66	0.706	102

Table 2.2. *Dipole moment, polarizability and bond energy from various chemical bonds [ROU 05]*

Finally, the vacuum, in which the dielectric constant is at its lowest ($\kappa = 1$), now constitutes an excellent dielectric. The development of porous materials (material skeleton containing pores) therefore represents a possible way to reduce the dielectric constant of a material.

2.3.1.3. *Deposition techniques*

In microelectronics, PECVD is the technique of choice for deposition of interconnect dielectrics. It enables thin layers, a few dozen nanometer to several micrometers thick, to be deposited with a high degree of uniformity. There are various types of PECVD and in the case of dielectric deposition for interconnects integrated according to a Damascene process (where the dielectric is deposited onto a flat substrate), the plasma is usually generated by a capacitively coupled radio frequency [COT 99]. This plasma technique is indeed simple to implement and has a lower cost per wafer compared with other PECVD deposition techniques. The reactor thus comprises two electrodes (see Figure 2.14).

The upper electrode, which is linked to a radiofrequency generator (generally at 13.56 Mhz), also acts as a gas supply showerhead (for precursors, carrier and cleaning gasses). The lower electrode is linked to the mass and also acts as a substrate carrier. An impedance matching box is inserted between the generator and the upper electrode in order to adjust the power transmitted according to the variation in impedance during deposition. The vacuum in the chamber is ensured using a primary pump (up to around a dozen torrs).

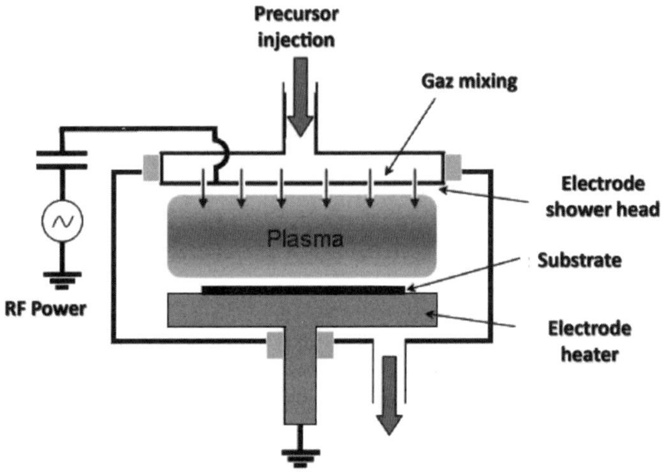

Figure 2.14. *Diagram of a capacitively coupled PECVD deposition reactor*

Compared with the traditional CVD, PECVD enables us to benefit from the energy contribution linked to the ionic and electronic bombardment and certain metastable species, as well as the electromagnetic radiation in the plasma. This enables us to bypass the activation energy from the chemical reactions and promotes movement of surface-absorbed species. For this reason, deposition can be at a lower temperature than in purely thermal CVD deposition. The precursor molecules are notably disassociated into radicals and ions that make the medium more reactive. Part of the energy is also used to form the surface film by heating the substrate. Numerous reactions may occur simultaneously, as shown in Figure 2.15.

Plasma being a complex medium outside thermodynamic equilibrium, numerous intermediary chemical reactions are possible in the gas phase (see Figure 2.15) and it is generally very difficult to predict exactly the chemical reactions leading to the deposition of thin layers (all the more so if one or more of the precursors used is complex). A wide range of materials can thus be obtained by varying the plasma power, pressure, flow and ratio of precursors, interelectrode spacing and the substrate temperature.

R1: Dissociation of molecules into radicals and ions.

R2: Diffusion of species toward the surface.

R3: Adsorption onto the surface.

R4: Migration onto the surface.

R5: Chemical sorption and film formation.

R6: Desorption of volatile species.

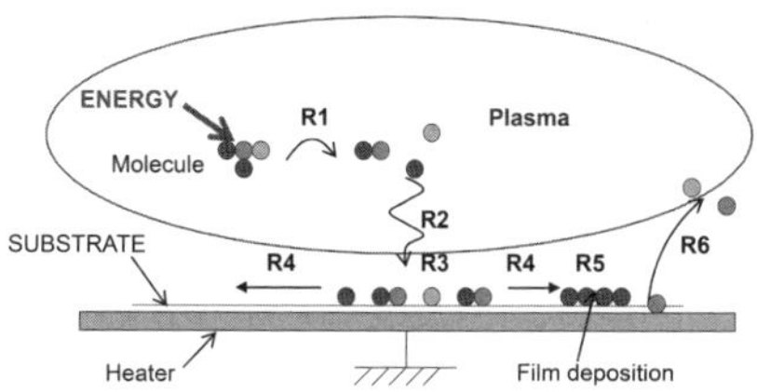

Primary reactions generated by electronic impact:	
ionization:	$e^- + A \rightarrow A^+ + 2e^-$
dissociative ionization:	$e^- + AB \rightarrow A^+ + B + 2e^-$
dissociative attachment:	$e^- + AB \rightarrow A^+ + B^- + 2e^-$
radiative attachment:	$e^- + A \rightarrow A^- + h\nu$
excitation:	$e^- + AB \rightarrow AB^*$
Excited species may later give rise to disexcitation reactions, with or without photon emissions: emissions:	
dissociation:	$AB^* \rightarrow AB + h\nu$
emission:	$AB^* \rightarrow A^* + B^*$
dissociative emission:	$AB^* \rightarrow A + B^* \rightarrow A + B + h\nu$
Secondary reactions involving electrons:	
Electron-ion recombination:	
dissociative:	$e^- + AB^+ \rightarrow A^* + B$
radiative:	$e^- + A^+ \rightarrow A^* + h\nu$
3-body:	$e^- + A^+ + B \rightarrow A + B$
Dielectric:	$e^- + A^+ \rightarrow A^{**} \rightarrow A^* + h\nu$
Excited molecule-electron reactions:	
dissociative attachment:	$AB^* + e^-_{lent} \rightarrow A^- + B$
disexcitation:	$A^* + e^- \rightarrow A + e$
Secondary reactions between two heavy particles:	
Neutral molecule-ion reactions:	
recombination:	$A^+ + BC \rightarrow AB^+ + C$
cluster formation:	$A^+ + B + M \rightarrow AB^+ + M$
charge transfer:	$A^+ + B \rightarrow A + B^+$
collisional detachment:	$AB^- + C \rightarrow AB + C + e^-$
associative detachment:	$A^- + B \rightarrow AB + e^-$
Neutral-excited molecule reactions:	
Penning ionization:	$A + B^* \rightarrow A^+ + B + e^-$
Penning dissociation:	$AB + C^* \rightarrow A + B + C$
Reactions between two ions:	
neutralization:	$A^+ + B^- \rightarrow A^* + B^*$
3-body:	$B^- + A^+ + M \rightarrow AB + M$
radiative neutralization:	$A^+ + B^- \rightarrow AB + h\nu$

Figure 2.15. *Various reactions which may be involved in a PECVD deposition [CASTEX]*

2.3.2. *Interline dielectrics*

2.3.2.1. *The traditional material: SiO_2*

Silica (SiO_2) is used as an interconnect dielectric in the case where signal propagation in an integrated circuit is not limited by interconnects (typically before 130 nm technology or in more recent higher level technologies). For this application, SiO_2 is generally deposited by PECVD from a silicon-containing precursor (either from the silane SiH_4 or from the tetraethylorthosilicate (TEOS, Si–$(OC_2H_5)_4$) and from an oxidant (generally O_2 or N_2O). With regard to silane-based deposition, the principal reaction is:

$$SiH_4\,(g) + 2O_2\,(g) \rightarrow SiO_2\,(s) + 2H_2O\,(g) \qquad \text{[Reaction 2.1]}$$

Numerous other reactions are possible and some active intermediary species are also formed in the gas phase (such as SiH_2, SiH_3 and SiH_3OH) [NGU 88, SMI 93]. It is thus necessary to make an appropriate choice of deposition conditions in order to avoid the formation of powder and particles and encourage deposition onto the substrate. As for the TEOS fragmentation, this occurs principally through removal of ethyl radicals [OKI 98]. Here again, numerous intermediary reactions are possible, depending on the deposition conditions [LEE 96, OKI 98]. Deposition of the TEOS-based SiO_2 is generally preferred in the case of conformal deposition or for applications requiring the filling of etched structures ("gap-fill" application, for example).

For optimal deposition conditions, the silicon oxides deposited by PECVD possess a low leakage current density (of the order of 10^{-11} A·cm^{-2} at 1 MV/cm) and a high breakdown field (greater than 6 MV/cm). Their dielectric constant is approaching $\kappa = 4.2$ (the dielectric constant of a thermal SiO_2 is 3.9). This difference is explained by the slightly lower density of the deposited oxides and by the presence of hydrogen in the PECVD oxide layers (1–2%).

2.3.2.2. *The first low* κ

Amorphous silicon fluoride oxide SiO_xF_y (denoted SiOF or fluorosilicate glass (FSG)) was the first low κ generation to replace SiO_2. Its use is found in 130 nm technologies; thin layers of FSG are prepared by adding SiF_4 to silane or TEOS, or even by adding a fluorocarbon precursor such as C_2F_6 to a TEOS-based oxide deposition process [SHA 95]. The addition of Si–F bonds that are less polarizable than Si–O bonds enables the value of the dielectric constant to be reduced. Indeed, the fluorine atoms exhibit strong electronegativity (3.98 on the Pauling scale) that enables electrons to be retained around the nucleus. The dielectric constant can thus be reduced to 3. However, for strong fluorine concentrations, the Si–F bond is no longer thermally stable, resulting in problems during the integration of these materials. For example, copper line corrosion or HF formation during wet (or chemical) etching or photoresist removal may then be observed [BAU 97]. FSG is therefore present in various interconnect technologies, but with generally low fluorine rate (<5%) and a dielectric constant of around 3.5.

For the following technology (90 nm), in an industrial context, FSG has been replaced by another modified silicon oxide: $SiO_xC_yH_w$ (denoted SiOCH and also known as organosilica glass (OSG)). This amorphous material is generally deposited by PECVD from an organosilicon-based precursor and O_2. The most widely used precursors are: trimethylsilane (3MS), tetramethylsilane (4MS), octamethylcyclotetrasiloxane (OMCTS) and dimethyldimethoxysilane (DMDMOS). The structures of these materials are made up of –Si–O–Si– bonds, in which a section of oxygen atoms is replaced by methyl groupings (see Figure 2.16). Incorporating the $–CH_3$ group enables the dielectric constant to be reduced as a result of the reduction in polarizability and the "free volume" generated around the methyls. Through this addition of $–CH_3$, the density changes from 2.2 for SiO_2 to values that are much lower, of the order of 1.2–1.4, for SiOCH. The dielectric constant value is dependent on the quantity of the incorporated $–CH_3$ groupings [LOB 94] and is close to 2.9 for industrial integrations of SiOCH.

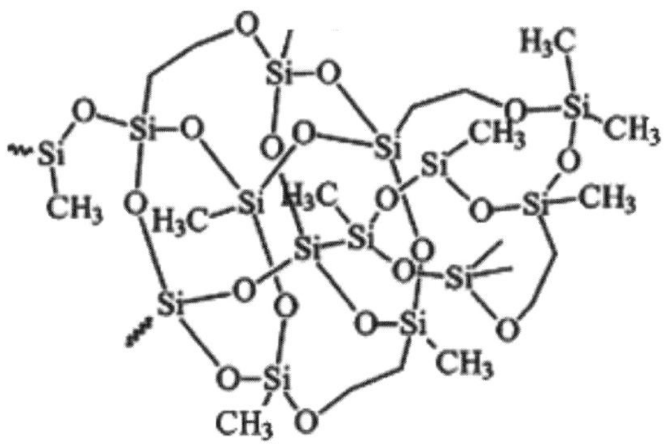

Figure 2.16. *Schematic representation of SiOCH [GRI 03]*

It even seems possible to reduce the value of the dielectric constant to 2.6 for SiOCH containing large quantities of methyl groupings. In order to obtain materials with so much $–CH_3$, one possible solution is to use siloxane precursors (i.e. that already contain Si–O bonds and do not require excessive amounts of an oxidizing gas to be added) [ROU 05]. However, one of the problems of SiOCH containing enormous quantities of $–CH_3$ is its lack of thermal stability [O'N 03]. Studies carried out with cyclical precursors (octamethylcyclopentasiloxane or decamethylcyclopentasiloxane) with the aim of reducing density still further, in particular through material cycle preservation, show that this approach is limited in reducing κ further [CAS 05].

Numerous other materials present dielectric properties that are compatible with interconnect specifications [MAE 03, VOL 10]. For example, numerous organic polymers have a low dielectric constant ($\kappa < 3.5$). They may be deposited by a spin coating provided that they are soluble in a solvent. Many review papers have been written that present those that are the best thermal stabilizers and that could be suitable candidates for integrated circuit interconnects [MOR 00, MAI 04]. These include aromatic or heteroaromatic (such as polyimides, polybenzoxazoles, polyquinolines and polyarylethers) and hydrocarbon (polynorbornenes, polyphenylenes, polyperfluorocyclobutanes, polyindanes and perfluorinated aliphatic) polymers. One of the polyphenylenes, available commercially from Dow Chemical as SiLK^{TM} (see Figure 2.17), has been integrated into 130 nm technology by IBM [TOW 97]. It is thermally stable up to 450°C and has a low dielectric constant ($\kappa = 2.7$).

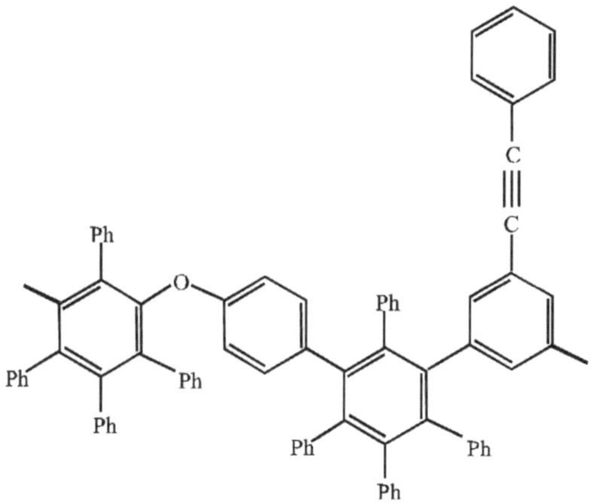

Figure 2.17. *SilK^{TM} (polyphenylene developed by Dow Chemical)*

Polysilsesquioxanes, which are organic–inorganic polymers with a formula $(R–SiO_{3/2})_n$, have also been studied as a low κ material. Within the R group, the two most tested configurations in microelectronics are poly-hydrogen-silsesquioxane (p-HSQ), in which R=H, and poly-methyl-silsesquioxane (p-MSQ), in which R=CH_3. p-MSQ has a dielectric constant of around 2.9, whereas that of p-HSQ varies between 3.0 and 3.2 [MAE 03].

These materials are also deposited by spin coating from a solution containing an organosilicon-based precursor in a solvent. These spin coating (organic polymers or sol-gel) approaches have not, however, been retained, and today the most used low κ material remains SiOCH deposited by PECVD.

2.3.2.3. *Introduction to porosity*

In order to further reduce the value of permittivity and obtain materials with very low dielectric constants (referred to in microelectronics as ultra low κ (ULK)), the principal method studied was to use an already low permittivity dielectric and introduce porosity to it (porosity being defined as the volumic fraction of pores over the total volume). Most studies have been undertaken on SiOCH-based skeletons. The first channels explored with a view to introducing nanometric pores into a thin dielectric layer have their origins in chemistry, with materials being deposited by spin coating from MSQ- or polymer-based skeletons [HED 98, MOR 00]. The background and the different approaches used are detailed in [VOL 10].

In 2001, Grill was the first to publish results indicating the possibility that porous SiOCH could be produced by PECVD [GRI 01]. The approach developed was called the porogen approach (also known as the substractive process or template method) and is a two-step technique. The first step consists of depositing a mixture of a precursor that will constitute the host matrix and a sacrificial organic molecule (called a porogen since it has its origin in the generation of pores). The thin layer thus obtained is sometimes called a “hybrid”. The material is then subjected to a thermal treatment that enables the sacrificial phase to be removed (see Figure 2.18). This heating step is generally accompanied by a film shrinkage linked to skeleton cross-linking or modification. Degradation of the porogens leads, under certain conditions, to the formation of porous thin layer with small pores (typically between 1 and 2 nm in diameter).

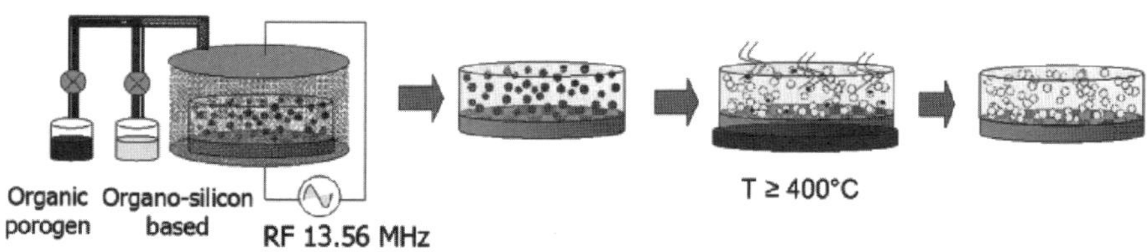

Figure 2.18. *Principle of the PECVD porogen approach*

Various organosilicon precursors have been studied in order to produce the matrix and various organic precursors to produce the porogen phase. Some molecule examples are shown in Table 2.3. From a practical point of view, one difficulty is the fact that most of these precursors are liquid at ambient temperatures, which necessitates the use of specific injectors and limits the choice of precursor chemicals possessing vaporization viscosities and temperatures that are compatible with this type of injection.

Di-ethoxy-methyl-silane (DEMS, also known as methyl-di-ethoxy-silane) is one of the reference precursors for this approach. Indeed, since organosilanes such as 3MS or 4MS have no Si–O bonds, a strong addition of oxygen is required for their use during the deposition step in order to form these bonds [ROU 06]. And yet, it is very difficult to incorporate organic porogens in a highly oxidant atmosphere. Use of 3MS or 4MS is not compatible with preserving porogen in the hybrid material. Therefore, we rather use molecules that already exhibit Si–O bonds. Moreover, Grill showed that it was harder to incorporate porogens from a chemical using a cyclical precursor, such as tetramethyl-cyclotetrasiloxane (TMCTS), whereas this incorporation was preferred for a precursor such as DEMS [GRI 08b].

The success of the process is also linked to the presence of a sufficiently mechanically rigid matrix. A more precise study of these materials shows that a SiOCH matrix is sufficiently mechanically rigid if it contains enough $Si–(O_{1/2})_4$ groupings. Below a certain percolation threshold, mechanical properties are too low to avoid a breakdown of the grid during the porogen removal step [BUR 04b]. Thus, a precursor leading to the creation of too large a quantity of $Si–(CH_3)_2$ bonds, to the detriment of $Si–(CH_3)_1$, does not enable porosity to be created due to the skeleton collapse during the post-treatment breakdown of the skeleton. Favennec showed that adding a low quantity of oxygen to the plasma enables the presence of $Si-(CH_3)_1$ bonds to be promoted, rendering very marginal the formation of $Si-(CH_3)_2$ bonds, which then enables the formation of porous materials [FAV 07].

The structure of porogen molecules generally presents chemical functions that are assumed to be reactive (epoxy based, double bond C=C, carbon cycle). These characteristics should enable incorporation during PECVD deposition and explain choices such as α-terpinene [VRT 03], cyclohexene oxide [FAV 05] or norbornadiene [GOU 07]. Finally, porogen, or rather the fragments that will be found in the film, should be thermally degradable at as low a temperature as possible.

Matrix precursors	Diethoxy-methyl-silane (DEMS)	Tetramethyl-cyclotetrasiloxane (TMCTS)
	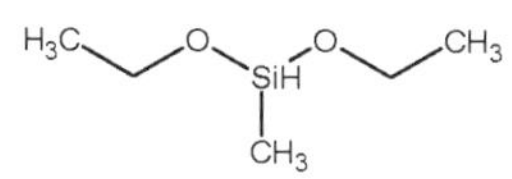	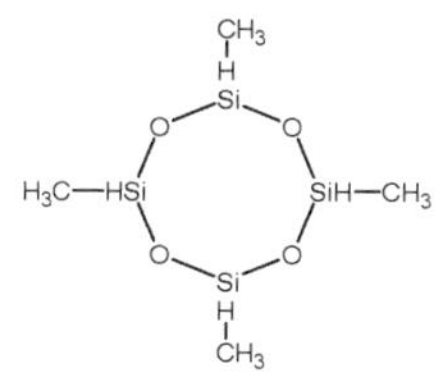
	Dimethyl-dioxiranyl-silane	
	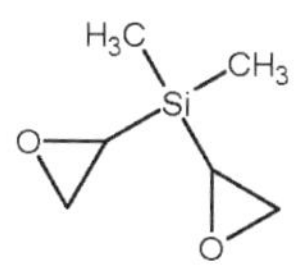	
Porogenic precursors	Norbornadiene (NBD)	α-terpinene (ATRP)
	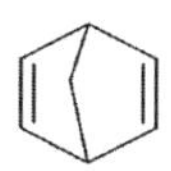	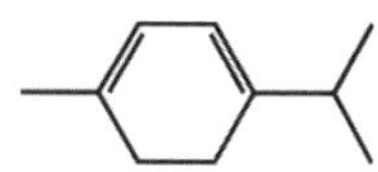
	Cyclopentene oxide (CPO)	Cyclohexene oxide (CHO)
	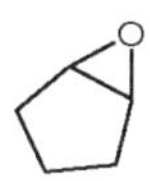	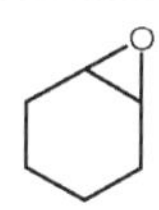

Table 2.3. *Examples of precursors used to carry out the PECVD porogen approach*

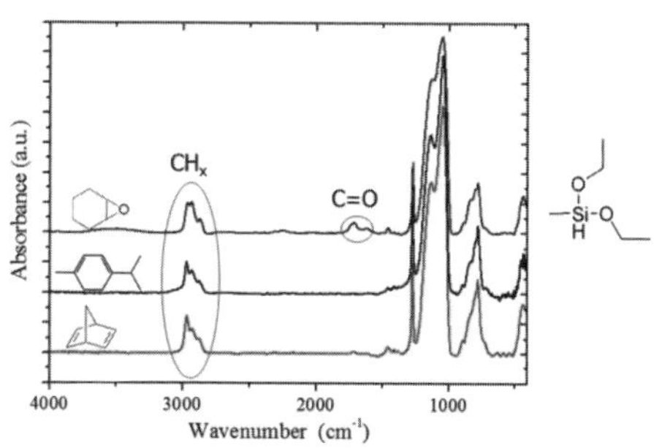

Figure 2.19. *IR spectrum of various SiOCH + porogen hybrids*

According to published studies, it thus seems possible to produce hybrids using PECVD with a large number of precursor couplings. Figure 2.19 shows the infrared spectra of various hybrids (DEMS + porogens): the porogen signature is generally characterized by the presence of C_xH_y bonds. In Figure 2.19, we observe the presence of C=O bonds, characteristic of the porogen used (cyclohexene oxide).

Porogen incorporation may be optimized by adjusting the deposition conditions. It is promoted at lower temperature and low power (pressure having a lesser impact on the incorporation but enabling variation in skeleton cross-linking). The quantity of porogens in the layers is directly linked to the porogen precursor/matrix precursor ratio [GOU 07]. Conversely, the way in which the porogens are incorporated into the SiOCH skeleton is still poorly understood, and porogen molecule size only seems to have a minor impact on the size of pores created [O'N 06]. The diagrammatic representation of the often proposed hybrid material shows a SiOCH skeleton with organic inclusions comprising multiple fragments from the porogen precursor used [O'N 06].

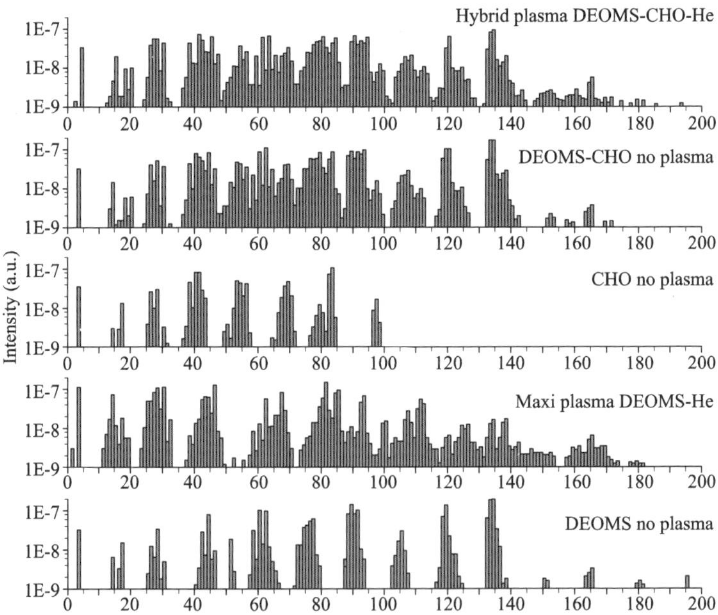

Figure 2.20. *Mass spectra obtained during DEMS- and CHO-based deposition by PECVD and examples of fragments likely to participate in deposition [CAS 08]*

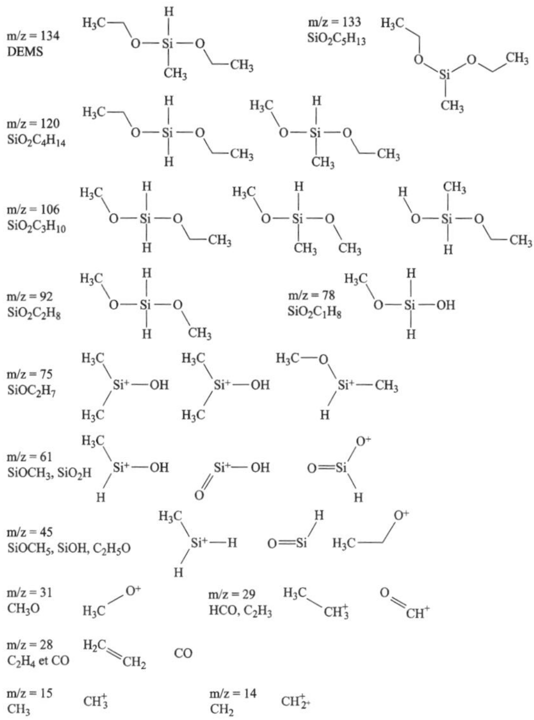

Figure 2.20. *(Continued) Mass spectra obtained during DEMS- and CHO-based deposition by PECVD and examples of fragments likely to participate in deposition [CAS 08]*

Few studies have been undertaken with a view to better understand this porogen segregation behavior in the "hybrid" thin layer. Indeed, it remains very difficult to study the chemical reactions occurring within the deposition of these materials [CAS 08]. Figure 2.20 illustrates this complexity for a deposition of DEMS and cyclo-hexene-oxide. The mass spectra are very complex and numerous fragments are observed at the deposition reactor output. Fragmentation of the DEMS proceeds for a significant part through the departure of methyl (Si–CH_3 rupture), ethyl (O–C_2H_5 rupture) and ethoxy (Si–OC_2H_5 rupture) groups, and produces numerous organic radicals [CAS 08]. These observations are in good agreement with the structures of the thin films obtained.

The second step of the porogen approach consists of removing the sacrificial organic phase that can be achieved using a thermal treatment (typically at 450°C for several hours). This treatment induces thermal degradation of the organic inclusions, leaving room for the free spaces (pores). This is generally accompanied by a reduction in the thickness of the thin layer. Numerous studies have been conducted in order to understand the modifications occurring on the skeleton. The principal chemical reaction 2.2 that occurs is a silanol bond condensation reaction:

–Si–OH + –Si–OH → –Si–O–Si– [Reaction 2.2]

This reaction is more or less preponderant according to the “hybrid” material deposition conditions, since it requires the presence of closely bonded Si–OH. We observe, for example, that materials deposited at 300°C present a very limited (< 5%) reduction of thickness, which is linked to the low quantity of silanols in the layers, whereas those deposited at lower temperatures reveal much higher levels of thickness reduction [GOU 10]. In order to optimize this step and reduce its temperature and duration, various alternative treatments have also been examined. The most advanced of these are UV- or electron beam-assisted thermal treatments [ITO 04, NAK 06, JOU 07]. These involve promoting porogen degradation in the assistant by energy radiation. Treatment times are thus reduced to several minutes (for thin layers of thickness < 600 nm) at a temperature of 400°C. Unlike thermal treatments, UV- or electron beam-assisted treatments are not totally selective and have also an impact on the carbon of the skeleton. We observe, for example, a reduction of the Si–CH_3 bonds in the material, which is worsened with treatment time. The principal chemical reactions that occur at this step [ITO 04, IAC 06, ZEN 07] are described by reactions 2.3–2.7:

– an Si–C bond “rupture” reaction that may lead to the formation of Si–H bonds:

–Si–CH_3 + eV → –Si* + CH_2(H) (g) [Reaction 2.3]

–Si–CH_3 + eV → –Si–H [Reaction 2.4]

– a cross-linking reaction 2.5 through Si–O–ethyl bonds:

–Si–O–CH_2–CH_3 + Si–CH_3 → Si–O* + –Si* + CH_3–CH_2–CH_3 (g) →
Si–O–Si + C_xH_y (g) [Reaction 2.5]

– reaction leading to the formation of –Si–C–Si bridges (presence of these bonds having not yet been clearly demonstrated):

–Si–CH_3 + H_3C–Si– → –Si–$(CH_2)_2$–Si– + H_2 (g) [Reaction 2.6]

$$-Si-CH_3 + H_3C-Si- \rightarrow -Si-CH_2-Si- + CH_4\,(g) \qquad \text{[Reaction 2.7]}$$

where X* represents an excited state of the corresponding X molecule.

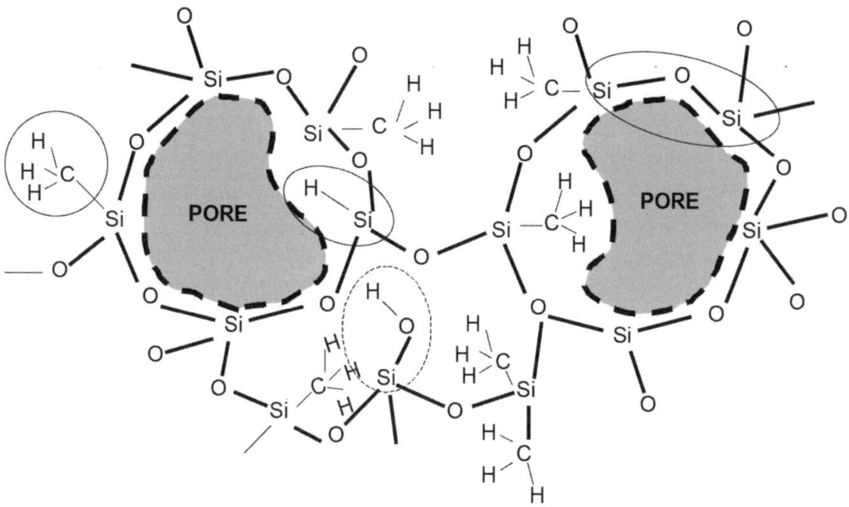

Figure 2.21. *Diagram of the film structure after removal of porogens by UV-assisted thermal treatment*

These assisted post-treatments also cause a more significant shrinkage in the skeleton than that observed with thermal treatment alone. Replacing Si–CH_3 by Si–H groupings encourages, for example, skeleton densification [ITO 04, ZEN 07]. The porosity introduced is thus lower, but the layer's mechanical properties are higher. In an industrial context, UV-assisted thermal heating (in 200–480 nm range) is preferred nowadays.

2.3.2.4. *Where are we going with the limitations encountered?*

The PECVD porogen approach is currently used in producing more advanced technological nodes (65 nm and below). According to the International Roadmap for Semiconductors, materials with low dielectric constants will probably be necessary in order to pursue miniaturization of integrated circuits [ITR 12]. The PECVD porogen approach enables the porosity, and thus the dielectric constant, to be varied over a wide range (from $\kappa = 3$ to $\kappa = 2$, see Figure 2.22).

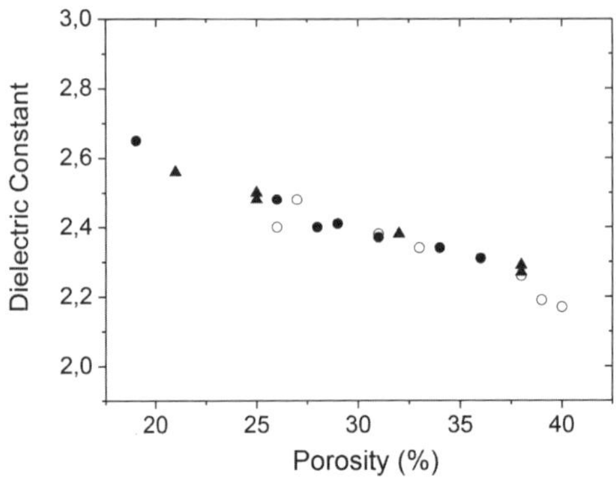

Figure 2.22. *Variations in the dielectric constant based on the porosity rate, for porous SiOCH layers [JOU 12]*

However, Favennec has shown that there is a porogen ratio threshold value in the hybrid material beyond that shrinkage of the dielectric constant is no longer possible [FAV 07]. This limitation is illustrated in Figure 2.23, which shows variations in porosity based on the rate of porogen injected into the reactor. For low ratios of porogen, material porosity increases significantly with the porogen ratio, but, past a limiting ratio, the increase in porogen does not result in an increase in porosity. At the same time, we note a strong increase in layer thickness reduction during the porogen removal stage. Above a "limit" porogen ratio, the matrix then has a tendency to collapse during the porogen removal stage. This behavior today limits the maximum amount of porosity that can be introduced into a porous SiOCH deposited using PECVD to a porosity ratio of close to 50%.

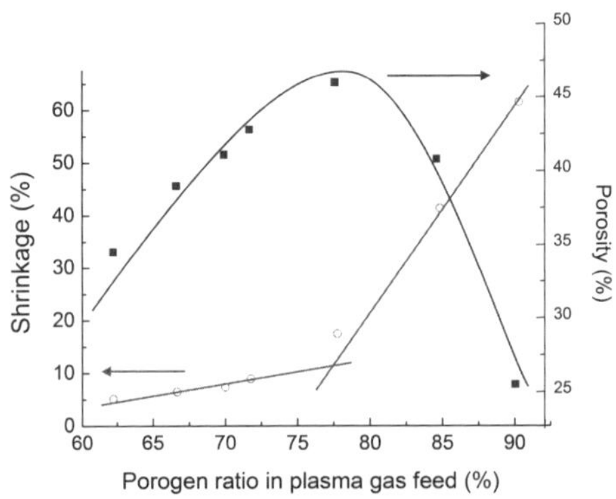

Figure 2.23. *Impact of porogen addition (in this example, NBD) on porosity creation and film compression*

Another limitation of porous thin layers is their poor elastic properties, which worsen when porosity increases. This is a real obstacle to their integration, notably during the CMP and packaging stages. In order to increase the elastic properties of SiOCH, there is a tendency to introduce Si–$(CH_2)_x$–Si bridges into the materials. The interest and feasibility of this approach has been demonstrated using solution approaches [DUB 05], but the transfer to PECVD remains difficult [GAT 09]. A final consideration concerns the presence of porogen residues in the layers. These are generally carbon residues containing a significant number of double bonds, which have a negative impact on dielectric properties [BAK 11, JOU 12]. Some complex post-treatment combinations (for example, a H_2–He plasma followed by UV treatment [URB 10]) are being studied with a view to limiting the presence of these residues.

In conclusion, the introduction of porosity into SiOCH has had a major impact on the domain of dielectrics in microelectronics. However, their use for interconnects has been accompanied by numerous integration problems linked to the presence of porosity and to their mediocre elastic properties. A number of these concerns are far from being resolved, which has, slowly, led industrialists to limit the porosity ratio of materials introduced into production (in 2012, integrated materials have a porosity ratio of approximately 20%). These porous materials are generally used in more complex stacks (associated with layers that are more dense) in order to enable their integration [GAL 08, GRI 08a]. Today, it is difficult to predict what will be used as the insulating dielectric for future generations of interconnects. Thus, alongside the porous SiOCH deposited using PECVD, other alternative approaches (such as polymers, zeolite and polycarbosilanes) continue to be tested [BAK 12].

2.3.3. *Barrier dielectrics*

2.3.3.1. *Brief history*

ILDs are not the only insulating materials to be found in interconnects of integrated circuits. Other dielectrics are used to satisfy specific integration functions:

1) etch stop layer function;

2) hard mask and CMP stop layer function;

3) in terms of integration, material must also be a barrier to diffusion of copper, where this metal is used as a conductor. The expression "barrier dielectric" is used to describe this.

The presence of these materials influences the global electrical properties of an interconnect level, and they must therefore have a low dielectric constant, while ensuring their function in integration. Silicon nitride deposited by PECVD is traditionally used as an etch stop layer. It is generally deposited from a silane, ammonia and nitrous mixture at a temperature of approximately 400°C [COL 93]. This material is stable, a barrier to copper diffusion, etch selective with regard to silicon oxides (SiO_2 or SiOCH) and compatible with the various stages of integration. However, its dielectric constant of between 6 and 8 is far too high, which strongly impairs the effective dielectric constant of an interconnect level. It has therefore had to be replaced in the new generations of integrated circuits.

2.3.3.2. *SiCH and SiCNH*

The main candidates are from the family of amorphous silicon carbons (a-SiCxHy or more simply SiCH). Intrinsic SiCH materials can be distinguished from those that are "doped", such as SiCNH (in reality, $SiC_xN_yH_z$). Initially, these materials were deposited from a mixture of silane and an organic precursor, but in microelectronics, organosilicon-based precursors have been favored for SiCH deposition due to their reactivity being lower than that of silane, thus enabling better control of the stoichiometry of the film deposited [LOB 94]. 3MS and 4MS are the most used precursors for the deposition of these materials (see Table 2.4).

During deposition of these two precursors, dilution is generally assured using helium or argon. As it is for SiOCH, it is possible to vary the dielectric constant of SiCH according to the deposition conditions and the precursor structures used [LOB 94]. Figure 2.24 illustrates an example of SiCH (κ = 4.8) structure [KIN 11]. The materials thus deposited have copper diffusion barrier properties similar to those of materials deposited using silane [LEE 01, CHI 03a]. Their etch rates are much lower than those of the ILDs derived from SiO_2, which makes them suitable as etch stop layers [POS 03]. This type of precursor enables the deposition of SiCH with a relatively low dielectric constant, depending on the deposition conditions (approximately 4–4.5).

Trimethylsilane (3MS)	Dimethylphenylsilane (DMPS)	Diphenylmethylsilane
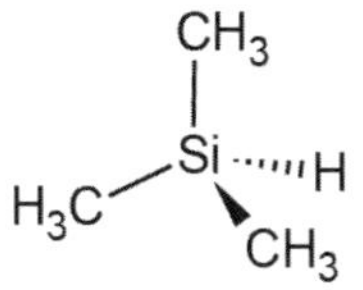	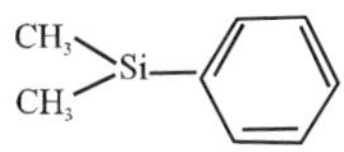	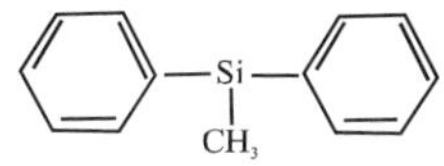
Tetramethylsilane (4MS)	1,1-divinylsilacyclopentane	Hexamethyldisilazane
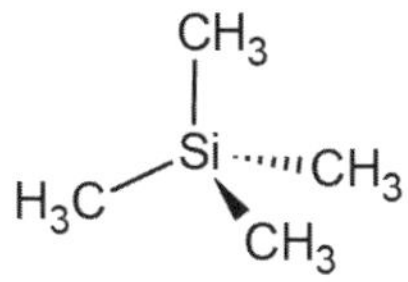	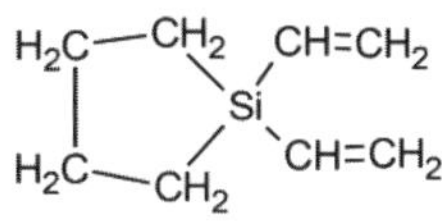	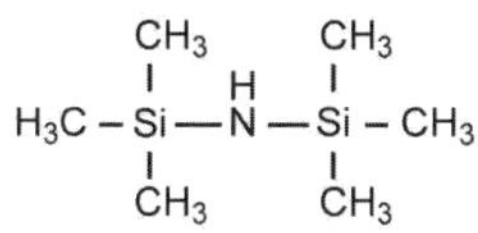

Table 2.4. *Examples of precursors used in SiCH and SiCNH deposition*

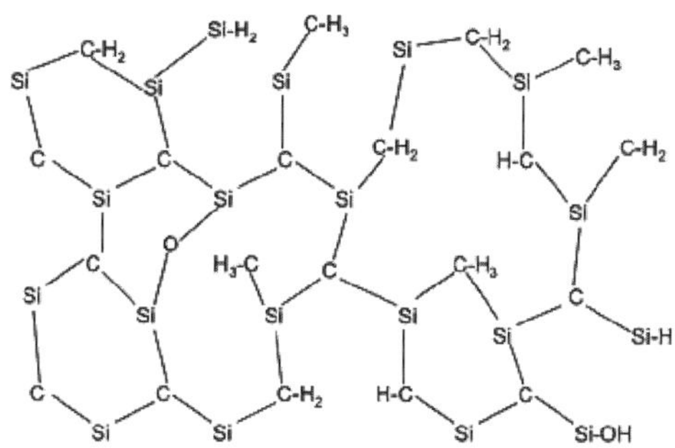

Figure 2.24. *Chemical structure of a SiCH (κ = 4.8) film, after exposure to air [KIN 11]*

SiCNH materials are generally deposited by PECVD, either from a mixture of 3MS and ammonia, or from a single hexamethyldisilazane (HMDSN) precursor. The 3MS + NH_3 mixture is the most widely used in industry, since it has the advantage of enabling better control of the quantity of nitrogen incorporated into the layer [WAN 04]. In addition, HMDSN is liquid at ambient temperature, which makes its use more complicated. Adding nitrogen into a SiCH dielectric layer leads to the formation of Si–N and N–H bonds, resulting in an increase in the dielectric constant ($\kappa \geq 5$).

Figure 2.25 compares the infrared spectra for SiCH and SiCNH deposited from 3MS, as well as the principal identifiable chemical bonds. Despite its dielectric constant being higher than that of SiCH, SiCNH is preferred for the production of 90, 65 and 45 nm technology circuits, since this material is less susceptible to adsorption of humidity, compared with SiCH deposited from 3MS [JOU 04a].

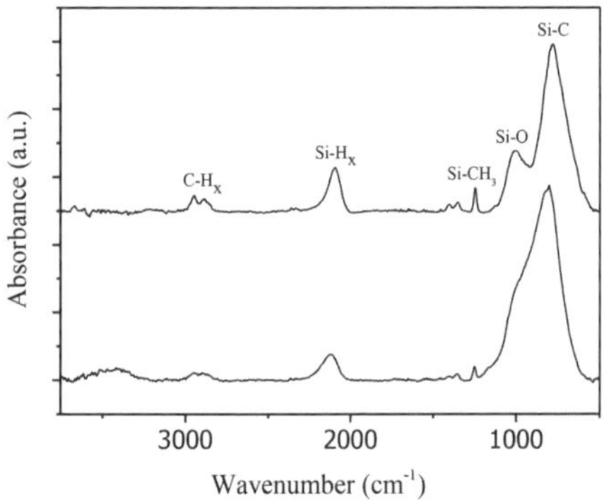

Figure 2.25. *Infrared spectra of thin layers of SiCH and SiCNH deposited by PECVD from 3MS*

However, in spite of the good behavior of SiCNH as a copper diffusion [CHI 03b, WAN 04] and humidity [KAL 02] barrier, the fact that its dielectric constant ($\kappa \approx 5$) is very much higher than that of ILD ($\kappa = 2.5$) gradually constitutes a handicap in improving circuit performance.

2.3.3.3. *"Advanced" dielectric barriers*

Several approaches are being considered with regard to the selection of dielectric barriers with low dielectric constants. The main approach consists of shrinking the density of thin layers by increasing the carbon ratio in the films. One initial method consists of using an organosilicon-based precursor with a significant carbon ratio. Nakashira, for example, has developed a SiCH material deposited from 1,1-divinylsilacyclopentane, which has a dielectric constant below 3.5 [NAK 08]. Favennec has studied the introduction of phenyl groupings using dimethylphenylsilane. They show that the dielectric constant of SiCH can be reduced to 3.2 and that materials deposited using this type of precursor are much more stable in storage than those deposited using only 3MS [FAV 06]. Remiat arrived at similar conclusions using diphenylmethyl-silane.

More recently, Charles-Alfred preferred adding a carbon precursor to 3MS in order to better control the proportion of carbon incorporated into the films [CHA 09a]. This approach turned out to be very promising, some films with a very low dielectric constant ($\kappa < 3.5$) having been obtained whichever organic precursor having been used (this study having compared propene, ethylene or toluene). Furthermore, they showed that incorporating phenyl rings does not bring about any particular advantage, with the reduction of the dielectric constant being principally due to the reduction in film density as a result of the incorporation of carbon (see Figure 2.26). Thus, it is possible to reduce the dielectric constant from 4.8 to 3.1 by increasing the C/Si ratio in films from 1 to 2.5 provided that "soft" plasma conditions (low power, high total and partial pressure from the carbon precursor) are used [CHA 09b].

As with ULK, this reduction in the dielectric constant is always accompanied by deterioration in the mechanical properties of the films. Various studies have also shown that reducing film density provides the disadvantage of degrading copper diffusion barrier properties. A compromise must therefore be found between a carbon ratio, which is high enough to reduce the dielectric constant, and a density, which is strong enough for the film to remain a suitable barrier.

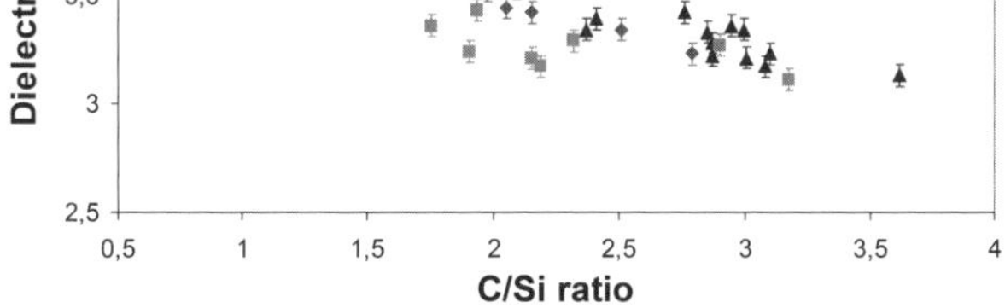

Figure 2.26. *Variation in dielectric constant κ based on incorporating carbon C for 3MS + organic precursor mixtures*

In order to rectify this problem, Bonilla [BON 08] successfully integrated a dual layer containing standard SiCNH (κ = 5.3) and a thicker layer of

SiCNH with permittivity $\kappa < 4$. The latter had a lower value of κ since the quantity of nitrogen incorporated was reduced and the quantity of carbon increased relative to the standard material as a result of the addition of an organic (non-precise) precursor into the plasma. The stack thus presents copper diffusion barrier properties and a constant $\kappa = 4.0$.

2.3.4. *Conclusion*

Today, the dielectric materials used in interconnects are principally organosilicon-based. We tend to increase the carbon ratio in order to shrink polarizability and density, with the final step being to introduce porosity. The preferred deposition technique used is PECVD. Here, the chemistry is complex, with numerous chemical reactions occurring in the plasma. The more advanced depositions require the use of co-deposition. The creation of porosity requires the use of a (UV-assisted thermal) posttreatment where the key words are cross-linking and condensation of silanols.

2.4. Deposition and properties of metal layers for interconnect structures

The deposition of metal layers is of critical significance in the manufacture of advanced interconnect structures. We will see that this affects not only the metal conductor, but also other connected layers necessary for devices to both function properly and be resistant to aging. Various deposition processes are used in order to manufacture a variety of structures, such as lines which are stacked and linked using contact holes (vias) tens of nanometers wide, or even through-silicon vias (TSVs) several tens of micrometers deep in the context of 3D integration.

First, we will describe the various stages involved in interconnect structure metallization. Indeed, these must follow increasingly demanding specifications. In order to satisfy these criteria, new materials and processes are continually being introduced. This approach necessitates that chemical considerations be increasingly taken into account at all levels. Quite clearly, the chemical nature of materials will have a large bearing on their functional properties; this is the case, for example, for copper diffusion barrier layers. Beyond this, interface chemistry plays a fundamental role in improving structural resistance to aging. We will see that it may be useful in the formation of true surface layers. Finally, no description of the benefits of chemistry in this domain would be complete without addressing some aspects of process engineering. By way of selected examples, we will illustrate a wide variety of processes either in existence or in the process of being developed, depending on whether the intention is to standardize, localize or even modulate depositions.

2.4.1. *The manufacture of interconnect structures*

During the 1990s, the introduction of copper to reduce electrical resistance in internal interconnects within circuits required a complete review of manufacturing diagrams and an introduction of new functional materials. Damascene architecture owed its success to the development of additional metal layer deposition as well as specific copper deposition processes. Today, these advancements are reused and reactualized for the manufacture of external interconnect structures, enabling multiple devices to be linked to each other, such as the TSVs in the so-called 3D circuit stack approach [LED 08].

The strategy remains the same for all of these approaches; the interconnect shapes (lines and vias) are inlaid into the substrate. These holes and trenches are filled with metal, with the excess being removed by a polishing step. This process necessitates the implementation of multiple successive depositions and treatments, which are detailed in the following section.

2.4.1.1. *Deposition of a copper diffusion barrier*

Copper is a metal that diffuses very easily in a number of materials, notably in silicon, where it creates a parasitic doping [IST 02]. While the diffusion is massive, some copper silicides may even be formed [NEW 82]. In both cases, the active parts of the circuits are destroyed. During the implementation of the copper, there is consequently a constant preoccupation with its confinement in interconnect structures, through the implementation of adequate barrier layers.

This is why the first stage of metallizing shapes into hollows consists of depositing such a layer. The most widely studied and used materials for this are tantalum, titanium and tungsten nitrides [MUR 95]. These thin layers (which may be up to several tens of nanometers, according to the structure) are generally formed using physical vapor deposition (PVD), but may also be formed by chemical methods (CVD, atomic layer deposition (ALD) or derived techniques).

2.4.1.2. *Deposition of a copper "seed layer"*

The deposition of copper takes place in two stages. First, a layer of metal (typically some tens of nanometers) is deposited. This film is known as a *seed layer*. PVD is the technique of choice for forming this layer [MUR 95]. However, chemical processes (CVD or wet line, as we will discuss later), are currently being used for TSV metallization.

2.4.1.3. *Filling the shapes with copper*

The filling of interconnect structures is then achieved by electrolytic copper deposition. This very long-standing technique has effectively been adapted, through sensible use of additives, to the problem of filling micrometric or submicrometric structures. Subtle mechanisms, which we will detail in section 4.4.2, effectively enable error-free filling of inlaid shapes to be assured. This discovery, credited to the teams at IBM at the beginning of the 1990s, has been the main enabler for the introduction of copper and Damascene architecture for the manufacture of interconnect structures [AND 98].

2.4.1.4. *Chemical mechanical polishing*

The various metal layers (barrier, seed and copper fill) not only fill the structures, but also cover the whole of the substrate. It is therefore necessary to eliminate excesses of copper above and between the shapes; this is achieved using chemical mechanical polishing. The substrate is friction polished over a polymer tissue, using the combined action of abrasive balls (made from silicon or aluminum) and chemical agents present in the polishing liquid. The nature and relative significance of the mechanical and chemical components are dependent on the type of material to be removed (chemical mechanical polishing processes enable numerous types of materials to be polished) [ZAN 04]. In the case of copper, the chemical component plays a major role. Indeed, metallic copper is too ductile to be efficiently tackled by the abrasive particles, which is why an oxidizing agent (most often hydrogen peroxide) is added to the polishing solution, in order to convert the metal into an oxide, which will be harder and more efficiently abraded.

2.4.1.5. *Encapsulation of structures*

At the end of the polishing stage, the interconnect lines and vias are perfectly formed and defined. However, a final stage is necessary in order to ensure that the copper is confined: the deposition of an encapsulation layer to act as a copper diffusion barrier. An insulating material such as silicon nitride or silicon carbonitride is generally used for this purpose.

2.4.1.6. *Copper-based interconnects: specifications*

In order to understand the reasons for choosing the materials and processes described earlier, but also and especially to explain researchers' and industrialists' motivations for developing new thin layers, we must recap and detail here some of the fundamental specifications described in section 2.1, to which copper interconnect structures must conform. The function of these structures is to distribute an electrical signal between the various circuit components. For this to happen, the lines and contact holes must offer the least resistance possible to the passage of the current. That may appear trivial, and we have already previously explained that it is because of this that copper has been retained as the favored metal conductor. However, we should not lose sight of this aspect insofar as other technical constraints that we will be describing, now generally result in choices that lead to an increase in the electrical resistance of metal structures.

The first of these constraints, which we tackled in the last section, is the requirement to guard against copper diffusion in the active zones of devices. Indeed, under the combined effect of heat (the temperature of a microprocessor during operation is approximately 100°C; similarly, manufacturing processes can reach temperatures of up to approximately 400°C) and intense electrical fields between adjacent lines (several MV/cm), copper may quite easily diffuse in the form of Cu^+ ions [MBR 86, IST 02]. As we have explained, this behavior is prevented by implementing barrier layers at various stages of the manufacturing process (before copper deposition, then after polishing in order to encapsulate the lines). Their effectiveness is dependent on their composition and microstructure, which must be perfectly controlled. At the same time, these layers must be as thin as possible, in order to limit their impact on the electrical resistance of the structures. In effect, the barrier materials are considerably more resistive than copper.

Finally, it is advisable to consider the durability of interconnects. Indeed, repeated thermal and electrical stress progressively degrades the circuits. Ageing, which currently limits device lifespans, affects the copper itself. Movements of material occur within the metal under the effect of temperature, combined either with mechanical stresses stored in the structures (known as stress migration), or with the strong cross currents (electromigration). In order to limit these behaviors, the addition of doping agents into the copper may be effective, since these contaminants stabilize the microstructure and block copper diffusion paths. In the case of electromigration, it is also known that the surfaces between the copper and the surrounding barrier layers play a critical role. Indeed, robust interfaces (with strong adhesion) limit the surface diffusion of copper and delay the formation of defects linked to electromigration [LAN 03]. In particular, the surface between the copper and the encapsulation layer plays a critical role in the formation and development of defects during the electromigration process.

The manufacture of advanced interconnect structures in microelectronics involves the implementation, as we have just seen, of multiple thin layers with well-specified characteristics. These different materials are carefully selected for their chemical properties, which dictate not only their intrinsic performance, but also that of the different surfaces created by stacking them. It is these aspects that we propose to tackle in the next two sections.

2.4.2. *The chemistry of materials and functional properties*

2.4.2.1. *Tantalum-based barriers*

As we have just explained, copper diffusion barrier layers play a critical role in interconnect quality. They must enable the copper to be effectively contained, while, at the same time, forming a good interface with it. Finally, they should be as conductive as possible so as to not impair the electrical performance too much.

The materials that best fulfill all of these conditions are the metal nitrides. Numerous compounds have been researched whereby the aluminum has been replaced by copper in interconnects, such as zirconium, tungsten, titanium and tantalum nitrides [TAK 00, KIZ 04]. Tantalum nitride has come to the fore for its barrier qualities and low resistivity. Its deposition using

PVD is straightforward and enables its composition and microstructure to be controlled [TSA 96]. As with the majority of nitrides, it forms a good interface with dielectric oxides, but a poor quality interface with copper, which is why an overlayer of Ta is added to it, which offers better performance in terms of its adhesion to copper and its electrical conductivity [STA 07].

2.4.2.2. *New barrier materials and adhesion promoters*

With the miniaturization of circuits, the volume occupied by the barrier further impairs the electrical conductivity of metal lines. Similarly, error-free filling of structures using copper is becoming increasingly difficult. For these two reasons, researchers are moving toward the development of new metal layers that are able to (1) act as copper diffusion barriers, (2) be conductive enough to initiate electrolytic copper deposition and (3) offer good adhesion with copper.

For this application, refractory metals (W, Pd, Pt, Ir, Os and Ru) have been principally considered, due to their thermal stability. However, most of these candidates do not fulfill all of the specifications. Tungsten can be corroded by immersion into the copper deposition electrolytes [POU 63]. Pd and Pt crystallize easily and form alloys with copper, which are in fact poor barriers. Similarly, the copper diffusion barrier properties of iridium and osmium remain hypothetical. Among these metals, ruthenium has retained the most interest. Numerous teams have studied its behavior as a diffusion barrier for copper, as well as an initiation layer for electrolytic deposition of it. It appears that this material is conductive enough to ensure uniform electrolytic deposition [ARM 11], but presents insufficient barrier properties [WOJ 12]. For this reason, efforts are turning toward Ru-based alloys. In particular, incorporating Ta (in nitride or carbon form) enables the barrier efficiency of the layers to be notably improved [FAN 11, CHA 11, WOJ 12, KIM 08]. However, this occurs to the detriment of their conductive properties and may affect the copper's nucleation, as well as its adhesion [CHA 11, WOJ 12]. Thus, the best compromise between these different properties should again be established. For completeness, we should mention another metal whose use is envisaged to the same degree as ruthenium. We refer to cobalt, which has also been combined with different bodies in order to improve their copper diffusion barrier properties [HEN 10a].

A more exploratory avenue of research is the use of organic layers to act as copper diffusion barriers and promoters of metal adhesion onto dielectric materials. These films, incorrectly called self-assembled monolayers (SAMs) by most authors, are, in reality, organic layers fixed onto the substrate by a specific chemisorption (or grafting, according to the authors) reaction [CAR 08, GAR 10a]. The main reaction used here is silanization, which comprises hydrolysis using residual water from the medium, then condensation of a silanol precursor onto an area comprising hydroxyl groupings [MUR 06, CAR 08]. This reaction has been conducted in wet [RAM 03] or supercritical [REB 09] liquid mode. The addition of copper by PVD [GAN 07], CVD [KIM 09, REB 10] or even "electroless" deposition [ASH 09, ARM 10] onto these layers has been carried out. A slowdown of copper diffusion as a result of these layers has been observed [CAR 08], as well as a reinforcement of the adhesion between the copper and the dielectric material [GAR 10b].

2.4.2.3. *The use of copper doping*

As mentioned in section 2.4.1.6, copper interconnects are subjected to ageing behaviors, which bring diffusion or self-diffusion of the metal into play. In order to limit these effects, one strategy that has been widely researched consists of doping the copper with a second metal, which acts as a copper diffusion and grain boundary path blocker. Indeed, during heating in the manufacture of structures, the doping element separates from the copper and segregates at the interfaces and the grain boundaries. This effect, as well as its benefit in containing the ageing of interconnects, has been demonstrated for doping metals such as Sn [KIM 05], Zr [IGA 98] and Al [MAE 08].

This approach was rationalized by Barmak who declared the criteria for choice of the second metal [BAR 06]. This metal must, in particular, be immiscible in copper and have a low reactivity with it (with no eutectic or intermetallic formation). Approximately 15 elements correspond to these criteria, among them are Pd, Al, Ag, Ti, In, Mg, Zn and Sn. The authors also added Mn, which does not quite satisfy all of the criteria, but which has moreover demonstrated satisfactory properties when added to copper (see section 2.4.3.2).

2.4.3. *The chemistry of interfaces*

2.4.3.1. *Adhesion properties*

The surface chemical properties of materials used in the manufacture of interconnect structures play a critical role in the quality of the interfaces created, which in turn influence circuit performance and resistance.

A first, well-documented example in print is that of barrier layers. Tantalum combines with the other metals mentioned to form barrier layers with a pronounced oxophilic characteristic, which signifies that a Ta or TaN layer is covered spontaneously and in a matter of seconds, through simple exposure to air, by a superficial oxide layer [POU 63]. This is why copper seed layer deposition must out of necessity be carried out immediately after deposition of the barrier layer, using the same equipment and without exposing it to air in the substrate. In the case of direct deposition of copper onto the barrier (see section 2.4.4.3), this exposure to air is inevitable; in this case, it is necessary to carry out a deoxidization treatment prior to the copper deposition [RAD 04, CHA 11].

A second example that provides a good illustration of the importance of interface quality is that of copper CVD. This deposition technique is, as we have mentioned above, intended for copper seed layer deposition. However, it has proved difficult for this technique to be accepted, as the layers formed are not sufficiently adherent. This poor adherence is attributed to interfacial fluoride contamination, which stems from the fluorinated copper precursors used during the reaction [MOT 00]. For this reason, the research effort is currently aimed toward the development of non-fluorinated precursors [AU 12].

2.4.3.2. *Self-forming barriers*

The knowledge and control of chemical processes extends well beyond simple interface engineering. Recently, the concept of self-forming barriers has emerged, where the aim is to retrospectively form the barrier layer, through the chemical reaction between the metal and its surrounding insulating material.

This approach follows on from research carried out on the effect of doping in copper outlined in section 2.4.2.3. It relies on the principle of a seed layer made of copper, doped (to several atomic percent) by an immiscible metal and deposited directly onto the insulating material (silica or silicate). The interconnect structure is subsequently filled with copper using electrolytic deposition and then annealed. During this thermal treatment, the doping metal demixes from the copper and moves toward the interface with the insulator, where it reacts to form a layer that acts as a copper diffusion barrier. This is a very elegant approach, since it enables the formation of very thin (of the order of a few nanometers) and perfectly covering barriers, solely on the structure walls (the barrier is absent from the base of the contact holes, where a Cu/Cu interface is formed). All of this contributes to the impact of the barrier on the structure conduction properties being limited.

This process has been demonstrated for metals such as Ti [LIU 05], Zr [LIU 02] and Mg [DIN 94, FRE 03]. However, manganese has received the most attention [KOI 07]. The nature of compounds formed at the interface with SiO_2 or SiOC have particularly been studied [SUN 11, HAN 07, WIL 10]. Manganese metal has been successfully integrated into Damascene structures [USU 06]; however, the consequence of copper doping is an increase in metal resistivity, which is why certain authors have proposed a hybrid approach, in which an interfacial layer of Ru is positioned between the copper and the dielectric material and doped with the reactive element [WOJ 12, HSU 11, HSU 12, TOK 10].

2.4.4. *The chemistry of metal deposition processes*

An essential contribution from chemists to the microelectronics industry is the development of manufacturing processes, some of which have become indispensable and which are benefitting from the unique characteristics of chemical reactions. An exhaustive review of these processes is largely outside the context of this section. However, in the following section, we will try to describe the main principles that govern the development and the implementation of these processes and to illustrate them using selected examples.

2.4.4.1. *The kinetics of reaction and multiscale control*

The manufacture of microelectronic devices is a highly parallel process. Tens, even hundreds of components are currently being manufactured at the same time over silicon substrates with a diameter of up to 300 mm. This implies that manufacturing processes must be succeeding in enabling controlled treatments simultaneously at macroscopic (uniform over the 700 cm² of the substrate) and microscopic, or even nanoscopic (each shape must be free from defect) scale. If we consider a chemical process, the crux of this problem is in controlling the kinetics of the reaction set into play at these various scales.

A chemical reaction generally comprises a large number of elementary steps, which are not strictly speaking limited to species. In particular, the mechanisms for transporting reagents and products of reactions to or from the treatment surface should be considered within the kinetic analysis of a process mechanism. Depending on the reaction conditions, one or more of these elementary steps may be kinetically limiting and regulate the reaction speed. This makes chemical processes much more versatile with regard to kinetic control, compared with physical processes. This variety of kinetic control modes is well illustrated by the diversity of deposition processes for metals used in interconnect structure manufacture.

2.4.4.2. *Background example: filling hollowed shapes using electrolytic copper deposition*

Electrolytic reactions are, by definition, conducted with the aid of an external current source. Copper deposition reaction 2.8 is the reduction of copper ions in metal based on:

$$Cu^{2+} + 2e^{-} \longrightarrow Cu^{0} \qquad \text{[Reaction 2.8]}$$

In this case, electrons are charged by an electrical generator. The current passes across the surface of the copper substrate (which acts as a cathode, the site of the reduction reaction), the electrolyte (by ionic migration) and then the surface of the copper anode (where the inverse copper dissolution reaction occurs).

It is important to note that in this type of process, the deposition rate $de(t)/dt$ (where $e(t)$ is the thickness of the deposition at time t) is directly linked to the current density $j(t)$ (the current per surface unit) according to Faraday's law equation [2.3]:

$$\frac{de}{dt}(t) = \frac{M_M}{\rho_M nF} j(t) \frac{de}{dt(t)} = \frac{M_M}{\rho_M nF} j(t) \qquad [2.3]$$

where M_M and ρ_M are the molar mass and density, respectively, of the metal M, *n* is the number of electrons exchanged (2 in the case, which is of interest to us) and F = 96,500 C/mol is the Faraday constant. At the microscopic scale, local variations in the current density will therefore be translated into local differences in deposition rate. Another consequence is that at the macroscopic scale, the (average) deposition rate is directly regulated by the applied current value, which is a measurement controlled by the operator.

2.4.4.2.1. At macroscopic scale: edge effect

In most industrial electrolytic copper deposition equipment, the silicon substrate (covered with a copper seed layer) is maintained using a boundary ring that additionally acts as an electrical contact on the copper surface, implying that the central section of the substrate is extended from these contact points. Hence, the current passing across the cell through this center zone must overcome resistance over and above that of the current passing through the boundary zone, that is the copper seed layer current (see Figure 2.27). In this configuration, the deposition rate therefore tends to be higher in the boundary zone of the substrate. This behavior is known as an edge effect [ARM 11].

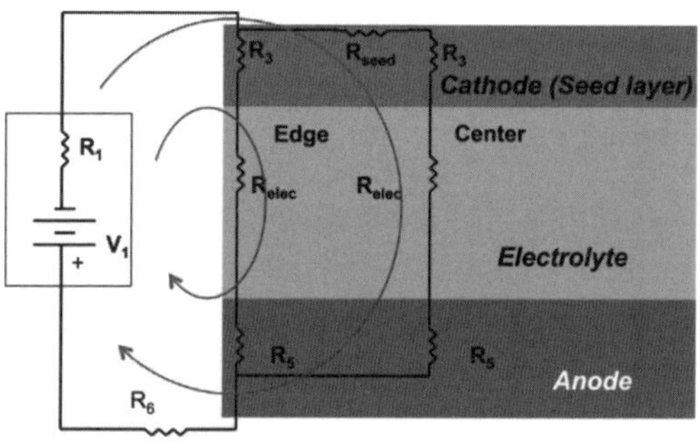

$$\frac{J_{center}}{J_{edge}} = \frac{R_{sum} + R_{elec}}{R_{sum} + R_{elec} + R_{seed}} < 1$$

Figure 2.27. *Equivalent circuit of a deposition cell, showing the effect of an Ohmic fall between the edge and center of the substrate, due to the seed layer resistance*

Various strategies have been implemented in order to counteract this effect. A first approach consists of increasing the resistance of the cell itself, in order to render the excess of resistance linked to the seed layer negligible. This has been achieved not only by adding an ion exchange membrane between the anode and the substrate, but also by modifying the formulation of deposition electrolytes. These electrolytes not only contain copper sulfate (source of copper ions), but also a strong concentration of sulfuric acid (up to 1M), which acts as a supporting electrolyte. By reducing the concentration of this compound, deposition baths have become much more resistive (while enabling a satisfactory deposition rate to be maintained). More recently, another approach has consisted of modifying the current distribution by using a segmented anode, with each segment being fed by an independent generator. Rebalancing the current between the various segments thus enables the natural disequilibrium between the edge and the center of the cell to be compensated for [ROU 04].

2.4.4.2.2. At microscopic scale: superconformal filling

The decisive factor that has lead to electrolytic deposition being chosen as the copper deposition technique for interconnect manufacture is, without doubt, its capacity for error-free filling of hollowed shapes. This skill has been made possible by the use of additives that have the ability to modulate the kinetics of the deposition reaction.

An electrolyte such as that described above (copper sulfate in an acidic medium) enables high-current density copper depositions (up to 50 mA/cm^2), to be produced, resulting in an increased rate (approximately 1 µm/min). However, the shape's topology tends to concentrate the lines of current onto the corners of the line or via. From Faraday's law, this signifies that the deposition will be quicker at this point. Conversely, the current density (and therefore the deposition rate) will be lower inside the shape. From this, a profile such as that described in Figure 2.28(b) is built up, which is said to be subconformal. Quite clearly, the thickness of the deposition results in the shape closing prematurely, before it has been filled. In the best case, the deposition is able to be rebalanced before becoming conformal (see Figure 2.28(a)). Even then, closure of the shape will occur by leaving small cavities at its center, resulting from the irregularity of the two sides that close on each other.

The only situation that enables the hollowed out shapes to be perfectly filled is when a deposition is superconformal (see Figure 2.28c), yet the teams at IBM demonstrated at the start of the 1990s that it was possible to force deposition to follow this type of profile, through the use of two counteracting additives: a deposition inhibitor on the one hand and an accelerator on the other hand [AND 98]. Since that time, these additives have been used to fill Damascene (submicronic) structures. Their mechanism of action was only however explained during the 2000s [MOF 05], and remains an actively researched topic to this day [ADO 11].

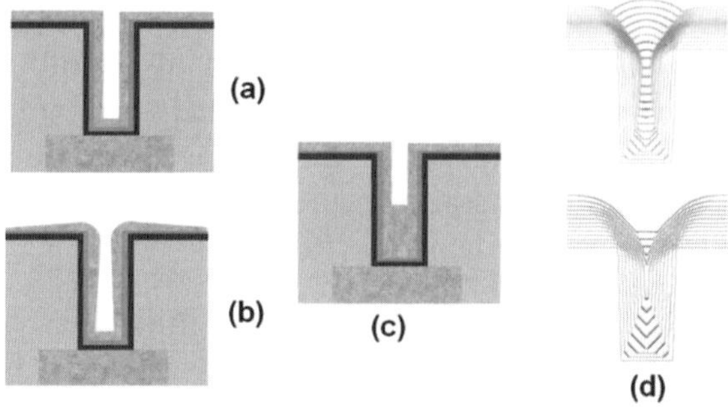

Figure 2.28. *Deposition modes: a) conformal, b) subconformal and c) superconformal. d) Example superconformal fill model, based on the CEAC model [MOF 05]*

The inhibitors used to fill Damascene shapes are polyethers such as polyethylene glycol, molecules with high molecular weights, which diffuse relatively slowly, but which are very rapidly adsorbed into the surface of copper. There they form an organic film that blocks the copper ions' access to the surface, preventing the charge transfer reaction from taking place. The result is a strong inhibition of the deposition reaction.

Accelerators are small sulfur molecules, which, unlike inhibitors, diffuse rapidly, but which are adsorbed slowly on the copper. The typical accelerator is bis-(3-sulfopropyl) disulfide (SPS). In spite of their name, these compounds cannot easily accelerate deposition on their own. On the contrary, they are able to force the desorption of inhibitors and to be adsorbed onto the copper in their place; indeed, their interaction with the metal is stronger than that of the inhibit. They enable inhibition to deposition to be partially relaxed by reactivating the charge transfer reaction locally.

Once the substrate has been immersed into the electrolyte, the additives are adsorbed onto the surface of the copper seed layer. The inhibitor is preferentially adsorbed, since it is faster in this respect than the accelerator [DAS 05]. However, the accelerator is adsorbed quite quickly during deposition and, surprisingly, is accumulated at the base of the shapes. This reactivates the deposition reaction, which remains largely inhibited over the remainder of the substrate. The deposition rate thus becomes more rapid, compared with the substrate surface, at the base of the shape; this corresponds to the researched superconformal deposition situation.

In order to understand which mechanism creates accelerator accumulation at the base of the shapes, Vereecken has studied the adsorption/desorption reactions of these additives, as well as their properties in relation to the complexation of cuprous and cupric ions [VER 05]. Approximately 20 processes have been identified, which all potentially play a part in the individual and synergetic action of these two additives, and which are dependent on factors such as copper concentration and the pH of the electrolyte. In particular, the chelatant action of the accelerator with respect to cuprous ions seems to play a fundamental role in the action of this additive.

However, the approach that has emerged as the most pertinent in terms of describing the accumulation effect of the accelerator has been developed by teams from the National Institute of Standards and Technology (NIST) [MOF 05]. Their model relies on the simple considerations set out above regarding the adsorption of two types of additives. Two further hypotheses are included:

– The two additives "float" on the growing copper surface. This hypothesis is supported by physicochemical analyses of the deposited copper, which contains only a very low amount of organic impurity.

– The accelerator, once adsorbed, does not move laterally over the surface. This hypothesis holds for Damascene shapes, which are filled in just a few seconds.

Initially, the additives are adsorbed onto the copper surface. The inhibitor is in the majority, but a small quantity of accelerator is adsorbed uniformly over the surface. During this phase, the deposition rate is the same throughout (conformal deposition). However, closure of the shape reduces the area of the surface base. The accelerator that had been adsorbed there thus reappears in concentrated form (there are as many adsorbed molecules, but over a reduced area). On the upper corners of the shape, the opposite happens; the area increases and the accelerator concentration is reduced (the newly created surface being preferentially re-covered by the inhibitor). The net result is therefore an increased accelerator concentration at the base of the shape, through the simple geometric effect of reducing the area on a concave surface during deposition. Incidentally, this has given its name to the model, known as curvature-enhanced accelerator coverage (CEAC), see Figure 2.28). This model is always relevant when explaining the triggering of superconformal filling in Damascene shapes, and is continually being improved in order to take account of effects such as those from the diffusion of the two additives (which is not taken into account for the initial model) [ADO 11].

With the development of 3D technologies that are based on the manufacture of TSVs that are tens of micrometers deep, the process of filling using copper has had to be reviewed. Indeed, two factors can complicate the task, compared with the submicronic shapes:

– The electrical resistance of the copper seed layer becomes sensitive between the top and bottom of the shape, resulting in a reduction in the current density at the TSV base.

– Hydrodynamic agitation of the bath is ineffective inside of the shape, the main mode of species (cupric ions) transportation is diffusion.

Both of these effects are combined in order to slow down the deposition at the base of the shape and to reinforce the subconformal characteristic of the deposition.

In the first generations of deposition electrolytes used for filling TSVs, the same types of additives as those described above have been used. These additives have resulted in a superconformal deposition, for which an example V characteristic profile is given in Figure 2.29(a). This profile is testament to a progressive increase in deposition kinetics deeper the via, which is characteristic of an acceleration of the deposition rate as a result of inhibitor depletion in the shape [LUH 09]. Indeed, the only diffusion from the additive, which is slow, is not enough to feed the base of the TSV. Thus, the superconformal profile is, in this case, the consequence of an inhibitor deficit in the shape. This type of mechanism limits the speed differential, which can be established between the base and the top of the shape, resulting in (1) slow filling (approximately 1 h for a TSV 5 μm in diameter and 50 μm deep) and (2) a significant deposition onto the field around the shapes (typically several μm), which must subsequently be eliminated using chemical–mechanical polishing.

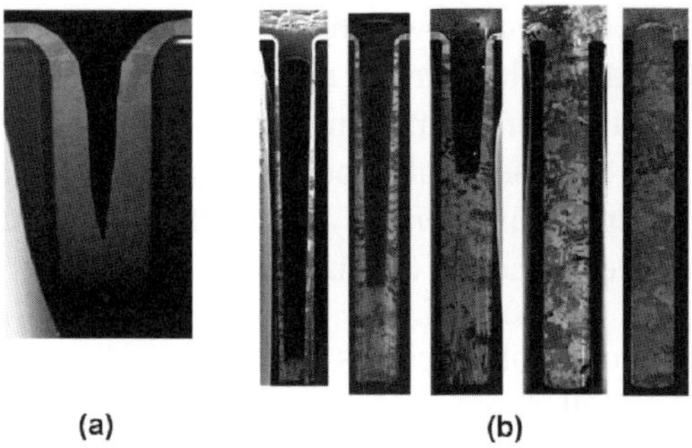

Figure 2.29. *a) Refill profile of a 5 × 17 μm TSV with a first-generation electrolyte. b) Execution of the filling of a 10 × 100 μm TSV with a later generation electrolyte*

In order to improve the kinetic behavior of the reaction, inhibitors have evolved. In current generations of electrolytes, other inhibitors are used. They inhibit deposition more effectively and are more strongly adsorbed on the surface of the metal (they are not, moreover, desorbed during deposition, but rather incorporated into the copper). Conversely, their adsorption kinetics are slower. This all provides them with a more irreversible behavior. In order to relax this strong inhibition, the copper deposition reaction must be strongly activated. When inhibition is increased, the deposition rate is therefore immediately very high and does not leave the inhibitor any time to be reabsorbed. During filling, this is expressed by an abrupt transition between the inhibited zones (water surface and via walls) and the activated zones (base of the vias). The growth profile is therefore very different, with vertical walls and a flat base, which very quickly "rises" toward the top of the via (see Figure 2.29(b)). The speed differential is at maximum between the base and top of the shape, which considerably reduces the time taken to fill and the thickness of the deposition over the field (of the order of 1–2 μm).

2.4.4.3. *Uniformly distributed deposition example: barrier and copper seed layer*

The physical deposition techniques usually used to deposit barrier and copper seed layers suffer from a low level of effectiveness over the vertical sections of substrates. On these surfaces, the quantity of metal deposited can be up to 10 times lower than on the field of the plate. With the advent of 3D integration, which requires holes of several tens of micrometers in depth to be metallized, these techniques will need to be replaced by more efficient processes for depositing the barrier and the seed layer. These processes should ideally deposit a very thin, conformal (i.e. of equal thickness across the whole shape) layer. Chemical depositions are particularly suited to this specification, since it is sufficient to ensure that the reaction speed is the same across the whole surface of the substrate.

The most studied class of processes for this is CVD and its derivatives. In these reactions, a metal precursor is decomposed onto the surface of the substrate, in a regime where the limiting step is the chemical reaction as such (and not the transportation of reactives or the reaction products). In most cases, this reaction is thermally activated and it is sufficient to ensure that the surface of the substrate has a homogeneous temperature. Table 2.5 provides a (non-exhaustive) list of the main metals and precursors studied for both the deposition of traditional materials (Ta, TaN, TiN and Cu) and also newly emerging metals (such as Ru). Another promising method used to form extremely thin films (of several atomic layers) is ALD (and its derivatives). Existing processes and associated precursors are also listed in Table 2.5. A more exhaustive review has been proposed by Profijt *et al.* [PRO 11].

Wet line processes are also an interesting alternative, notably for metallizing TSVs, where economics plays a preponderant role. Some electroless chemical reactions have been studied for this purpose. Barrier deposition by this method is a very active subject [SHA 03], notably with Ni-based [CHO 08, INO 11, TRU 09, YOS 07] rather than Co-based [EIN 05] alloy deposition. Conversely, studies appear to be less promising for copper deposition using this type of reaction [SHA 03]. In order to form the seed layer, electrolytic processes are favored. However, they need to enable continuous deposition onto resistive layers, such as TaN/Ta, which necessitates the replacement of the acid electrolytes used in filling with low-conductivity electrolytes, in an alkaline and chelatant medium [HAU 05, ROU 07, KIM 07]. Another, more marginal method consists of using a complementary electrografting reaction in order to assist the copper plating reaction [TRU 09]. This elegant approach makes good use of the specific properties of electrografting reactions, which have a low sensitivity to Ohmic voltage drop effects [HAU 03].

Material	Technique	Precursor: Name	Precursor: Formula	Ref.
Co(P)	CVD	$Co_2(CO)_8$		[HEN 10] [HEN 10b]
Co(W)	ALD	bis(cyclopentadienyl) cobalt(II) bis(cyclopentadienyl) tungsten(IV) dihydride	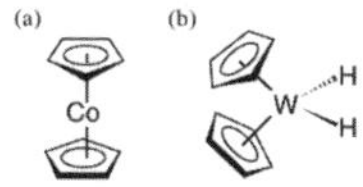	[SHI 12]
Cu	CVD	CupraSelect®	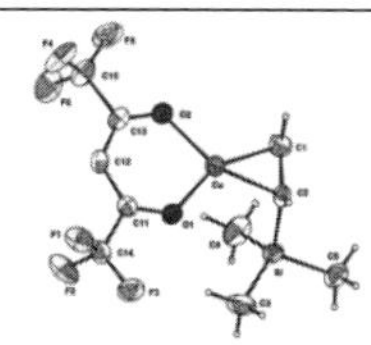	[VOS 00] [WEI 00]
		{2-$(Me_3Si)_2$ C(Cu)C_5H_4N}$_2$	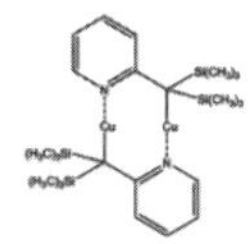	[YE 05]
		GigaCopper	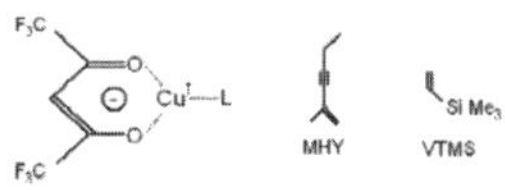	[JOU 02]
Cu	ALD	Cu(I) N,N′-di-sec-butylacetamidinate	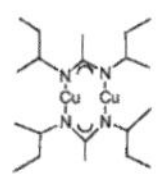	[LI 06]
		$[(^nBu_3P)_2Cu(acac)]$	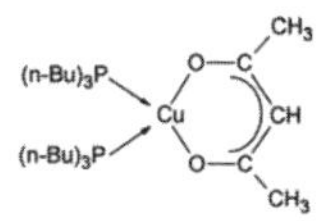	[WAE 09]
Mn oxide	CVD	bis(ethylcyclopentadienyl) manganese	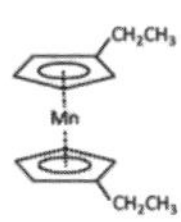	[DIX 11]
Ru	ALD	IMBCHRu, EBCHRu, EBECHRu	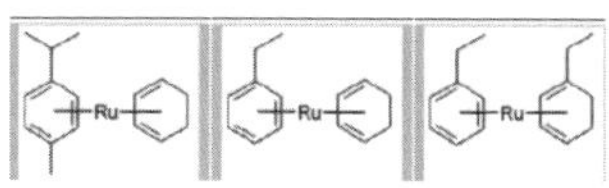	[KIM 11]
		$Ru(EtCp)_2$		[CHO 10]

Table 2.5. *Non-exhaustive list of precursors for CVD or ALD deposition of barrier and copper seed layers*

TaN	ALD	PDMAT	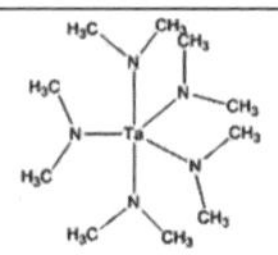	[VAN 11]
		TBTDET	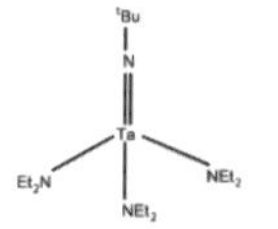	[ZHA 05]
TiN	CVD	TDMAT		[MOT 00]
		TDEAT		[LIU 98]
WCN	ALD	WF_6, NH_3, TEB		[BEY 02]

Table 2.5. *(Continued) Non-exhaustive list of precursors for CVD or ALD deposition of barrier and copper seed layers*

2.4.4.4. *Localized deposition example: copper line encapsulation layers*

As we saw in section 2.4.1.6, the interface between the copper and the encapsulation layer is the main weak point of structures, with regard to electromigration behavior. In order to limit this behavior and increase the lifespan of interconnects, it is proposed that this metal/insulator interface be replaced by a more robust metal/metal interface. However, this approach assumes that it is possible to localize the deposition of the new barrier precisely above the copper lines. Indeed, the presence of conductive material between adjacent lines would result in the formation of short-circuits that would compromise the operation of the device. It is theoretically possible to arrive at this result by using a supplementary lithographic process, which would superpose this barrier layer onto the metal concerned. However, the high cost associated with the technical problems of alignment make this approach impractical.

The solution used to resolve this problem takes advantage of the unique possibilities offered by the kinetic control of chemical reactions. Indeed, the kinetics of reactions such as CVD or electroless depositions can be regulated so that deposition is only activated over certain selected metallic surfaces [GOS 06]. Electroless deposition reactions proceed depending on the reduction in metallic cations, similar to reaction [2.8]. The metal which is principally used for this reaction is cobalt [ALM 07, GAM 06, OLI 10], but nickel has also been widely studied [CHO 08, JOU 04b]. For electroless depositions, the electrons necessary to form the metal are provided by a chemical species, directly onto the surface of the substrate. The deposition reaction is thus fed by a simultaneous oxidation reaction of a reducing agent such as sodium hypophosphite or dimethylamine borane (DMAB) (see Figure 2.30). Yet, it is possible to find operatory conditions (pH, temperature, the addition of stabilizing agents) in which this reaction occurs extremely slowly: we then refer to a metastable medium. The species lack the "boost" necessary in order to react, which is provided by certain catalytic metallic surfaces.

Compound	Role	Example : CoWB
Metal salt	Metal source	Cobalt sulfate $CoSO_4$, $7H_2O$
Complexing agents		Sodium citrate $C_6H_5Na_3O_7$, $2H_2O$ Malic acid $C_4H_6O_5$
Reducing agent	Electron donnor	Morpholine borane $C_4H_9ON\text{-}BH_3$
Refractory metal salt	Codeposition (alloy)	Sodium tungstate Na_2WO_4, $2H_2O$
Additives	Stabilizer	Surfacants
Acid/base	pH control	TMAH

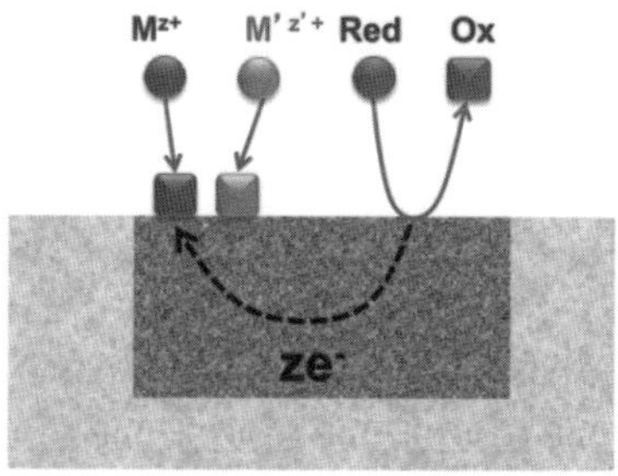

Figure 2.30. *Typical composition of an electroless deposition bath, principle diagram of the deposition reaction*

The catalytic properties of certain metals have been understood and used for more than a century [FRE 11]. Interaction of reactants with these surfaces translated into a significant increase in speed and efficiency of chemical reaction. For a given reaction, the effectiveness of a catalyst is dependent on its nature as well as on the state of its surface. In the example that is of interest to us, it is important to note that the deposited metal catalyzes the deposition reaction. This is why those reactions we also called auto catalytic. This also implies that the reaction, once initiated, continues while the substrate is immersed in the bath.

We therefore need to initiate this reaction selectively at the surface of the copper lines. There are two scenarios, depending on the nature of the reducing agent. For DMAB and its derivatives, copper is a good reaction catalyst [OHN 85]. Metal deposition (Co or Ni) is therefore spontaneously initiated over the surface of the metal lines. However, these redactor agents are particularly powerful, so it is difficult to eliminate all undesirable nucleation of substrate defects that will inevitably be present, as well as those in the reactor itself. These baths are therefore particularly unstable and difficult to control within a stabilized industrial process. For this reason, sodium hydrophosphite baths, which are not as powerful, are preferred in an industrial context. Deposition baths are more stable, but this is offset by the fact that copper is no longer a reaction catalyst. These baths require that more effective catalysts, such as palladium, are implemented [OHN 85]. Fortunately, this metal is more noble than copper. During a simple immersion of the substrate in a palladium salt solution, a redox reaction occurs, where the copper is oxidized into cupric ions in solution form, whereas the palladium ions are reduced to metal on the surface of the copper. This is referred to as surface activation using palladium.

The processes currently being studied at an industrial level for the electroless deposition of cobalt-based barriers are therefore divided into several steps:

– cleaning of the copper surface (notably, the elimination of chemical mechanical polishing residues);

– activation by palladium;

– electroless deposition of the barrier *per se*;

– eventually, specific cleaning to eliminate parasitic depositions.

These processes are still maturing at an industrial level. However, their performances are astonishing in terms of thickness control at the silicon plate scale (see Figure 2.31). Indeed, the deposition rate is constant over time and uniquely set by the operating conditions; if these are perfectly controlled and homogeneous over the whole surface of the plate, a single deposition time is sufficient to control the deposited thickness, to the nanometer. The aim of the current processes is also to attain the extremely demanding levels of selectivity of this application. The least parasitic or even very localized deposition, which short circuits two adjacent lines, leads to an unacceptable degradation in performance of the circuits (see Figure 2.32) [OLI 08].

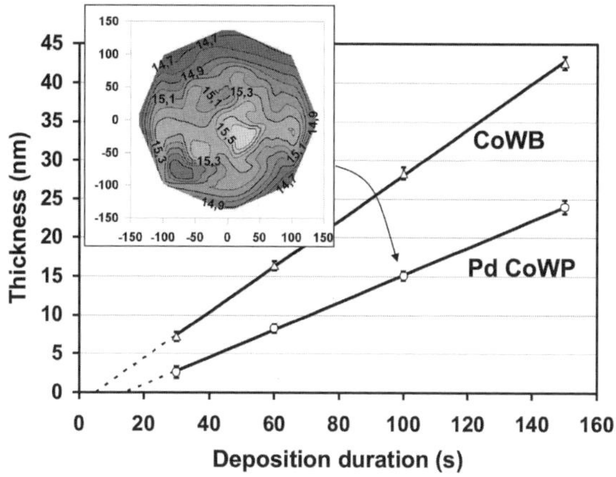

Figure 2.31. *Typical kinetic of an electroless deposition and control of the uniformity of a 15 nm deposition on a silicon plate [OLI 10]*

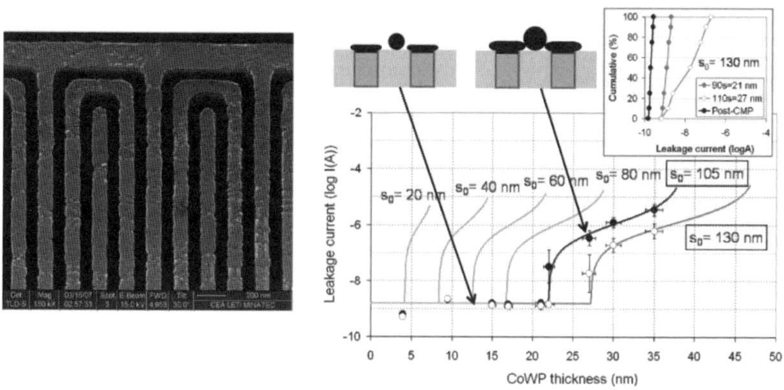

Figure 2.32. *a) Selective deposition of 15 nm CoWP on comb/coil structures and b) illustration of the catastrophic increase in leakage current with the formation of bridging defects between adjacent lines*

A concurrent method concerns localized deposition using CVD. In this case, a reactivity differential between the metallic zones and the insulator also enables the deposition to be localized. For instance, among the approaches studied are the selective depositions of cobalt and its alloys [HEN 10b], Mn [AU 10] or the localized formation of CuSiN [GOS 06].

2.4.5. *Conclusion*

The metal layer stacks used for manufacturing advanced interconnect structures are becoming increasingly complex. New materials and processes have been introduced in order to respond to increasingly demanding specifications. Within this evolution, the understanding and control of chemical behaviors are growing in importance, for several reasons.

First, the intrinsic chemical properties of bodies guide the choice of compounds for producing different operating layers. In particular, the miscibility or otherwise of metals, their tolerance to contaminants, dictates the properties of different barrier layers and the effectiveness of such copper doping in reinforcing structure resistance.

Next, the chemistry of surfaces and interfaces enables the design of adherent stacks, which are necessary for robust structures to be obtained. Hereafter, interfacial reactions may even enable true interfacial functional layers, such as copper diffusion barriers, to be formed *in situ*.

Finally, the engineering of chemical processes enables manufacturing methods to be developed, which take advantage of the unique properties of chemical reactions with respect to deposition rate control. Operating modes can be modulated at will, according to whether a uniform and homogeneous (barrier, seed layer) or localized (electroless barrier) deposition is needed, or whether the requirement is to vary the growth rate for different sections of the substrate (superconformal refilling).

2.5. Cleaning process for copper interconnects

2.5.1. *Introduction*

As we have already seen, the performance of integrated circuits is continually improving, mainly due to reductions in transistor sizes. According to Moore's law, performance is doubling every 18 months. Interconnect systems that enable the signal to be distributed have had to follow this evolution by limiting, as far as possible, propagation delays, expressed by the product *RC*, where *R* is the resistance of the interconnect and *C* is the cumulative capacity between the different dielectric levels. To reduce this product *RC*, copper has replaced aluminum (resistivity below 30%: (ρ(Al) =2.66 $\mu\Omega.cm^{-2}$) and ρ (Cu)= 1.67 $\mu\Omega.cm^{-2}$) and dense SiO_2 has been advantageously replaced by various dielectrics with lower permittivities (fluorosilicate and dense, then porous organosilicate) [BAK 07].

Introducing copper into the microcircuit manufacturing process has modified the integration schema developed for aluminum interconnects. Before the difficulties in simply etching the copper material, the problem was reported for the etching of the dielectric material. This integration, which is commonly known as "Damascene", requires chemical mechanical polishing (CMP) steps for the removal of surplus copper and thus to obtain a flat surface. Various cleaning steps have been introduced such as (1) post-CMP cleaning, which eliminates CMP particles and residues, and (2) post-etching cleaning of the grooves, which removes the copper oxide from the base of the via, the sputtered metallic copper on the dielectric sides (see Figure 2.33(a)) and the polymer etching residues (see Figure 2.33(b)). These planarization and cleaning processes, produced in wet conditions, are able to generate corrosion of the copper or other metals present [MIL 01] in the structure.

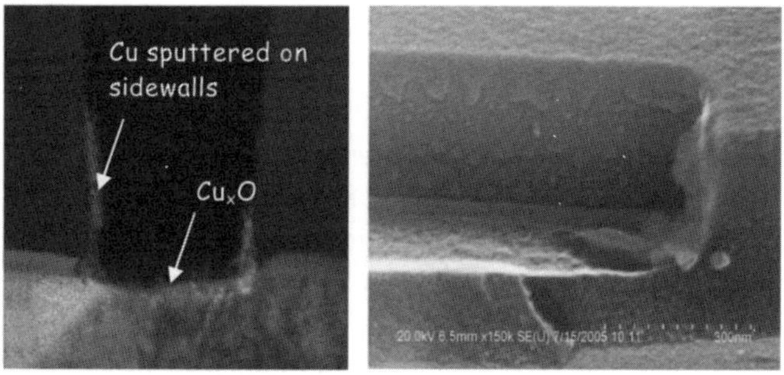

Figure 2.33. *a) SEM image of polymer residues on the sides of dielectrics observed after etching using carbofluorinated plasma and b) TEM-EELS image after via etching*

In addition to the processes of thin layer deposition elaboration described in section 2.4 and used in back-end, such as electrolytic copper, or even electroless (without current) deposition, of self-aligning CoWP barriers, electrochemistry is proving to be a useful diagnostic tool in microelectronics. Notably, this tool enables the identification and quantification or corrosion behaviors that may generate major defects, which would be fatal for an electronic circuit.

In this section, which discusses electrochemical and microelectronic applications, we will succinctly introduce various types of corrosion that the microelectronics industry faces experientially, and we will detail the electrochemical methods used to diagnose corrosion by way of an example, namely the cleaning carried out after the etching of vias.

2.5.2. *Impact of corrosion in microelectronics*

The impact of corrosion may vary on the basis of its nature.

2.5.2.1. *Generalized corrosion*

In the case of generalized corrosion, it may create an increase in roughness, since it takes place preferentially at the grain boundaries of the metal. This type of corrosion has been observed at the end of the chemical–mechanical copper planarization step (see Figure 2.34(a)). The solution used to planarize the structures contains an oxidizing agent, usually peroxide, a chelatant such as citric acid and a corrosion inhibitor such as benzotriazole. Peroxide, which is a powerful oxidant, enables a copper oxide layer to be formed, which is then removed through a mechanical abrasive action, to reveal particles of SiO_2 contained in the polishing solution. The organic acid enables chelation of the metallic elements and the corrosion inhibitor ensures the integrity of the copper structures [BER 07]. Other additives, known as surfactants, make the process more stable, or, rather, less sensitive to oxygen or light. The choice and proportion of each one of its constituents enables the oxidation reaction, complexation and protection kinetics to be adjusted so that multimaterial structures can be polished and planarized, without degradation of the copper lines, whose densities may vary on the same plate. Some electrochemical studies have been carried out in order to simulate as far as possible what happens during the CMP step. Thus, polarization curves $E=f(I)$ have been recorded with polishing solutions by applying a mechanical pressure onto the copper rotating disk electrodes [CAR 95]. More recently, triboelectrochemical studies have been carried out on patterned substrates [JOO 12] so that CMP can be better understood and optimized. Adding an additive which uses oxygen, also known as an "oxygen scavenger", is also a solution for reducing the oxygen concentration dissolved into the solution and to reduce or even suppress the effects of corrosion.

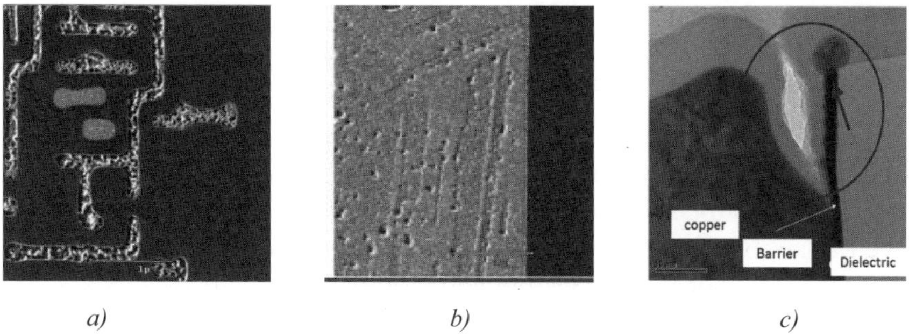

Figure 2.34. *Corrosion observed following CMP: a) copper line generalized corrosion, b) pitting corrosion and c) dissolution of the copper line by galvanic coupling with Ta/TaN copper diffusion barrier*

2.5.2.1.1. Photo-corrosion

The post-CMP cleaning stage may also be crucial, although post-CMP cleaning solutions are extremely diluted. These solutions mostly comprise organic acids and corrosion inhibitors. The task of this cleaning step is to remove the remaining slurry particles, as well as the metallic contamination. During this step, it has been possible to identify photo-corrosion behaviors, specifically the appearance of a potential gradient generated through the illumination of adjacent p–n junctions (see Figure 2.35). The potential difference generated may reach 600 mV.

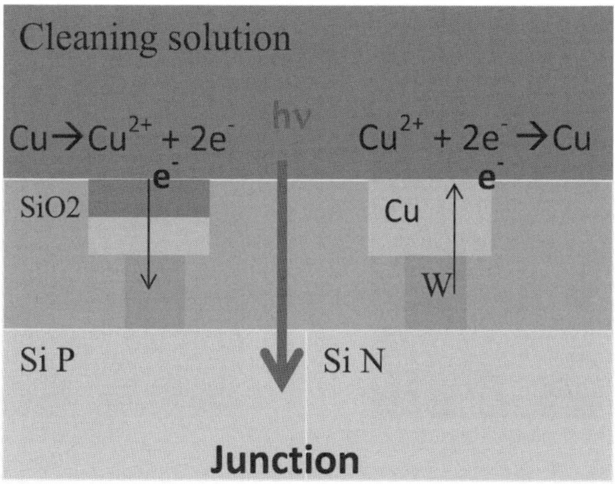

Figure 2.35. *Basic photo-corrosion diagram*

Everything happens as if a potential difference had been applied to two copper electrodes immersed in an electrolyte. The first electrode acts as an anode, where the oxidation and, in particular, copper oxidation reactions take place, and the second electrode acts as a cathode, where the reduction and copper ion (where present in the cleaning solution) reduction reactions take place. Thus, photo-corrosion may lead to certain lines completely dissolving, as well as an overdeposition of copper at other points (see Figure 2.36). The photo-corrosion signature is visible under an optical microscope; some tests for accelerated photo-corrosion are able to quickly distinguish between cleaning solutions of different compositions (see Figures 2.37a and 2.37b).

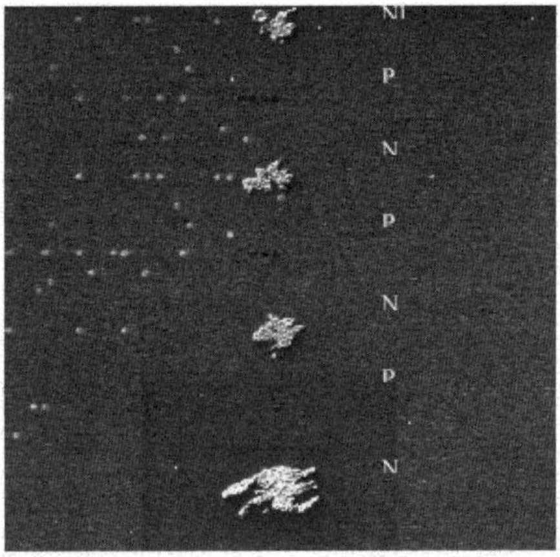

Figure 2.36. *Scanning electron microscope image of photo-corrosion behavior inducing copper dendrite formation and localized dissolution*

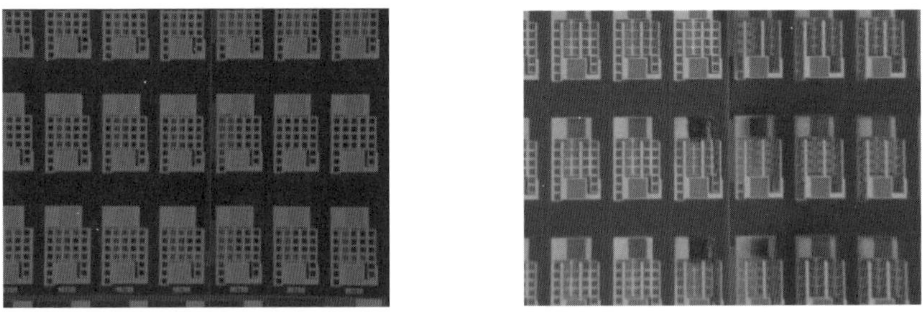

a) b)

Figure 2.37. *Accelerated photo-corrosion tests: immersion for 1 h a) in an optimized cleaning solution under illumination and b) in distilled water*

2.5.2.1.2. Pitting corrosion

In the case of pitting corrosion, the material may be locally "pitted" as a result of a local more significant reagent concentration (see Figure 2.34(b) after CMP). This type of corrosion is particularly observed on aluminum–copper blocks, which are still used in packaging. This corrosion is induced by rupture of the passive Al_2O_3 film under wet conditions and in the presence of halides. The metal becomes fragile and defects form on the aluminum blocks. The composition of the cleaning solution is essential for conservation of the integrity of the structures. Generalized corrosion is significant in the case of sulfuric acid, rapid etching of the metal takes place and pitting corrosion has been highlighted in oxalic acid. Glycolic acid, on the other hand, is an interesting candidate, forming a protective film on the surface of the metal by adsorption. The electrochemical study, which has enabled these conclusions to be drawn, is the subject of [BIL 10].

2.5.2.2. *Galvanic corrosion*

Finally, with galvanic corrosion, one of the stacking materials may be completely dissolved. This case has been observed during the CMP step (see Figure 2.34(c)). The bringing together of two metallic materials with different properties, in this example Cu and Ta, immersed in a conductive solution, the polishing solution, creates a short-circuit that generates an electrical current. An increase in corrosion in the least noble material is observed, and a shrinking or suppression of corrosion is observed in the most noble material. Assiongnon [ASS 06] enhanced this behavior and identified the galvanic corrosion between Cu and Ta, due to electrochemical methods such as the Tafel method, electrochemical impedance spectroscopy and the direct measurement of galvanic corrosion current (zero resistance ameter (ZRA)). These methods will be explained in detail later.

This type of corrosion has also been suspected during CoWP or CoWB self-aligning barrier dissolution during post-etching via cleaning. These barriers have been developed for the 90 nm technology node in order to improve the interconnect reliability. These barriers are situated at the base of the via structure, between two lines. They may disappear completely during post-etching cleaning unless some precautions are taken. Figure 2.38 highlights complete dissolution of the material during cleaning carried out with standard acid solutions [CHH 06]. A detailed electrochemical study has enabled the identification of the type of corrosion, galvanic or generalized, which is preponderant in the dissolution of the barrier based on the constituents of the cleaning solution [PER 07]. Electrochemical techniques have demonstrated that galvanic corrosion is preponderant in a solution of diluted HF, whereas generalized corrosion of cobalt is responsible for dissolution of the barrier in diluted organic acid and HF solutions. The addition of BTA, which is well known for its copper corrosion inhibitor properties, is the only additive tested that enables generalized and galvanic corrosion in cobalt to be limited (see Figure 2.39). In the case of generalized corrosion in cobalt, the cathode reaction, in this case the reduction of oxygen, is the limiting step [BIL 09]. Consequently, elimination of oxygen in the reactional medium induces an associated oxidation barrier and, therefore, a barrier to the corrosion of the affected material. Manufacturers are working on various chamber configurations in "single wafer" equipment, where one wafer is treated at a time; some propose nitrogen blanket formation above the wafer over which the cleaning chemical is dispensed, whereas others propose systems that are completely inert, with oxygen concentrations below 100 ppm. Adjusting the composition of the cleaning solution and configuring the equipment are, therefore, critical for the implementation of an effective cleaning process, which ensures preservation of the integrity of the structure and its dimensions.

Figure 2.38. *Complete corrosion of the CoWP barrier following standard cleaning*

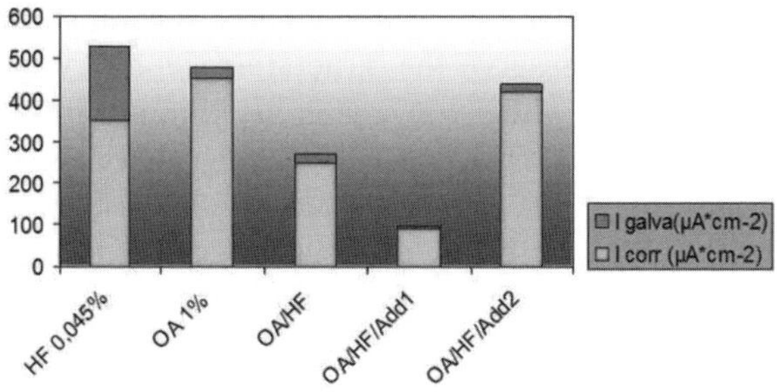

Figure 2.39. *Influence of cleaning solution composition on the generalized corrosion of cobalt and on galvanic coupling with copper*

2.5.3. *Electrochemical diagnostic tools*

To considerably reduce the defectivity and electrical performance losses created by corrosion behaviors, some systematic electrochemical studies have been carried out on some representative samples in order to determine the reactional mechanisms that play an important role. Electrochemical methods provide data on the influence of each of the reagents added to the cleaning solution compositions and enable such solution formulations to be optimized without plate consumption. Optimized solutions are later clean room tested on industrial equipment. Consequently, traditional electrochemical methods enable us to adjust technological processes not only by limiting the costs linked to clean room infrastructures, but also by reducing the consumption of silicon wafers.

The most used electrochemical methods are the drawing of Tafel lines from polarization curves (the corrosion technique traditionally used), which enable the material etch rate to be defined, and impedance spectroscopy, in order to better comprehend the interface mechanisms. The combination of conventional electrochemical methods (such as abandon potential measurement and cyclic voltammetry) with the quartz crystal microbalance device provides supplementary information on gain or loss of material mass in the medium studied.

2.5.3.1. *Etch rate*

The criterion used to qualify a cleaning process is the etch rate of materials that are supposed to be in contact with the cleaning chemistry. The material may be uniformly lightly etched in order to be cleaned from any organic or metallic contamination, of the order of several Ångströms, but the cleaning step must also preserve the structure's geometry and dimensions. While the etch rate for dielectric materials cannot be captured using electrochemical techniques, as they are insulating materials, the etch rate of metallic materials such as copper is accessible using electrochemical techniques. There are two principal electrochemical methods used to obtain this measurement: the Tafel method and the quartz crystal microbalance.

2.5.3.1.1. Tafel method

The etch rates of a material in different mediums can be determined, for example from polarization curves (see Figure 2.40) by extrapolation of Tafel lines. Anode/cathode Tafel lines describe the anode and cathode limits of the Butler–Volmer general equation [BARD]. This equation describes the charge transfer kinetic and links the current density to the potential of the electrode. This equation is consisted of two terms, which are associated with anode and cathode behaviors, respectively. The intersection of the anode and cathode branches gives the corrosion current value and the corrosion potential. Faraday's law, which links the current to the number of oxidized (or reduced) moles, enables the etch rate (i.e. the thickness of material etched per minute) to be calculated, according to equation [2.4]:

$$\text{Etched thickness (nm)} = \frac{I_{\text{Corr}} M t_{\text{Clean}}}{F z \rho} \qquad [2.4]$$

where I_{corr} is the corrosion current, M is the molar mass of the etched material, t_{clean} is the cleaning time, F is the Faraday number, z is the number of electrons exchanged during the redox reaction and ρ is the density of the material.

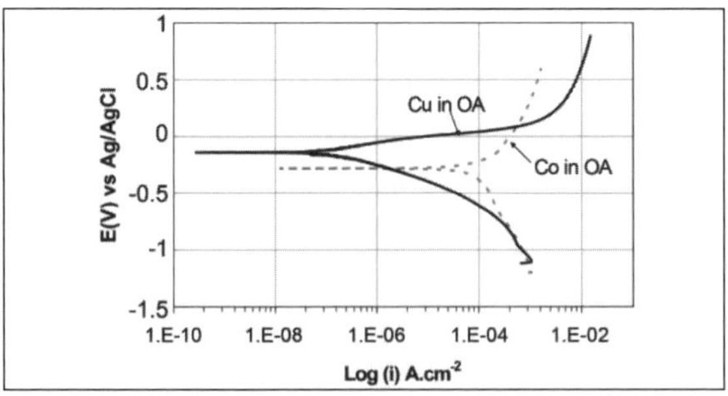

Figure 2.40. *Example of Cu and Co polarization curves for determining the galvanic corrosion and coupling currents (OA = organic acid, in this case Glycolic acid 1% wt T 25°C 1,000 rpm)*

2.5.3.1.2. Direct measurement with a quartz crystal microbalance

Another more direct technique consists of measuring the loss of mass of material immersed into a solution with a quartz crystal microbalance. Thus, the frequency of the device varies linearly with the variation in mass of the electrode, according to Sauerbrey's law, which is represented according to equation [2.5]:

$$\Delta f = \frac{2 f_0^2 \Delta m}{A(\rho_q \mu_q)} \qquad [2.5]$$

where Δf is the frequency variation of the device, Δm is the variation in mass of the electrode, f_0 is the natural frequency of the quartz, A is the specific area of the quartz, ρ_q is the density of the quartz and μ_q is the shear modulus of the quartz.

Specifically, a piezoelectric device can be associated to electrochemical techniques. For example, having effected a deposition of copper onto the surface of one of the two Pt metallizations of the quartz, the quartz is immersed into the cleaning solution. The potential and the mass of the copper electrode thus formed are recorded on the basis of time. The variation in mass quickly becomes linear, which enables the etch rate of the copper to be determined, expressed in A/min. This method has enabled the etch rates for various acids at different concentrations to be compared and an etch rate for copper oxide to be measured. The conclusions of these tests have enabled changes of oxalic acid for glycolic acid for pre-industrial lines to be validated and have enabled the initial glycolic acid concentration to be reduced from 4% to 1%, since the etch rate is identical for any concentration ranging between 1% and 4% (see Figure 2.41). This electrochemical study has been followed by a single wafer validation. Equivalent tendencies have been observed in equipment, but with appreciably different etch rates. The problem with this method is linked to the availability of suitable material depositions.

Other electrochemical methods, such as cyclic voltammetry, combined with quartz crystal microbalance (QCM), may provide information about reactional mechanisms. The chelation of copper and its oxide using different organic acids [PER 06] have thus emphasized the singular behavior of oxalic acid, which is traditionally used as a chelatant during the post-etch cleaning stage. Indeed, during the oxidation of copper in 3.5% oxalic acid solution, first a loss and then an increase in mass is observed, which indicates that cleaning begins with a slight dissolving of copper, followed by the formation of a passivation layer, which ensures the integrity of the structure (see Figure 2.42).

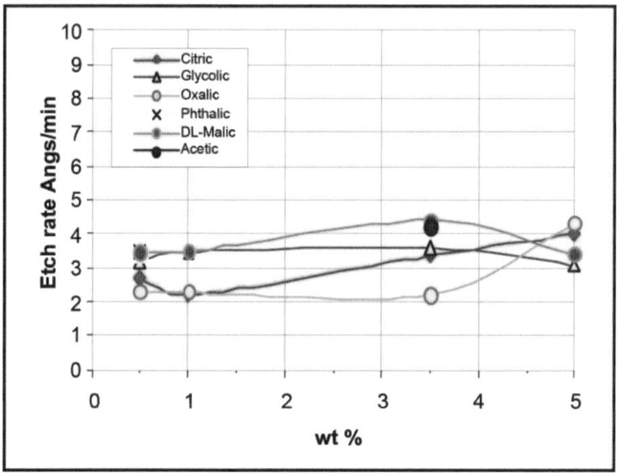

Figure 2.41. *Etch rate of Cu based on the mass concentration of acid determined using QCM*

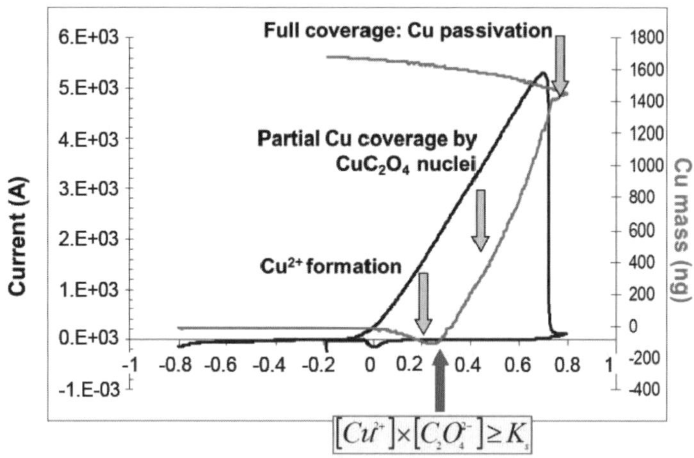

Figure 2.42. *Cyclic voltammetry and quartz crystal microbalance recorded for a copper electrode in 3.5% oxalic acid. Copper passivation mechanism in oxalic acid*

2.5.3.1.3. Quantification of galvanic coupling

The polarization curve of two metals also enables us to access the value for the galvanic corrosion current. Indeed, as one acts as an anode and the other as a cathode, the intersection of the anode and cathode branches enables the potential and the corrosion current generated by this bimetallic coupling to be determined. The galvanic copper–cobalt coupling has been determined using interpolation of their polarization curves (see Figure 2.40) [BIL 09a]. Another method consists of directly measuring the coupling current by using a dual electrode system containing the two metals from the galvanic coupling in the relevant solution. This technique is known as ZRA.

2.5.3.2. *Identification of interface mechanisms*

The understanding of mechanisms, which take place at the interface between solid and liquid during the cleaning and CMP stages, can also be studied using electrochemical impedance spectroscopy. This technique consists of enforcing a sinusoidal perturbation to a voltage or current signal and recording the current or voltage response. The complex impedance resulting from the U/I ratio is recorded on the basis of frequency. The graphical representation often used is the Nyquist diagram, where the imaginary part of the complex number is represented on the *y*-axis and the real part is represented on the *x*-axis. The semi-circles obtained are the signature of the events that take place at the interface. By analogy with electrical circuits, the interface is modeled using RLC, which enables characteristic measurements such as charge transfer resistance and the dual layer thickness to be calculated. This method has also good performance in the identification of adsorption behaviors and in the release of corrosion sub-products. This extremely attractive technique, through the *in situ* analysis of the interface and its ease of implementation, is, however, difficult to exploit. Some physicochemical characterizations are necessary in order to validate models that exist to define the equivalent RLC circuit. For example, the study of the impact of adding BTA to glycolic acid for post-etch via cleaning has emphasized the etch action of the organic acid for which an inductive loop, which is characteristic of dissolution product release, is observed (see Figure 2.43). The BTA forms a protective layer over the copper and the signature of this passivation is the second capacitive loop. Associating these two chemical agents in the same solution produces antagonistic effects: glycolic acid etches the passivant Cu(I)-BTA layer formed over the copper; the two behaviors of corrosion and passivation occur concurrently [BIL 09a].

In general, electrochemistry seems to be an indispensable tool for developing wet microelectronic and nanotechnological processes. Its capacity for examination of any type of interface is a promising future within high added value technologies, such as the rise in electronic devices associated with biological functions.

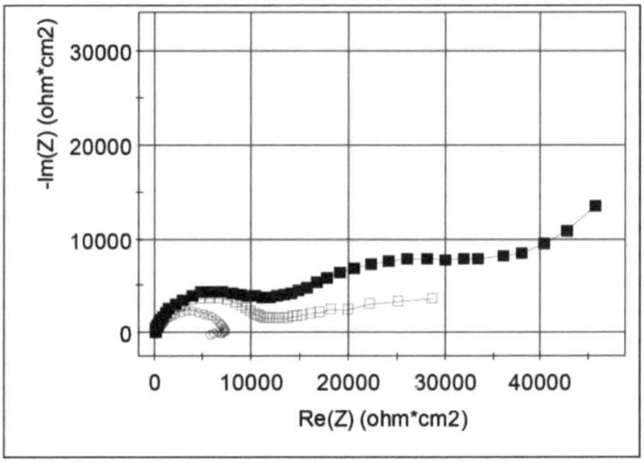

Figure 2.43. *Nyquist diagram obtained for a copper electrode in different solutions: μ: 1% glycolic acid, ν: 0.1 % BTA and ρ: 1% glycolic acid–0.1% BTA mixture*

2.5.4. *Equipment for cleaning copper interconnections*

With the introduction of low-k dielectric materials for copper dual damascene interconnection of semiconductor circuits, cleaning processes not only have to provide efficient removal of etching residues, but also need to either passivate or remove sidewall damage that is left by the plasma etching process [BRO 09]. Typically, formulated multicomponent chemical solutions are focused on passivating the low-k damage and preserving the etched pattern dimensions. Dilute HF offers a lower cost process by removing the damaged layer. If a dilute HF cleaning approach is pursued, a negative bias in pattern dimensions is required to account for the removal of low-k material from the trench and via sidewalls, which would ultimately lead to wider Cu lines and vias.

Formulated chemical solutions often include a passivating component to prevent loss of copper at the bottom of the etched via. As previously described, when dissimilar metals are present, such as when self-aligned electromigration barriers are used, galvanic corrosion can occur. With dilute

HF solutions, copper loss and galvanic corrosion can be significantly reduced or eliminated by removing most of the dissolved oxygen from the solution. Studies show that dissolved oxygen is a key component in the galvanic corrosion mechanism and its removal disrupts that corrosion pathway [BIL 09b].

Removal of oxygen from water and dilute HF solutions is fairly straightforward using commercially available contactors or complete systems [GOT 11]. Removal of dissolved oxygen to below the 100 ppb level appears to be sufficient to reduce the copper loss to less than 0.4Å/min and suppresses galvanic corrosion effects when a CoWP self-aligned barrier is used on top of the copper lines [LAU 09a]. Implementation of a deoxygenated process in the treatment of semiconductor circuitry requires not only deoxygenating the dilute HF treatment solution, but also maintaining the deoxygenated state throughout the time that the solution contacts the surface of the substrate. Atmospheric oxygen diffuses through polymers typically used in equipment construction and can reoxygenate the solution [GOT 11]. Also, atmospheric oxygen can easily be redissolved if the solution is dispensed on the substrate in an atmospheric environment [LAU 09a]. Continuous deoxygenation and treatment in a purged, controlled circulating system is necessary to achieve corrosion-free cleaning results.

The most common equipment for cleaning copper interconnection in advanced manufacturing is the single-wafer spin type (Figure 2.44(a)). A chemical is dispensed onto the spinning wafer from a nozzle or from a spraybar. If a nozzle is used, it is often mounted on a swingarm and scanned over the spinning wafer surface while a chemical is dispensed (Figure 2.44(b)). In the case of the spraybar, it is fixed above the spinning wafer and extends from the center to the edge of the wafer (Figure 2.44(c)). In both cases, modification is required to conduct the cleaning process in a low-oxygen environment, which is required to prevent the corrosion of copper and self-aligned metal barriers.

Different approaches have been used to achieve a reduced oxygen environment in single-wafer processing tools. In one case, a modification to the typical open-cup chamber is made by adding a so-called "shield-plate" that is held in close proximity to the surface of the spinning wafer, creating a narrow gap through which the deoxygenated chemical and an inert gas, such as dry nitrogen, is flowed from the center toward the edge of the wafer, as shown in (Figure 2.45) [IMA 09]. Optimization of gap, gas flow rate, dispense rate of chemical solution and spin speed of the substrate is necessary to prevent atmospheric oxygen from diffusing at the outside edge of the substrate.

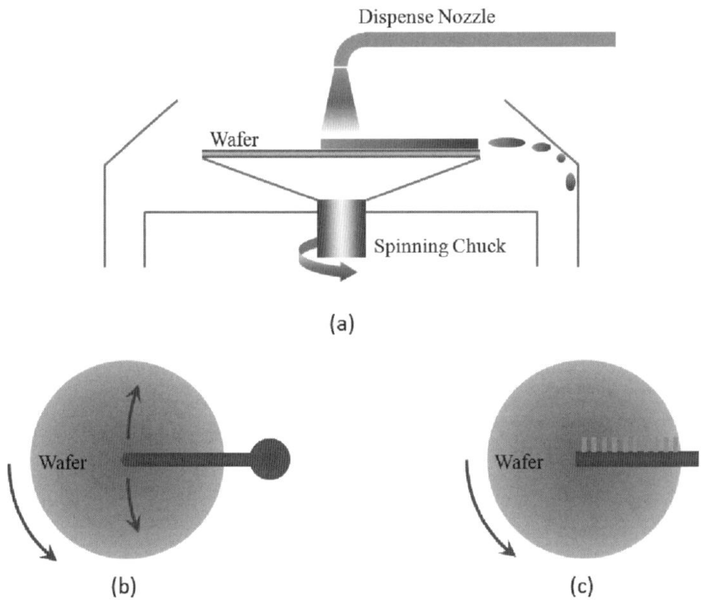

Figure 2.44. *a) Simple schematic of single-wafer cleaning chamber. b) Top view of swingarm scanning nozzle chemical dispense. c) Top view of fixed spraybar chemical dispense*

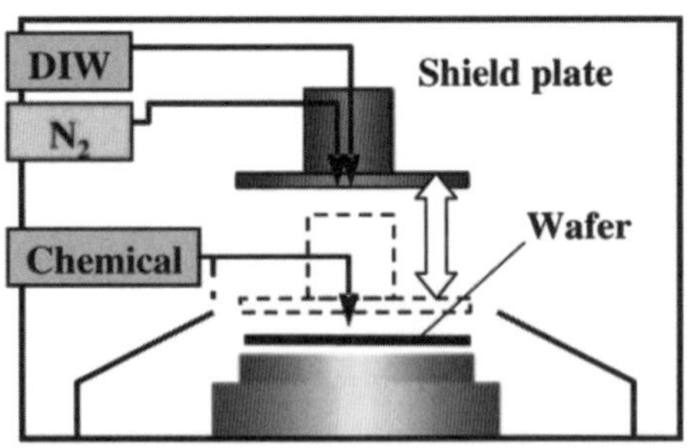

Figure 2.45. *Single wafer cleaning chamber with shield plate to provide purged environment near wafer surface [IMA 09]*

In another approach, a closed chamber is created by providing cover over the top of the chamber that is sealed at the outer edge. Inert gas is flowed into the chamber from the center of the cover and then exits at the edge of the wafer and into the chamber exhaust (Figure 2.46). Additional inert gas is flowed into the chamber at the outside edge and prevents atmospheric oxygen from reentering the chamber and reoxygenating the solution [LAU 09b].

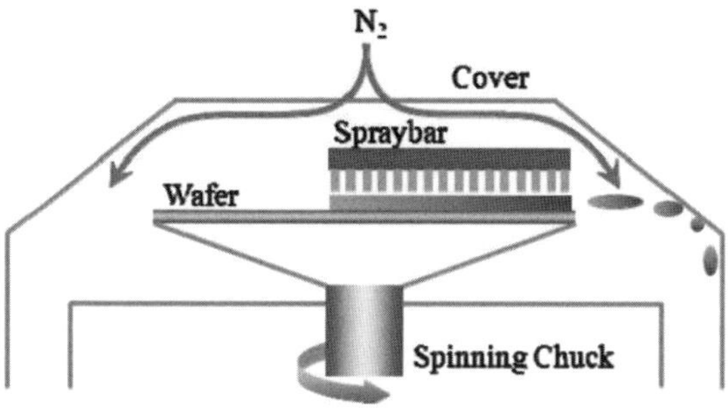

Figure 2.46. *Single-wafer cleaning chamber with sealed cover and fixed spraybar to provide purged environment*

An illustration of the importance of both deoxygenating the cleaning solution and purging the process chamber is shown in Figure 2.47 [LAU 09a]. In this figure, three different combinations of chemicals and environments are compared for their effect on copper etching during the dispense of dilute HF at the center of the wafer in a closed chamber. When the HF solution is not deoxygenated, but the chamber is purged and filled with nitrogen, copper etching is high at the center of the wafer where the chemical is first dispensed. However, as the chemical flows across the wafer, dissolved oxygen diffuses out into the nitrogen ambient, leading to decreased copper etch toward the edge of the wafer (Figure 2.47(a)). When the dilute HF solution is deoxygenated, but the chamber is filled with atmospheric gases, copper etching is suppressed at the center of the wafer. In this case, as the chemical flows across the wafer surface, oxygen is reabsorbed from the atmospheric gases in the chamber, leading to increased copper etch toward the edge of the wafer (Figure 2.47(b)). When both the HF solution is deoxygenated and the chamber is purged and filled with nitrogen, copper etching is suppressed across the entire wafer surface (Figure 2.47(c)).

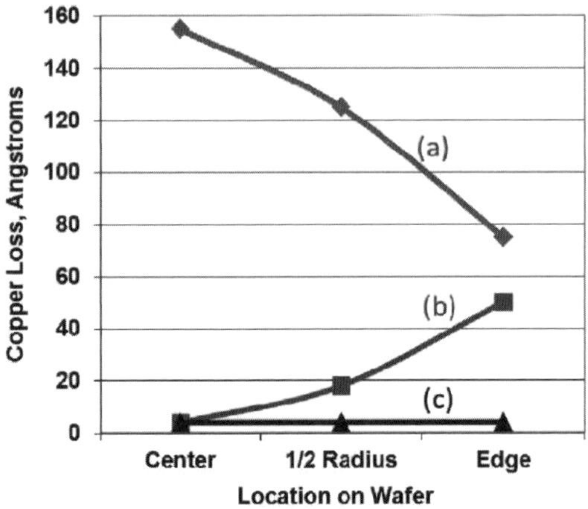

Figure 2.47. *Measured copper loss during 5-min center dispense of 0.1% dilute HF solution under conditions where a) the solution is saturated with atmospheric O_2 and the chamber is purged with N_2, b) the solution is deoxygenated to <100 ppb O_2 and the chamber is filled with ambient atmosphere, and c) the solution is deoxygenated and the chamber is purged [LAU 09a]*

Organic acid has been added to the deoxygenated dilute HF for cleaning copper interconnections [PER 06]. An organic acid, such as glycolic and citric acid, can act as a chelating agent, helping to keep dissolved copper in solution. When using a TiN metal hard mask for patterning low-k dielectric, addition of organic acid to the dilute HF solution can help prevent redeposition of dissolved copper on the TiN surface. A properly designed circulation system provides an easy way for organic acid to be added to the dilute HF solution.

To maintain a repeatable cleaning process on all wafers, it is necessary to continuously measure and carefully control the temperature and concentration of the solution in the circulation loop which supplies chemical solution to several single-wafer cleaning chambers simultaneously. HF concentrations in the range of 500–1,500 ppm have been used with organic acid concentration of up to 1% by weight. Typically, concentration is controlled to +/–1% of set point. Solution temperature in the range of 25–60°C has been used. Typically, the temperature is controlled to +/–0.5°C or better. Carefully designed filling, circulating and heating hardware is required to maintain this level of control while dispensing and refilling and blending the chemical solution during wafer cleaning operations.

Although the discussion to this point has focused on dilute HF solutions, formulated multicomponent cleaning solutions are also used in cleaning copper interconnections. Similar equipment, as shown in Figure 2.44, is also used for dispensing formulated chemical solutions. Due to the higher cost, after a formulated chemical solution is dispensed on a wafer surface, it is usually collected and reused by adding it back into the circulating chemical solution. Since this may bring contaminants from the wafers into the circulating cleaning solution, advanced filtration techniques are used to keep the circulating solution free of particles. Also, *in situ* concentration monitors for key components of the formulated multicomponent solution are used to ensure that the reused, circulating solution stays within prescribed tolerances and to determine when a completed chemical solution changeout is required.

In summary, design and operation of the cleaning equipment is important to ensure that the desired chemical reactions do occur on the wafer surface and the undesirable chemical reactions are suppressed.

2.6. General conclusions and perspectives

These pages have presented how chemical phenomena and their control have been involved in the development and implementation of current interconnects in ULSI devices: we can take, for example, the establishment of filling methods based on the electrodeposition Cu. These methods have allowed the development of damascene integration. Similarly, optimizing insulating film properties requires a detailed understanding of the chemical reactions involved in the removal of porogen. Finally, the control of surface treatments can be only done by taking into account the interactions between chemical solutions and materials involved.

Although the Damascene Cu technology developed at the end of the 20th Century might be considered as being relatively recent, its limitations have already been identified. Thus, in 1998, a first study presented experimental proof of what was going to be called the "Cu scaling effect" [CHE 98]. In this study, the authors reported a notable increase in resistivity when the width of lines decreased. Later, this effect was confirmed by numerous groups [KUA 00, HIN 01, WEN 01, STE 02, GUI 03]. From 2001, it was integrated into ITRS as one of the main problems of interconnect technology [INT 00].

The reasons for the increase in resistivity should be found in the physics of the last century. For a copper line with a critical size less than 100 nm, a bamboo-type granular structure and a grain size proportional to the line size are experimentally observed [STE 05, BRA 10]. At these dimensions, the distances between grain boundaries and walls are of the order of magnitude of the mean free path of the electrons (38 nm at ambient temperature). Thereby, diffusion and reflection of the electronic wave function at the interfaces occurs, resulting in an increase in resistivity. The behavior itself was observed at the start of the 19th Century in silver films [THO 01]

With regard to the insulator, the integration of more porous films into interconnects is, today, still a difficult task. Porosity itself is the principal cause of this difficulty:

– Because of the low number of strong bonds, films are mechanically poor, if not mediocre. This impacts heavily the integrity of the structure during the chemical mechanical polishing and packaging stages.

– The vast majority of films used have open porosity: gas or liquid diffusion within the film is greatly facilitated. The dielectric can be easily degraded during etching, cleaning, polishing and stripping. It will lose its ability to insulate and its low dielectric constant. The metal may diffuse in the insulator during deposition, leading to an increase in leakage currents.

Significant work on elementary processes and the setting up of innovative integration strategies have enabled a certain number of difficulties to be overcome. To realize the work, readers must delve into the conference proceedings of IITC, Advanced Metallization Conference (AMC) and Materials for Advanced Metallization (MAM).

With regard to reliability, while a downturn in the increase in current densities has been observed in recent years, some 1×10^6 A/cm² is still expected by the 2020s. While Cu is intrinsically reliable, reliabilities studies have shown that, by reducing the line size, the microstructure of Cu plays a major role in resistance to electromigration [AR 03, HU 07]; the Cu finds a large number of diffusion paths within the increased grain boundary surfaces. The activation energy of the diffusion at the grain boundaries is less than that of the diffusion in the volume, as shown at the start of this chapter. Later works show the importance of local microstructures on "early failure". Strong integration density makes this a crucial point. Several solutions aiming to block diffusion paths have been introduced into industrial integration diagrams and have been presented in section 2.4: the covering of the top of the line with an alloy which ensures a more resistant interface (such as barrierless CoWP, CVD Co deposition and CuSiN treatment) or the insertion of alloy elements (such as Al, Mn and Ag). Some of these solutions are nowadays industrialized. It should be noted that increased reliability occurs to the detriment of line resistance and therefore performance.

The continual reduction in dimensions and the pursuit of Moore's law remains a challenge for interconnects. New evolutions or revolutions are predicted.

Some projections can be attempted and the future may come down to three main points:

– Pursuing yet more research into performance, using technologies similar to those currently available, is still possible. The most promising route is the integration of air instead of insulating materials [GRA 08].

– It is possible to foresee a slow evolution of technologies with a view to making them more reliable (such as work on integration of the dielectric and encapsulation of metals), more cost-effective (work on low-cost processes, such as electrochemical processes) or more compatible with a sustainable future (such as reduction in polishing effluents and elimination of rare earths)

– Finally, a profound revolution may emerge which passes through new paradigms.

Let us concentrate on the latter option.

Some studies are highly advanced on specific points: optical or wireless interconnects for connections between chip blocks. As during the introduction of Cu, the solution could come from materials, such as carbon nanotubes (CNT), graphene, topological insulators or high-temperature superconductors. These could be used in conjunction with other state variables (such as spin wave and optical). Signal propagation speed, consumption and integration are critical points to consider. Note that only C(CNT, graphene) has been the subject of detailed and continuous evaluations [DIJ 10a, DIJ 10b]. We can notice the Fujitsu effort, an effort prolonged within Tsukuba community, which has enabled good quality CNT vias to be obtained [NIH 12]. Associating graphene and CNT still appears to be a potential solution. There are still a large number of points to be tested, the most critical of which seem to be electrical contact management, the localization and control of large-scale film and tube quality. Cleaning, chemical self-positioning and catalytic control are some problems to be considered.

Ultimately, numerous solutions could come from the world of chemistry and these could subsequently be applied to the world of micro- and nanotechnologies.

2.7. List of Abbreviations

BEOL:	Back End Of the Line
CMP:	Chemical Mechanical Polishing
ILD:	InterLine Dielectric
CVD:	Chemical Vapor Deposition
PVD:	Physical Vapor Deposition
PECVD:	Plasma Enhanced Chemical Vapor Deposition
TEOS:	TetraEthylOrthoSilicate
FSG:	Fluoro Silica Glass
OSG:	Organo Silica Glass
3MS:	TriMethylSilane
4MS:	TetraMethylSilane
OMCTS:	OctaMethylCycloTetraSiloxane
DMDMOS:	DiMethylDiMethoxyOxySilane
PHSD:	PolyHydrogenSilsesQuioxane
PMSQ	PolyMethylSilsesQuioxane
DEMS:	DiEthoxyMethylSilane
TMCTS:	TetraMethylCycloTetraSiloxane
CPO:	CycloPentene Oxide
CHO:	CycloHexene Oxide
NBD:	NorBornaDiene
ATRP:	AlphaTeRPinene
DMPS:	DiMethylPhenylSilane
HMDSN:	HexaMethylDiSilazaNe
TSV:	Through-Silicon Via
SAM:	Self Assembled Monolayer
SPS:	SulfoPropyldiSulfide

CEAC:	Curvature-Enhanced Accelerator Coverage
NIST:	National Institute of Standards and Technology
LETI:	*Laboratoire d'Electronique des Technologies de l'Information*
DMAB:	DiMethylAmine Borane
ZRA:	Zero Resistance Ameter
QCM:	Quartz Crystal Microbalance
BTA:	BenzoTriAzole
AMC:	Advanced Metallization Conference
MAM:	Materials for Advanced Metallization
CNT:	Carbon NanoTube

2.8. Bibliography

[ADO 11] ADOLF J., LANDAU U., "Predictive analytical fill model of interconnect metallization providing optimal additives concentrations", *Journal of the Electrochemical Society*, vol. 158, no. 8, pp. D469–D476, 2011.

[AIN 72] AINSLIE N.G., D'HEURLE F.M., WELSS O.C., "Coating, mechanical constraints, and pressure effects on electromigration", *Applied Physics Letters*, vol. 20, no. 4, pp. 173–174, 1972.

[ALM 07] ALMOG R.O., SVERDLOV Y., GOLDFARB I., *et al.*,"CoWBP capping barrier layer for sub 90 nm Cu interconnects", *Microelectronic Engineering*, vol. 84, pp. 2450–2454, 2007.

[AND 98] ANDRICACOS P., UZOH C., DUKOVIC J., *et al.*, "Damascene copper electroplating for chip interconnections", *IBM Journal of Research and Development*, vol. 42, pp. 567–574, 1998.

[ARM 10] ARMINI S., CARO A., "Materials engineering for future interconnects: 'Catalyst-free' electroless Cu deposition on self-assembled monolayer alternative barriers", *Journal of the Electrochemical Society*, vol. 157, pp. D74–D80, 2010.

[ARM 11] ARMINI S., TOKEI Z., VOLDERS H., *et al.*, "Impact of terminal effect on Cu electrochemical deposition: filling capability for different metallization options", *Microelectronic Engineering*, vol. 88, pp. 754–759, 2011.

[ARN 03] ARNAUD L., BERGER T., REIMBOLD G., "Evidence of grain-boundary versus interface diffusion in electromigration experiments in copper damascene interconnects", *Journal of Applied Physics*, vol. 93, no. 1, pp. 192–204, 2003.

[ARN 07] ARNAL V., FARCY A., AIMADEDDINE M., *et al.*, "Materials and processes for high signal propagation performance and reliable 32 nm node BEOL", *Proceedings of the International Interconnect Technology Conference (IITC)*, Piscataway, NJ, pp. 1–3, 2007.

[ASH 09] ASHER T., INBERG A., GLICKMAN E., *et al.*, "Formation and characterization of low resistivity sub-100 nm copper films deposited by electroless on SAM", *Electrochimica Acta*, vol. 54, pp. 6053–6057, 2009.

[ASS 06] ASSIONGBON K.A., EMERY S.B., GORONTLA V.R.K., *et al.*, "Electrochemical impedance characteristics of Ta/Cu contact regions in polishing slurries used for chemical mechanical planarization of Ta and Cu: considerations of galvanic corrosion", *Corrosion Science*, vol. 48, no. 2, pp. 372–388, 2006.

[AU 10] AU Y., LIN Y., KIM H. *et al.*, "Selective chemical vapor deposition of manganese self-aligned capping layer for Cu interconnections in microelectronics", *Journal of the Electrochemical Society*, vol. 157, no. 6, pp. D341–D345, 2010.

[AU 12] AU Y., WANG Q.M., LI H. *et al.*, "Vapor deposition of highly conformal copper seed layers for plating through-silicon vias (TSVs)", *Journal of the Electrochemical Society*, vol. 159, pp. D382–D385, 2012.

[BAK 07] BAKLANOV M., GREEN M., MAEX K., *"Dielectric films for advanced microelectronic"*, John Wiley and Sons, West Sussex, England, 2007.

[BAK 11] BAKLANOV M., ZHAO L., VAN BESIEN E., *et al.*, "Effect of porogen residue on electrical characteristics of ultra low-k materials", *Microelectronic Engineering*, vol. 88, no. 6, pp. 990–993, 2011.

[BAK 12] BAKLANOV M., "Advanced Interconnects for ULSI Technology", BAKLANOV M., HO P., EHRENFRIED ZSCHECH E. (eds), John Wiley & Sons, Chichester, 2012.

[BAL 04] BALDACCI A., RIVERO C., GERGAUD P., *et al.*, "Stresses in copper blanket films and damascene lines: measurements and finite element analysis", *34th European Solid-State Device Research Conference (ESSDERC)*, Leuven, Belgium, pp. 105–108, 2004.

[BAR 06] BARMAK K., CABRAL J., RODBELL K.P., *et al.*, "On the use of alloying elements for Cu interconnect applications", *Journal of Vacuum Science, Technology B*, vol. 24, pp. 2485–2498, 2006.

[BAU 97] BAUD L., PASSEMARD G., GOBIL Y., *et al.*, "Integration of a stack of two fluorine doped silicon oxide thin films with interconnect metallization for a sub-0.35 μm inter-metal dielectric application", *Microelectronic Engineering*, vol. 37, pp. 261–269, 1997.

[BER 01] BERGER T., Analyse du comportement d'interconnexions damascènes en cuivre testées en électromigration, Doctoral Thesis, INSA Lyon, 2001.

[BER 07] BERNARD M.C., GABRIELLI C., JOIRET S., *et al.*, "Investigations on the corrosion of copper patterns in the course of the post-CMP cleaning of integrated electronic microcircuits in oxalic acid aqueous solutions", *Electrochimica Acta*, vol. 53, no. 3, pp. 1325–1335, 2007.

[BEY 02] BEYER G., SATTA A., SCHUHMACHER J., *et al.*, "Development of sub-10-nm atomic layer deposition barriers for Cu/low-k interconnects", *Microelectronic Engineering*, vol. 64, pp. 233–245, 2002.

[BIL 04] BILODEAU S.M., BOROVIK A.S., EBBING A.A., *et al.*, "Chemical Routes to Improved Mechanical Properties of PECVD Low K Thin Films", *MRS Proceedings*, 812, F6.18, 2004, DOI:10.1557/PROC-812-F6.18.

[BIL 06] BILODEAU S.M., CHEN P., GIANNETTO W., *et al.*, "Porogen precursors for ULK (K < 2.2) PECVD dielectrics", *Proceedings of the Advanced Metallization Conference*, p. 367, 2006.

[BIL 09a] BILOUK S., NOGUEIRA R.P., BROUSSOUS L., *et al.*, "Electrochemical behavior of copper and cobalt in post-via-etch cleaning solutions", *Microelectronic Engineering*, vol. 86, no. 10, pp. 2038–2044, 2009.

[BIL 09b] BILOUK S., PERNEL C., BROUSSOUS L., *et al.*, "Electrochemical behaviour of cobalt in post-via etch cleaning solutions", *Solid State Phenomena*, vol. 145–146, pp. 343–346, 2009.

[BIL 10] Thesis Work presented the 2010/03/30 untitled "Corrosion du cuivre et des alliages CoWP et AlCu dans les microprocesseurs 32nm et les imageurs: vers des solutions de nettoyage optimisées".

[BLA 06] BLAMPEY B., GALLITRE M., FLECHET B., *et al.*, "Delay and crosstalk on future 32 nm node interconnects: impact of ULK-air-gap architecture", *IEEE Workshop on Signal Propagation on Interconnects*, pp. 75–78, 2006.

[BLE 66] BLECH I.A., SELLO H., "The failure of thin aluminum current-carrying strips on oxidized silicon", *IEEE 5th Annual Symposium on the Physics of Failure in Electronics*, Piscataway, NJ, pp. 496–505, 1966.

[BLE 76a] BLECH I.A., HERRING C., "Stress generation by electromigration", *Applied Physics Letters*, vol. 29, no. 3, pp. 131–133, 1976.

[BLE 76b] BLECH I.A., "Electromigration in thin aluminum films on titanium nitride", *Journal of Applied Physics*, vol. 47, no. 4, pp. 1203–1208, 1976.

[BON 08] BONILLA G., CLEVENGER L.A., HU C.K., *et al.*, "Advanced bilayer low-k dielectric cap for reliable high performance Cu-Low-k interconnects in ULSI devices", *Proceedings of the Advanced Metallization Conference*, pp. 673–679, 2008.

[BRA 10] BRANDSTETTER S., RAUCH E.F., CARREAU V., *et al.*, "Pattern size dependence of grain growth in Cu interconnects", *Scripta Materialia*, vol. 63, no. 10, pp. 965–968, 2010.

[BRO 09] BROUSSOUS L., PUYRENIER W., REBISCOUL D., *et al.*, "Porous low-k wet etch in HF-based solutions: focus on cleaning process window, pore-sealing and κ recovery", *Solid State Phenomena*, vol. 145–146, pp. 295–302, 2009.

[BUR 03] BURKEY D.D., GLEASON K.K., "Structure and mechanical properties of thin films deposited from 1,3,5-trimethyl-1,3,5-trivinylcyclotrisiloxane and water", *Journal of Applied Physics*, vol. 93, no. 9, pp. 5143–5150, 2003.

[BUR 04a] BURKEY D.D., GLEASON K.K., "Temperature-resolved Fourier transform infrared study of condensation reactions and porogen decomposition in hybrid organosilicon porogen films", *Journal of Vacuum Science Technology A*, vol. 22, pp. 61–70, 2004.

[BUR 04b] BURKEY D.D., GLEASON K.K., "Organosilicon thin films deposited from cyclic and acyclic precursors using water as an oxidant", *Journal of the Electrochemical Society*, vol. 151, p. F105, 2004.

[CAR 95] CARPIO R., FARKAS J., JAIRATH R., "Initial study on copper CMP slurry chemistries", *Thin Solid Films*, vol. 266, no. 2, pp. 238–244, 1995.

[CAR 06] CARRON D., http://en.wikipedia.org/wiki/Integrated_circuit, 2006.

[CAR 08] CARO A.M., MAES G., BORGHS G., *et al.*, "Screening self-assembled monolayers as Cu diffusion barriers", *Microelectronic Engineering*, vol. 85, no. 10, pp. 2047–2050, 2008.

[CAS 05] CASTEX A., FAVENNEC L., JOUSSEAUME V., *et al.*, "Study of plasma mechanism of hybrid a-SiOC:H low-k film deposition from decamethylcyclopentasiloxane and cyclohexene oxide", *Microelectronic Engineering*, vol. 82, no. 3–4, pp. 416–421, 2005.

[CAS 08] CASTEX A., JOUSSEAUME V., DEVAL J., *et al.*, "Ultra low κ films by using a plasma enhanced chemical vapor deposition porogen approach: study of the precursor reaction mechanisms", *Journal of Vacuum Science Technology A*, vol. 26, pp. 1343–1354, 2008.

[CHA 02] CHANDRA G., KAPUR P., SARASWAT K.C., "Scaling trends for the on chip power dissipation", *Proceedings of the IEEE International Interconnect Technology Conference (IITC)*, pp. 170–172, 2002.

[CHA 09a] CHARLES-ALFRED C., JOUSSEAUME V., GRANIER A., "Plasma enhanced chemical vapor deposition of a-SiC:H for advanced microelectronics interconnections", *Proceedings of the 19th International Symposium on Plasma Chemistry (ISPC)*, pp. 356–359, 2009.

[CHA 09b] CHARLES-ALFRED C., JOUSSEAUME V., GRANIER A., "Impact of organic precursors on the deposition of advanced a-SiC:H low-k copper diffusion barrier layer", *Proceedings of the Advanced Metallization Conference (AMC)*, pp. 381–386, 2009.

[CHA 11] CHAKRABORTY T., GREENSLIT D., EISENBRAUN E.T., "Nucleation and growth characteristics of electroplated Cu on plasma enhanced atomic layer deposition-grown RuTaN direct plate barriers", *Journal of Vacuum Science, Technology B: Microelectronics and Nanometer Structures*, vol. 29, no. 3, pp. 030605–8, 2011.

[CHE 98] CHEN F., GARDNER D., "Influence of line dimensions on the resistance of Cu interconnections", *IEEE Electron Device Letters*, vol. 19, no. 12, pp. 508–510, 1998.

[CHH 06] CHHUN S., GOSSET L.G., MICHELON J., *et al.*, "Cu surface treatment influence on Si adsorption properties of CuSiN self-aligned barriers for sub-65 nm technology node", *Microelectronic Engineering*, vol. 83, no. 11–12, pp. 2094–2100, 2006.

[CHI 03a] CHIANG C.C., WU Z.C., WU W.H., *et al.*, "Physical and barrier properties of plasma-enhanced chemical vapor deposited a-SiC:H films from trimethylsilane and tetramethylsilane", *Japanese Journal of Applied Physics*, vol. 42, pp. 4273–4277, 2003.

[CHI 03b] CHIANG C.C., CHEN M.C., KO C.C., *et al.*, "Physical and barrier properties of plasma-enhanced chemical vapour deposited α-SiCN:H films with different hydrogen contents", *Japanese Journal of Applied Physics*, vol. 42, pp. 5246–5250, 2003.

[CHO 08] CHOU Y.-H., SUNG Y., BAI C.-Y., *et al.*, "Effects of molybdate concentration on the characteristics of Ni-Mo-P diffusion barriers grown by nonisothermal electroless deposition", *Journal of the Electrochemical Society*, vol. 155, no. 9, pp. D551–D557, 2008.

[CHO 10] CHOI B.H., LIM Y.H., LEE J.H., *et al.*, "Preparation of Ru thin film layer on Si and TaN/Si as diffusion barrier by plasma enhanced atomic layer deposition", *Microelectronic Engineering*, vol. 87, pp. 1391–1395, 2010.

[COL 93] COLTER T.J., CHAPPLE-SOKOL J., "High quality plasma-enhanced chemical vapour deposited silicon nitride films", *Journal of the Electrochemical Society*, vol. 140, no. 7, pp. 2071–2075, 1993.

[COT 99] COTE D.R., NGUYEN S.V., STAMPER A.K., *et al.*, "Plasma-assisted chemical vapour deposition on dielectric thin films for ULSI semiconductor circuits", *IBM Journal of Research Development*, vol. 43, pp. 5–38, 1999.

[DAS 05] DA SILVA S., MOURIER T., HAUMESSER P.H., *et al.*, "Gap fill enhancement with medium acid electrolyte for the 45 nm node and below", *Proceedings of the Advanced Metallization Conference (AMC)*, pp. 513–517, 2005.

[DEC 07] DECORPS T., Barrières auto alignées déposées par voie electroless pour la fabrication des interconnexions au nœud 32 nm, Doctoral Thesis, 2007.

[DIJ 10a] DIJON J., FOURNIER A., SZKUTNIK P.D., *et al.*, "Carbon nanotubes for interconnects in future integrated circuits: the challenge of the density", *Diamond and Related Materials*, vol. 19, no. 5–6, pp. 382–388, 2010.

[DIJ 10b] DIJON J., OKUNO H., FAYOLLE M., *et al.*, "Ultra-high density carbon nanotubes on Al-Cu for advanced vias", *Proceedings of the International Electronic Devices Meeting (IEDM),* p. 33.4, 2010.

[DIN 94] DING P., LANFORD W.A., HYMES S., *et al.*, "Oxidation resistant high conductivity copper films", *Applied Physics Letters*, vol. 64, no. 21, pp. 2897–2899, 1994.

[DIX 11] DIXIT V.K., NEISHI K., AKAO N., *et al.*, "Structural and electronic properties of a Mn oxide diffusion barrier layer formed by chemical vapor deposition", *IEEE Transactions on Device and Materials Reliability*, vol. 11, no. 2, pp. 295–302, 2011.

[DUB 05] DUBOIS G., MAGBITANG T., VOLKSEN W., *et al.*, "New spin-on oxycarbosilane low-k dielectric materials with exceptional mechanical properties", *Proceedings of the International Interconnect Technology Conference (IITC)*, pp. 226–228, 2005.

[EDE 97] EDELSTEIN D., HEIDENREICH J., GOLDBLATT R., *et al.*, "Full copper wiring in a sub-0.25 µm CMOS ULSI technology", *Proceedings of the International Electron Devices Meeting*, pp. 773–776, 1997.

[EIN 05] EINATI H., BOGUSH V., SVERDLOV Y., *et al.*, "The effect of tungsten and boron on the Cu barrier and oxidation properties of thin electroless cobalt-tungsten-boron films", *Microelectronic Engineering*, vol. 82, no. 3–4, pp. 623–628, 2005.

[FAN 11] FANG J.S., LIN J.H., CHEN B.Y., *et al.*, "Ultrathin Ru-Ta-C barriers for Cu metallization", *Journal of the Electrochemical Society*, vol. 158, no. 2, pp. H97–H102, 2011.

[FAR 07] FARCY A., GALLITRE M., ARNAL V., *et al.*, TORRES J., "Impact of process parameters on circuit performance for the 32 nm technology node", *Microelectronic Engineering*, vol. 84, no. 11, pp. 2738–2743, 2007.

[FAV 05] FAVENNEC L., JOUSSEAUME V., ROUESSAC V., *et al.*, "Ultra low K PECVD porogen approach: matrix precursors comparison and porogen removal treatment study", *Proceedings of the Materials Research Society Symposium*, vol. 863, pp. 49–54, 2005.

[FAV 06] FAVENNEC L., JOUSSEAUME V., ZENASNI A., *et al.*, "New low κ a-SiC:H dielectric barrier for advanced interconnects", *Proceedings of the International Interconnect Technology Conference (IITC)*, pp. 110–112, 2006.

[FAV 07] FAVENNEC L., JOUSSEAUME V., GERBAUD G., *et al.*, "Ultralow-k using a plasma enhanced chemical vapor deposition porogen approach: matrix structure and porogen loading influences", *Journal of Applied Physics*, vol. 102, no. 6, pp. 064107–9, 2007.

[FLO 02] FLORE J.A., CHASON E., CAMMARATA R.C., *et al.*, "Physical origins of intrinsic stresses in Volmer-Weber thin films", *MRS Bulletin*, vol. 27, no. 1, pp. 19–25, 2002.

[FRE 03] FREDERICK M.J., GOSWAMI R., RAMANATH G., "Sequence of Mg segregation, grain growth, and interfacial MgO formation in Cu-Mg alloy films on SiO_2 during vacuum annealing", *Journal of Applied Physics*, vol. 93, pp. 5966–5972, 2003.

[FRE 11] FREYSCHLAG C.G., MADIX R.J., "Precious metal magic: catalytic wizardry", *Materials Today*, vol. 14, pp. 134–142, 2011.

[FUJ 84] FUJIMOTO F., OOTUKA A., KOMAKI K., *et al.*, "Hydrogen content in a-SiC: H films prepared by plasma enhanced of silane and methane or ethylene", *Japanese Journal of Applied Physics*, vol. 23, pp. 810–814, 1984.

[GAL 08] GALL S., EUVRARD C., SHHUN S., *et al.*, "Investigation of ULK (κ = 2.5) damage by direct CMP process for 45 nm technology node", *Proceedings of the Materials Research Society Symposium*, vol. 23, pp. 115–121, 2008.

[GAM 06] GAMBINO J., WYNNE J., GILL J., *et al.*, "Self-aligned metal capping layers for copper interconnects using electroless plating", *Microelectronic Engineering*, vol. 83, pp. 2059–2067, 2006.

[GAN 07] GANDHI D.D., LANE M.W., ZHOU Y., *et al.*, "Annealing-induced interfacial toughening using a molecular nanolayer", *Nature*, vol. 447, pp. 299–302, 2007.

[GAR 10a] GARG S., SINGH B., TEKI R., *et al.*, "Hydrophobic fluoroalkylsilane nanolayers for inhibiting copper diffusion into silica", *Applied Physics Letters*, vol. 96, no. 14, pp. 143121–3, 2010.

[GAR 10b] GARG S., JAIN A., KARTHIK C., *et al.*, "Metal-dielectric interface toughening by molecular nanolayer decomposition", *Journal of Applied Physics*, vol. 108, no. 3, pp. 034317–4, 2010.

[GAT 09] GATES S.M., DUBOIS G., RYAN E.T., *et al.*, "Adjusting the skeleton and pore structure of porous SiCOH dielectrics", *Journal of the Electrochemical Society*, vol. 156, no. 10, pp. G156–G162, 2009.

[GOS 06] GOSSET L.G., CHHUN S., GUILLAN J., *et al.*, "Self aligned barrier approach: overview on process, module integration and interconnect performance improvement challenges", *Proceedings of the International Interconnect Technology Conference (IITC)*, pp. 84–86, 2006.

[GOT 11] GOTTSCHALK C., BRAMMER U., LE TIEC Y., "Deoxification of liquids for advanced semiconductors processing", in HATTORI T., RUZYLLO J., MERTENS P., NOVAK R. (eds), *ECS Transactions, Semiconductor Cleaning Science and Technology 12*, vol. 41, no. 5, pp. 287–294, 2011.

[GOU 07] GOURHANT O., JOUSSEAUME V., FAVENNEC L., *et al.*, "Extendibility of the PECVD porogen approach for ULK materials", *Proceedings of the Materials Research Society Symposium*, vol. 990, pp. B06–06, 2007.

[GOU 10] GOURHANT O., GERBAUD G., ZENASNI A., *et al.*, "Crosslinking of porous SiOCH films involving Si–O–C bonds: impact of deposition and curing", *Journal of Applied Physics*, vol. 108, pp. 124105–9, 2010.

[GRA 08] GRAS R., GAILLARD F., BOUCHU D., *et al.*, "300 mm multi level air gap integration for edge interconnect technologies and specific high performance applications", *International Interconnect Technology Conference (IITC)*, pp. 196–198, 2008.

[GRI 99] GRILL A., PATEL V., "Low dielectric constant films prepared by plasma-enhanced chemical vapor deposition from tetramethylsilane", *Journal of Applied Physics*, vol. 85, no. 6, pp. 3314–3318, 1999.

[GRI 01] GRILL A., PATEL V., "Ultralow-k dielectrics prepared by plasma-enhanced chemical vapour deposition", *Applied Physics Letters*, vol. 79, no. 6, pp. 803–805, 2001.

[GRI 03] GRILL A., PATEL V., "Structure of low dielectric constant to extreme low dielectric constant SiCOH films: Fourier transform infrared spectroscopy characterization", *Journal of Applied Physics* vol. 94, 10, pp. 6697–6707, 2003.

[GRI 07] GRILL A., "Low and ultralow dielectric constant films prepared by plasma enhanced chemical vapor deposition", in BAKLANOV M., GREEN M., MAEX K. (eds), *Dielectric Films for Advanced Microelectronics*, John Wiley & Sons, West Sussex, England, pp. 1–32, 2007.

[GRI 08a] GRILL A., EDELSTEIN D., LANE M., *et al.*, "Interface engineering for high interfacial strength between SiCOH and porous SiCOH interconnect dielectrics and diffusion caps", *Journal of Applied Physics*, vol. 103, no. 5, pp. 054104–6, 2008.

[GRI 08b] GRILL A., PATEL V., "Ultralow dielectric constant pSiCOH films prepared with tetramethylcyclotetrasiloxane as skeleton precursor", *Journal of Applied Physics*, vol. 104, no. 2, pp. 024113–9, 2008.

[GRI 09] GRILL A., "Porous pSiCOH ultralow-k dielectrics for chip interconnects prepared by PECVD", *Annual Review of Material Research*, vol. 39, pp. 49–69, 2009.

[GUE 07] GUEDJ C., CLARET N., ARNAL V., *et al.*, "Evidence for 3-D/2-D transition in advanced interconnects", *IEEE Transactions on Devices and Materials Reliability*, pp. 64–68, 2007, DOI: 10.1109/TDMR.2007.902172.

[GUI 03] GUILLAUMOND J.F., ARNAUD L., MOURIER T., *et al.*, "Analysis of resistivity in nano-interconnect: full range (4.2–300 K) temperature characterization", *Proceedings of the IEEE International Interconnect Technology Conference (IITC)*, pp. 132–134, 2003.

[HAN 07] HANEDA M., IIJIMA J., KOIKE J., "Growth behavior of self-formed barrier at Cu-Mn/SiO_2 interface at 250–450 degrees C", *Applied Physics Letters*, vol. 90, no. 25, pp. 252107–3, 2007.

[HAU 03] HAUMESSER P.H., GIBLAT F., AMEUR S., *et al.*, "Electro-grafting: a new approach for copper seeding or direct plating", *Advanced Metallization Conference (AMC)*, pp. 575–581, 2003.

[HAU 05] HAUMESSER P.H., ROULE A., MAITREJEAN S., *et al.*, "Seed enhancement: a bridging technology", *Future Fab*, vol. 19, pp. 81–83, 2005.

[HED 98] HEDRICK J.L., MILLER R.D., HAWKER C.J., *et al.*, "Templating nanoporosity in thin film dielectric insulators", *Advanced Materials*, vol. 10, no. 13, pp. 1049–1053, 1998.

[HEN 10a] HENDERSON L.B., EKERDT J.G., "Effect of phosphorus and carbon incorporation in amorphous cobalt films prepared by chemical vapor deposition", *Journal of the Electrochemical Society*, vol. 157, no. 1, pp. D29–D34, 2010.

[HEN 10b] HENDERSON L.B., EKERDT J.G., "Chemically capping copper with cobalt", *Microelectronic Engineering*, vol. 87, no. 4, pp. 588–592, 2010.

[HIN 01] HINODE K., HANAOKA Y., TAKEDA K., *et al.*, "Resistivity increase in ultrafine-line copper conductor for ULSIs", *Japanese Journal of Applied Physics Part 2-Letters*, vol. 40, no. 10B, pp. L1097–L1099, 2001.

[HOM 03] HOMMA Y., FUKUSHIMA K., KONDO S., *et al.*, "Effects of mechanical parameters on CMP characteristics analyzed by two-dimensional frictional-force measurement", *Journal of the Electrochemical Society*, vol. 150, no. 12, pp. G751–G757, 2003.

[HSU 11] HSU K.-C., PERNG D.-C., WANG Y.-C., "Robust ultra-thin RuMo alloy film as a seedless Cu diffusion barrier", *Journal of Alloys and Compounds*, vol. 516, pp. 102–106, 2011.

[HSU 12] HSU K.-C., PERNG. D.-C., YEH Y.-B., *et al.*, "Ultrathin Cr added Ru film as a seedless Cu diffusion barrier for advanced Cu interconnects", *Applied Surface Science*, vol. 258, no. 18, pp. 7225–7230, 2012.

[HU 95] HU C.-K., RODBELL K.P., SULLIVAN T.D., *et al.*, "Electromigration and stress-induced voiding in fine Al and Al-alloy thin-film lines", *IBM Journal of Research and Development*, vol. 39, no. 4, pp. 465–497, 1995.

[HU 07] HU C.-K., GIGNAC L., BAKER B., *et al.*, "Impact of Cu microstructure on electromigration reliability", *IEEE International Interconnect Technology Conference (IITC)*, pp. 93–95, 2007.

[HUA 00] HUANG W.H., RAGHAVAN S., FANG Y., *et al.*, "Electrochemical behavior of copper in tetramethyl ammonium hydroxide based solution", *Proceedings of the Materials Research Society Symposium*, vol. 566, p. 161, 2000.

[IAC 06] IACOPI F., TRAVALY Y., EYCKENS B., *et al.*, "Short-ranged structural rearrangement and enhancement of mechanical properties of organosilicate glasses induced by ultraviolet radiation", *Journal of Applied Physics*, vol. 99, p. 053511, 2006.

[IGA 98] IGARASHI Y., ITO T., "Electromigration properties of copper-zirconium alloy interconnects", *Journal of Vacuum Science, Technology B: Microelectronics and Nanometer Structures*, vol. 16, no. 5, pp. 2745–2750, 1998.

[IMA 09] IMAI M., YAMASHITA Y., FUTATSUKI T., *et al.*, "Effect of dissolved oxygen on Cu corrosion in single wafer cleaning process", *Japanese Journal of Applied Physics*, vol. 48, no. 4, pp. 04C023-01–04C023-4, 2009.

[INO 11] INOUE F., SHIMIZU T., YOKOYAMA T., *et al.*, "Formation of electroless barrier and seed layers in a high aspect ratio through-Si vias using Au nanoparticle catalyst for all-wet Cu filling technology", *Electrochimica Acta*, vol. 56, pp. 6245–6250, 2011.

[IST 02] ISTRATOV A.A., WEBER E.R., "Physics of copper in silicon", *Journal of the Electrochemical Society*, vol. 149, no. 1, pp. G21–G30, 2002.

[ITR 00] International Technology Roadmap for Semiconductors (ITRS), http://www.itrs.net, 2000.

[ITR 12] International Technology Roadmap for Semiconductors (ITRS), http://www.itrs.net, 2012.

[ITO 04] ITO F., TAKEUCHI T., HAYASHI Y., "Improvement of mechanical properties of porous SiOCH films by post-cure treatments", *Proceedings of the Advanced Metallization Conference (AMC)*, pp. 32–33, 2004.

[JEN 94] JENG S.P., HAVEMANN R.H., CHANG M.C., "Process integration and manufacturability issues for high performance multilevel interconnect", *Materials Research Society Proceeding*, vol. 337, p. 25, 1994, DOI: 10.1557/PROC-337-25.

[JOO 12] JOO S., LIANG H., "Tribo-electrochemical characterization of copper with patterned geometry", *Microelectronic Engineering*, vol. 98, pp. 12–18, 2012.

[JOU 02] JOULAUD M., ANGEKORT C., DOPPELT P., *et al.*, "Evaluation of (hfac) Cu (MHY) for Cu CVD", *Microelectronic Engineering*, vol. 64, pp. 107–115, 2002.

[JOU 04a] JOUSSEAUME V., ROCHAT N., FAVENNEC L., *et al.*, "Mechanical stress in PECVD a-SiC:H: aging and plasma treatments effects", *Proceedings of the Materials Research Society Symposium*, vol. 7, pp. 301–305, 2004.

[JOU 04b] JOULAUD M., HAUMESSER P.H., JAMMER W., *et al.*, "Electroless NiMo-P films as capping layers for copper interconnects", *Advanced Metallization Conference (AMC)*, pp. 815–821, 2004.

[JOU 07] JOUSSEAUME V., ZENASNI A., FAVENNEC L., *et al.*, "Comparison between e-beam and ultraviolet curing to perform porous SiOC:H", *Journal of the Electrochemical Society*, vol. 154, no. 5, pp. G103–G109, 2007.

[JOU 12] JOUSSEAUME V., GOURHANT O., GONON P., *et al.*, "Dielectric constant of porous ultra low-k thin films", *Journal of the Electrochemical Society*, vol. 159, no. 5, pp. G49–G55, 2012.

[KAL 02] KALTENPOTH G., SIEBERT W., STUBHAN F., *et al.*, "Moisture barrier properties of plasma enhanced chemical vapor deposited SiC_xN_y films on polyethylene naphthalate sheets and epoxy molding compound", *Surface Coating Technology*, vol. 161, pp. 96–101, 2002.

[KIM 05] KIM H., KOSEKI T., OHBA T., *et al.*, "Process design of Cu(Sn) alloy deposition for highly reliable ultra large-scale integration interconnects", *Thin Solid Films*, vol. 491, no. 1–2, pp. 221–227, 2005.

[KIM 07] KIM S., DUQUETTE D.J., "Morphology control of copper growth on TiN and TaN diffusion barriers in seedless copper electrodeposition", *Journal of the Electrochemical Society*, vol. 154, no. 4, pp. D195–D200, 2007.

[KIM 08] KIM S.-W., KWON S.-H., JEONG S.-J., *et al.*, "Improvement of copper diffusion barrier properties of tantalum nitride films by incorporating ruthenium using PEALD", *Journal of the Electrochemical Society*, vol. 155, no. 11, pp. H885–H888, 2008.

[KIM 09] KIM A.R., PARK H.J., JEONG K.H., *et al.*, “Effects of iodine on the initial growth of MOCVD Cu on MPTMS monolayer surface at a low temperature of 110 degrees C”, *Thin Solid Films*, vol. 517, no. 14, pp. 3827–3830, 2009.

[KIM 11] KIM S.H., “Low temperature atomic layer deposition of Ru thin films with enhanced nucleation using various Ru(0) metallorganic precursors and molecular O_2”, *ECS Transciations*, vol. 41, pp. 19–23, 2011.

[KIN 11] KING S.W., FRENCH M., BIELEFELD J., *et al.*, “Fourier transform infrared spectroscopy investigation of chemical bonding in low-k a-SiC:H thin films”, *Journal of Non-Crystal Solids*, vol. 357, no. 15, pp. 2970–2983, 2011.

[KIT 86] KITTEL C., *Introduction to Solid State Physics*, John Wiley & Sons Inc, New York, 1986.

[KIZ 04] KIZIL H., STEINBRUCHEL C., “TiN, TaN and W_xN as diffusion barriers for Cu on SiO_2: capacitance-voltage, leakage current, and triangular-voltage-sweep tests after bias temperature stress”, *Thin Solid Films*, vol. 449, no. 1–2, pp. 158–165, 2004.

[KOI 07] KOIKE J., WADA M., TAKAHASHI S., *et al.*, “Semiconductor device with a barrier films which contains manganese”, US Patent 7304384 B2, 4 December 2007.

[KUA 00] KUAN T.S., INOKI C.K., OEHRLEIN G.S., *et al.*, “Fabrication and performance limits of sub-0.1 μm Cu interconnects”, *Proceedings from MRS Spring Meeting*, vol. 612, p. D7.1.1, 2000, DOI:10.1557/PROC-612-D7.1.1.

[LAN 03] LANE M.W., LINIGER E.G., LLOYD J.R., “Relationship between interfacial adhesion and electromigration in Cu metallization”, *Journal of Applied Physics*, vol. 93, no. 3, pp. 1417–1421, 2003.

[LAU 09a] LAUERHAAS J., “Suppression of galvanic corrosion in advanced BEOL integration”, *Solid State Technology*, vol. 52, no. 9, pp. 12–15, 2009.

[LAU 09b] LAUERHAAS J.M., COLLINS J.D., GAST T.A., *et al.*, “Tools and methods for processing microelectronic workpieces using process chamber designs that easily transition between open and closed modes of operation”, US Patent 20090280235, 12 November 2009.

[LED 05] LEDUC P., SAVOYE M., MAITREJEAN S., *et al.*, “Understanding CMP-induced delamination in ultra low-k/Cu integration”, *IEEE International Interconnect Technology Conference (IITC)*, pp. 209–211, 2005.

[LED 08] LEDUC P., DI CIOCCIO L., CHARLET B., *et al.*, "Enabling technologies for 3D chip stacking", *VLSI Technology, Systems and Applications*, pp. 76–78, 2008, DOI: 10.1109/VTSA.2008.4530806.

[LEE 01] LEE S.G., KIM Y.J., LEE S.P., *et al.*, "Low dielectric constant 3MS α-SiC:H as Cu diffusion barrier layer in Cu dual damascene Process", *Japanese Journal of Applied Physics*, vol. 40, no. 4B, pp. 2663–2668, 2001.

[LEE 96] LEE J.H., KIM D.S., LEE Y.H., "Room temperature deposition of silicon dioxide films by ion-assisted plasma enhanced chemical vapour deposition", *Journal of the Electrochemical Society*, vol. 143, no. 4, pp. 1443–1451, 1996.

[LI 06] LI Z., RAHTU A., GORDON R.G., "Atomic layer deposition of ultrathin copper metal films from a liquid copper (I) amidinate precursor", *Journal of the Electrochemical Society*, vol. 153, no. 11, pp. C787–C794, 2006.

[LI 07] LI Y., *Microelectronic Applications of Chemical Mechanical Planarization,* John Wiley & Sons, 2007.

[LIU 02] LIU C.J., CHEN J.S., "Low leakage current Cu(Ti)/SiO_2 interconnection scheme with a self-formed TiO_x diffusion barrier", *Applied Physics Letters*, vol. 80, pp. 2678–2680, 2002.

[LIU 05] LIU C.J., CHEN J.S., "High-temperature self-grown ZrO_2 layer against Cu diffusion at Cu(2.5 at. % Zr)/SiO_2 interface", *Journal of Vacuum Science, Technology B*, vol. 23, pp. 90–95, 2005.

[LIU 98] LIU X., LU Y.Z., GORDON R.G., "Improved conformality of CVD titanium nitride films", *MRS Proceedings*, vol. 555, p. 135, 1998, DOI:10.1557/PROC-555-135.

[LOB 94] LOBODA M.J., SEIFFERLY J.A., DALL F.C., "Plasma-enhanced chemical vapour deposition of a-SiC:H films from organosilicon precursors", *Journal of Vacuum Science Technology A*, vol. 12, no. 1, pp. 90–96, 1994.

[LUH 09] LUHN O., RADISIC A., VEREECKEN P.M., *et al.*, "Changing superfilling mode for copper electrodeposition in blind holes from differential inhibition to differential acceleration", *Electrochemical and Solid-State Letters*, vol. 12, no. 5, pp. D39–D41, 2009.

[MAE 03] MAEX K., BAKLANOV M.R., SHAMIRYAN D., *et al.*, "Low dielectric constant materials for microelectronics", *Journal of Applied Physics*, vol. 93, no. 11, pp. 8793–8841, 2003.

[MAE 08] MAEKAWA K., MORI K., SUZUMURA N., *et al.*, "Impact of Al in Cu alloy interconnects on electro and stress migration reliabilities", *Microelectronic Engineering*, vol. 85, no. 10, pp. 2137–2141, 2008.

[MAI 04] MAIER G., "The search for low κ and ultra low k dielectrics: how far can you get with polymers?", *Electrical Insulation Magazine*, vol. 20, pp. 6–17, 2004.

[MCB 86] MCBRAYER J.D., SWANSON R., SIGMON T., "Diffusion of metals in silicon dioxide", *Journal of the Electrochemical Society*, vol. 133, p.1242, 1986.

[MCB 86] MCBRAYER J.D., SWANSON R.M., SIGMON T.W., "Diffusion of metals in silicon dioxide", *Journal of the Electrochemical Society*, vol. 133, no. 6, pp. 1242–1246, 1986.

[MCG 10] MC GAHAY V., "Porous dielectrics in microelectronic wiring applications", *Materials*, vol. 3, pp. 536–562, 2010.

[MER 03] MERCADO L.L., KUO S.-M., GOLDBERG C., *et al.*, "Impact of flip-chip packaging on copper/low-k structures" *IEEE Transactions on Advanced Packaging*, vol. 26, no. 4, pp. 433–440, 2003.

[MIL 01] MILLER A.E., FISCHER P.B., FELLER A.D., *et al.*, "Chemically induced defects during copper polish", *Proceedings of the International Interconnect Technology Conference (IITC)*, pp. 143–145, 2001.

[MOF 05] MOFFAT T.P., WHEELER D., EDELSTEIN M.D., *et al.*, "Superconformal film growth: mechanism and quantification", *IBM Journal of Research and Development*, vol. 49, no. 1, pp. 19–36, 2005.

[MOR 97] MORAND Y., LERME M., PALLEAU J., *et al.*, "Copper integration in self aligned dual damascene architecture", *Proceedings of the VLSI Conference*, pp. 31–32, 1997.

[MOR 00] MORGEN M., RYAN E.T., ZHAO J.H., *et al.*, "Low dielectric constant materials for ULSI interconnects", *Annual Review Materials Science*, vol. 30, pp. 645–680, 2000.

[MOT 00] MOTTE P., PROUST M., TORRES J., *et al.*, "TiN-CVD process optimization for integration with Cu-CVD", *Microelectronic Engineering*, vol. 50, no. 1–4, pp. 369–374, 2000.

[MUR 06] MURTHY B.R., YEE W.M., KRISHNAMOORTHY A., *et al.*, "Self-assembled monolayers as Cu diffusion barriers for ultralow-k dielectrics", *Electrochemical and Solid State Letters*, vol. 9, no. 7, pp. F61–F63, 2006.

[MUR 95] MURARKA S.P., HYMES S.W., "Copper metallization for ULSI and beyond", *Critical Reviews in Solid State and Materials Sciences*, vol. 20, no. 2, pp. 87–124, 1995.

[MUR 97] MURARKA S.P., "Multilevel interconnections for ULSI and GSI era", *Materials Science and Engineering R, Reports*, vol. 19, no. 3–4, pp. 87–151, 1997.

[NAK 06] NAKAO S.-I., USHIO J., OHNO T., *et al.*, "UV/EB cure mechanism for porous PECVD/SOD low-k SiCOH materials", *Proceedings of the International Interconnect Technology Conference (IITC)*, pp. 66–68, 2006.

[NAK 08] NAKAHIRA J., INAISHI Y., NAKAO S., *et al.*, "Novel oxygen-free barrier SiC film (κ < 3.5) with high etching selectivity", *Proceedings of the Advanced Metallization Conference*, pp. 551–557, 2008.

[NEW 82] NEWMAN R.C., "Defects in silicon", *Reports on Progress in Physics*, vol. 45, pp. L1163–1210, 1982.

[NGU 88] NGUYEN S.V., "Plasma-assisted chemical vapour deposition", in SCHUEGRAF K.K. (ed.), *Handbook of Thin-Film Deposition Processes and Techniques*, Park Ridge, NJ, pp. 112–141, 1988.

[NIH 12] NIHEI M., SATO M., "Carbon Nanotubes for Interconnects", in BAKLANOV M.R., HO P.S., ZSCHECH E. (eds), *Advanced Interconnects for ULSI Technology*, Wiley, 2012.

[NIX 89] NIX W.D., "Mechanical-Properties of thin films", *Metallurgical Transactions a-Physical Metallurgy and Materials Science*, vol. 20, no. 11, pp. 2217–2245, 1989.

[NIX 09] NIX W.D., "Exploiting new opportunities in materials research by remembering and applying old lessons", *MRS Bulletin*, vol. 34, no. 2, pp. 82–91, 2009.

[OHN 85] OHNO I., WAKABAYASHI O., HARUYAMA S., "Anodic oxidation of reductants in electroless plating", *Journal of the Electrochemical Society*, vol. 132, no. 10, pp. 2323–2330, 1985.

[OKI 98] OKIMURA K., MAEDA N., "Dissociation processes in plasma enhanced chemical vapour deposition of SiO_2 films using tetraethoxysilane", *Journal of Vacuum Science Technology A*, vol. 16, no. 6, pp. 3157–3163, 1998.

[OLI 08] OLIVIER S., DECORPS T., BERNARD M., *et al.*, "Physical investigation of the impact of electrolessly deposited self-aligned caps on insulation of copper interconnects", *Microelectronic Engineering*, vol. 85, no. 10, pp. 2051–2054, 2008.

[OLI 10] OLIVIER S., DECORPS T., CALVO-MUNOZ M.L., *et al.*, "Inhomogeneous nucleation and growth of palladium and alloyed cobalt during self-aligned capping of advanced copper interconnects", *Thin Solid Films*, vol. 518, no. 17, pp. 4773–4778, 2010.

[O'N 03] O'NEILL M.L., VRTIS R.N., VINCENT J.L., *et al.*, "Optimized materials properties for organosilicate glasses produced by plasma-enhanced chemical vapour deposition", *Technology and Reliability for Advanced Interconnects and Low κ Dielectrics Symposium, Materials Research Society*, vol. 766, pp. 321–326, 2003.

[O'N 06] O'NEILL M.L., HAAS M.K., PETERSON B.K., *et al.*, "Impact of pore size and morphology of porous organosilicate glasses on integrated circuit manufacturing", *Proceedings of the Materials Research Society Symposium*, vol. 914, p. 3, 2006, available at http://dx.doi.org/10.1557/PROC-0914-F01-02.

[ORA 04] ORAIN S., BARBE J.-C., FEDERSPIEL X., *et al.*, "FEM-based method to determine mechanical stress evolution during process flow in microelectronics, application to stress-voiding", *Proceedings of the 5th International Conference EuroSimE, Thermal and Mechanical Simulation and Experiments in Microelectronics and Microsystems,* pp. 47–52, 2004.

[PER 06] PERNEL C., FARKAS J., LOUIS D., "Copper in organic acid based cleaning solutions", *Journal of Vacuum Science and Technology B*, vol. 24, no. 5, pp. 2467–2471, 2006.

[PER 07] PERNEL C., BILOUK S., NOGUEIRA R.P., *et al.*, "Impact of post via clean solution on self aligned barriers: spontaneous corrosion and galvanic coupling", *Proceedings of the 24th Advanced Metallization Conference (AMC)*, 2007.

[POS 03] POSSEME N., CHEVOLLEAU T., JOUBERT O., *et al.*, "Etching mechanisms of low-k SiOCH and selectivity to SiCH and SiO_2 in fluorocarbon based plasmas", *Journal of Vacuum Science Technology B*, vol. 21, no. 6, pp. 2432–2440, 2003.

[POU 63] POURBAIX M., *Atlas d'équilibres électrochimiques*, Gauthier-Villars & Cie, Paris, 1963.

[PRO 11] PROFIJT H.B., POTTS S.E., VAN DE SANDEN M.C.M., *et al.*, "Plasma-assisted atomic layer deposition: basics, opportunities, and challenges", *Journal of Vacuum Science & Technology A: Vacuum, Surfaces, and Films*, vol. 29, no. 5, pp. 050801-1–050801-26, 2011.

[RAD 04] RADISIC A., OSKAM G., SEARSON P.C., "Influence of oxide thickness on nucleation and growth of copper on tantalum", *Journal of the Electrochemical Society*, vol. 151, no. 6, pp. C369–C374, 2004.

[RAM 03] RAMANATH G., CUI G., GANESAN P.G., *et al.*, "Self-assembled subnanolayers as interfacial adhesion enhancers and diffusion barriers for integrated circuits", *Applied Physics Letters*, vol. 83, no. 2, pp. 383–385, 2003.

[REB 09] REBISCOUL D., PERRUT V., RENAULT O., *et al.*, "Alkoxysilane layers deposited by SC CO_2 process on silicon oxide for microelectronics applications", *Journal of Supercritical Fluids*, vol. 51, no. 2, pp. 287–294, 2009.

[REB 10] REBISCOUL D., PERRUT V., MOREL T., *et al.*, "Alkoxysilane layers compatible with copper deposition for advanced semiconductor device applications", *Langmuir*, vol. 26, no. 11, pp. 8981–8987, 2010.

[ROU 04] ROULE A., MOREL T., HAUMESSER P.H., *et al.*, "Copper metallization of advanced interconnects: strategies for beyond 65 nm", *Advanced Metallization Conference 2004 (AMC)*, pp. 341–346, 2004.

[ROU 05] ROUESSAC V., FAVENNEC L., REMIAT B., *et al.*, "Precursor chemistry for ULK CVD", *Microelectronic Engineering*, vol. 82, no. 3–4, pp. 333–340, 2005.

[ROU 07] ROULE A., AMUNTENCEI M., DERONZIER E., *et al.*, "Seed layer enhancement by electrochemical deposition: the copper seed solution for beyond 45 nm", *Microelectronic Engineering*, vol. 84, no. 11, pp. 2610–2614, 2007.

[RYN 91] RYNDERS S.W., SCHEELINE A., BOHN P.W., "Structure evolution in a-SiC:H films prepared from tetramethylsilane", *Journal of Applied Physics*, vol. 69, no. 5, pp. 2951–2960, 1991.

[SHA 95] SHAPIRO M.J., NGUYEN S.V., MATSUDA T., *et al.*, "CVD of fluorosilicate glass for ULSI applications", *Thin Solid Films*, vol. 270, no. 1–2, pp. 503–507, 1995.

[SHA 03] SHACHAM-DIAMAND Y., INBERG A., SVERDLOV Y., *et al.*, Electroless processes for micro-and nanoelectronics", *Electrochimica Acta*, vol. 48, no. 20–22, pp. 2987–2996, 2003.

[SHI 07] SHINRIKI M., TAJIMA N., HAMADA T., *et al.*, "Si-C_2H_4-Si network formation for improvement of PECVD SiOCH film properties", *Proceedings of the Advanced Metallization Conference (AMC),* pp. 339–344, 2007.

[SHI 12] SHIMIZU H., SAKODA H., SAKODA K., *et al.*, "Atomic layer deposited Co (W) film as a single-layered barrier/liner for next-generation Cu-interconnects", *Japanese Journal of Applied Physics*, vol. 51, no. 5, pp. 05EB02–7, 2012.

[SIN 02] SINGH R.K., BAJAJ R., "Advances in chemical-mechanical planarization", *Materials Research Society Bulletin*, vol. 27, pp. 743–751, 2002.

[SMI 93] SMITH D.L., "Controlling the plasma chemistry of silicon nitride and oxide deposition from silane", *Journal of Vacuum Science Technology A*, vol. 11, no. 4, pp. 1843–1850, 1993.

[STA 07] STANGL M., FLETCHER A., ACKER J., *et al.*, "Influence of Ta-based diffusion barriers on the microstructure of copper thin films", *Journal of Electronic Materials*, vol. 36, no. 12, pp. 1625–1629, 2007.

[STE 02] STEINHOGL W., SCHINDLER G., STEINLESBERGER G., *et al.*, "Size-dependent resistivity of metallic wires in the mesoscopic range" *Physical Review B*, vol. 66, no. 7, pp. 0754141–4, 2002.

[STE 05] STEINHOGL W., SCHINDLER G., STEINLESBERGER G., *et al.*, "Comprehensive study of the resistivity of copper wires with lateral dimensions of 100 nm and smaller", *Journal of Applied Physics*, vol. 97, no. 2, pp. 023706–7, 2005.

[SUN 11] SUN H., QIN X., ZAERA F., "Chemical nature of the thin films that form on SiO_2/Si (100) surfaces upon manganese deposition", *The Journal of Physical Chemistry Letters*, vol. 2, no. 20, pp. 2525–2530, 2011.

[TAK 00] TAKEYAMA M.B., NOYA A., SAKANISHI, "Diffusion barrier properties of ZrN films in the Cu/Si contact systems", *Journal of Vacuum Science, Technology B: Microelectronics and Nanometer Structures*, vol. 18, no. 3, pp. 1333–1337, 2000.

[THO 01] THOMSON J.J., "On the theory of electric conduction through thin metallic films", *Proceedings of the Cambridge Philosophical Society*, vol. 11, pp. 120–122, 1901.

[TOK 10] TOKEI Z., CROES K., BEYER G.P., "Reliability of copper low-k interconnects", *Microelectronic Engineering*, vol. 87, no. 3, pp. 348–354, 2010.

[TOW 97] TOWNSEND P.H., MARTIN S.J., GODSCHALX J., *et al.*, "Silk polymer coating with low dielectric constant and high thermal stability for ULSI interlayer dielectric", *Proceedings of the Materials Research Society Symposium*, vol. 476, p. 9, 1997, DOI:10.1557/PROC-476-9.

[TRU 09] TRUZZI C., RAYNAL F., MEVELLEC V., "Wet-process deposition of TSV liner and metal films", *Proceedings of the EEE International Conference on 3D System Integration*, pp. 1–6, 2009.

[TSA 96] TSAI M., SUN S., TSAI C., *et al.*, "Comparison of the diffusion barrier properties of chemical-vapor-deposited TaN and sputtered TaN between Cu and Si", *Journal of Applied Physics*, vol. 79, pp. 6932–6938, 1996.

[URB 10] URBANOWICZ A.M., VANSTREELS K., VERDONCK P., *et al.*, "Improving mechanical robustness of ultralow-k SiOCH plasma enhanced chemical vapor deposition glasses by controlled porogen decomposition prior to UV-hardening", *Journal of Applied Physics*, vol. 107, no. 10, pp. 104122–7, 2010.

[USU 06] USUI T., NASU H., TAKAHASHI S., *et al.*, "Highly reliable copper dual-damascene interconnects with self-formed $MnSi_xO_y$ barrier layer", *IEEE Transactions on Electron Devices*, vol. 53, no. 10, pp. 2492–2499, 2006.

[VAN 11] VAN ELSHOCHT S., DELABIE A., DEWILDE S., *et al.*, "ALD barrier deposition on porous low-k dielectric materials for interconnects", *ECS Transactions*, vol. 41, no. 2, pp. 25–32, 2011.

[VEN 97] VENKATESAN S., GELATOS A.V., HISRA S., *et al.*, "A high performance 1.8 V, 0.20 nm CMOS technology with copper metallization", *Proceedings of the International Electron Devices Meeting*, pp. 769–772, 1997.

[VER 05] VEREECKEN P.M., BINSTEAD R.A., DELIGIANNI H., *et al.*, "The chemistry of additives in damascene copper plating", *IBM Journal of Research and Development*, vol. 49, no. 1, pp. 3–18, 2005.

[VOL 10] VOLKSEN W., MILLER R.D., DUBOIS G., "Low dielectric constant materials", *Chemical Review*, vol. 110, no. 1, pp. 56–110, 2010.

[VOS 00] VOSS S., GANDIKOTA S., CHEN L.Y., *et al.*, "Chemical studies of CVD Cu deposited on Ta and TaN barriers under various process conditions", *Microelectronic Engineering*, vol. 50, no. 1–4, pp. 501–508, 2000.

[VRT 03] VRTIS R.N., O'NEILL M.L., VINCENT J.L., *et al.*, "Plasma enhanced chemical vapor deposition of porous organosilicate glass ILD films with $\kappa < 2.4$", *Proceedings of the Materials Research Society Symposium*, vol. 766, p. 259, 2003.

[WAE 09] WAECHTLER T., OSWALD S., ROTH N., *et al.*, "Copper oxide films grown by atomic layer deposition from Bis (tri-n-butylphosphane) copper (I) acetylacetonate on Ta, TaN, Ru, and SiO_2", *Journal of the Electrochemical Society*, vol. 156, no. 6, pp. H453–H459, 2009.

[WAN 04] WANG Y.H., MOITREYEE M.R., KUMAR R., *et al.*, "A comparative study of low dielectric constant barrier layer, etch stop and hard mask films of hydrogenated amorphous Si-(C,O,N)", *Thin Solid Films*, vol. 460, no. 1–2, pp. 211–216, 2004.

[WEI 00] WEISS K., RIEDEL S., SCHULZ S.E., *et al.*, "Development of different copper seed layers with respect to the copper electroplating process", *Microelectronic Engineering*, vol. 50, pp. 433–440, 2000.

[WEN 01] WEN W., MAEX K., "Studies on size effect of copper interconnect lines", *Proceedings of the 6th International Conference on Solid-State and Integrated-Circuit Technology*, vol. 1, pp. 416–418, 2001.

[WIL 93] WILSON S.R., TRACY C.J., FREEMAN J.L.J., *Handbook of Multilevel Metallization for Integrated Circuits*, Noyes Publication, Park Ridge, NJ, 1993.

[WIL 10] WILSON C.J., VOLDERS H., CROES K., *et al.*, "In situ X-ray diffraction study of self-forming barriers from a Cu-Mn alloy in 100 nm Cu/low-k damascene interconnects using synchrotron radiation", *Microelectronic Engineering*, vol. 87, no. 3, pp. 398–401, 2010.

[WOJ 12] WOJCIK H., KALTOFEN R., MERKEL U., *et al.*, "Electrical evaluation of Ru-W (-N), Ru-Ta (-N) and Ru-Mn films as Cu diffusion barriers", *Microelectronic Engineering*, vol. 92, pp. 71–75, 2012.

[YE 05] YE D.-X., CARROW B., PIMANPANG S., *et al.*, "Evaluation of a novel Cu (I) precursor for chemical vapor deposition", *Electrochemical and Solid-State Letters*, vol. 8, no. 7, pp. C85–C88, 2005.

[YOO 04] YOON B.U., KONDO S., TOKITOH S., *et al.*, "Characterization of patterned low-k film delamination during CMP for the 32 nm-node Cu/ultra low-k (κ = 1.6–1.8) integration", *Proceedings of the 7th IEEE Annual International Interconnect Technology Conference (IITC)*, Burlingame, CA, pp. 239–241, 2004.

[YOS 07] YOSHINO M., MASUDA T., YOKOSHIMA T., *et al.*, "Electroless diffusion barrier process using SAM on low-k dielectrics", *Journal of the Electrochemical Society*, vol. 154, no. 3, pp. D122–D125, 2007.

[ZAN 04] ZANTYE P.B., KUMAR A., SIKDER A.K., "Chemical mechanical planarization for microelectronics, applications", *Materials Science and Engineering R*, vol. 45, pp. 89–220, 2004.

[ZEN 07] ZENASNI A., JOUSSEAUME V., HOLLIGER P., *et al.*, "The role of ultraviolet radiation during ultralow κ films curing: strengthening mechanisms and sacrificial porogen removal", *Journal of Applied Physics*, vol. 102, no. 9, p. 094107, 2007.

[ZHA 05] ZHAO X., MAGTOTO N.P., KELBER J.A., "Chemical vapor deposition of tantalum nitride with tert-butylimino tris (diethylamino) tantalum and atomic hydrogen", *Thin Solid Films*, vol. 478, no. 1–2, pp. 188–195, 2005.

Chapter 3

The Chemistry of Wet Surface Preparation: Cleaning, Etching and Drying

3.1. Introduction

This chapter covers the main aspects of the surface preparation that are used in current industry and research laboratories. Wet cleaning is a key driver since advanced technologies request new answers with decreasing technology nodes to less than 20 nm. Over the last 10 years, the specification list given to the cleaning experts is receiving new criteria and, today, there are key focuses concerning material consumption or loss, as well as selectivity between exposed materials. New materials have been introduced to address advanced nodes (see Chapters 1 and 2 covering front end of the line and back end of the line materials). In successive parts, this chapter describes the cleaning, the etching and the final rinsing-drying sequence. Mechanisms, chemical reactions, impact of concentration, temperature and time are reported to allow a better understanding of the wet surface preparation challenges.

Chapter written by Yannick LE TIEC and Martin KNOTTER.

3.2. Cleaning

3.2.1. *Ammonium hydrogen peroxide mixture (APM)*

3.2.1.1. *Brief history and application*

In 1970, Kern and Poutinen presented to the semiconductor community a wet cleaning technology that was able to remove contaminants from a silicon surface [KER 70]. Since then, it has been widely used in the industry and thoroughly investigated [OHM 90]. The first step in the RCA clean is Standard Clean 1 (SC1) or more descriptively named ammonium hydrogen peroxide mixture (APM). It is a mixture of two commercially available chemicals in water, that is, 25–30% ammonium hydroxide (NH_4OH) in water and 29–31% hydrogen peroxide (H_2O_2) in water. The SC1 composition is expressed as a ratio x:y:z, whereby x is the volume of ammonium hydroxide solution, y the volume of hydrogen peroxide solution and z the volume of water. In manufacturing, a wide variety of mix ratios are in use ranging from 1:1:5 (most concentrated); 1:20:100 (reduced ammonia concentration); to 1:1:200 (most diluted version). The temperature for use is from 20°C to 80°C with most common 50–60°C.

The original intent of APM was to remove organic contamination. Kern optimized the mixture and operating temperature for the removal of bees wax that he used as part of his cleaning challenge. He worked with a 1:1:5 mix ratio to be used at 80°C. His choice for these chemicals was quite obvious:

– hydrogen peroxide is a strong oxidant that is able to oxidize material and make it more soluble. However, due to a high pH, it is not strong enough to oxidize organic matter. It leaves no residues on the wafer as it evaporates or decomposes in harmless products: water and oxygen;

– ammonia solution is known for its degreasing abilities (it is used around the house for degreasing surfaces prior to painting) and unlike tetramethylammonium hydroxide or potassium hydroxide it leaves no residues as it evaporates (boiling point is at –33°C).

Nowadays, its application has changed. Its main objective is to remove particles and regrow a chemical oxide layer on silicon surfaces. Over the years, there have been some reasons to re-evaluate the concentration and operation temperature. For the reduction in cost, the concentration has been reduced. Furthermore, engineers become concerned about the amount of silicon that can be lost during a cleaning step and APM etches silicon surfaces. By reducing the operating temperature and adapting the composition, the amount of silicon loss could be reduced.

3.2.1.2. *Chemical mechanism*

APM is known to be a good particle removal solution, a property that is obtained as a result of the combined action of silicon oxide growth and etching of this newly formed chemical oxide. The etching compound in the chemical mixture is OH^-, which is formed by the reaction of NH_3 with water (see reaction 3.1):

$$NH_3 + H_2O \Leftrightarrow NH_4^+ + OH^- \qquad \text{[Reaction 3.1]}$$

The reaction product is a silicate (see reaction 3.2):

$$OH^- + SiO_2 \Leftrightarrow SiO_3H^- \qquad \text{[Reaction 3.2]}$$

This latter reaction is a very slow process step, because the breaking of the Si-O bonds to extract one molecule out of the SiO_2 matrix is an energetically unfavorable process. Reported etch rates of SiO_2 in APM are 0.1 nm/min at 60°C and 0.37 nm/min at 85°C [STO 99a]. The etch rate of SiO_2 is controlled by the OH^- concentration, which depends on the square root of the NH_3/H_2O_2 ratio [VER 96]. Silicon is etched much faster in ammonium hydroxide. The reason for this is that it is easier to break a Si atom out of a Si matrix than out of a SiO matrix comparing the bond energy differences: Si-Si = 327 kJ/mol and Si-O = 810 kJ/mol.

A summary of possible reactions at the surface of a silicon wafer is given in Figure 3.1. If bare silicon is exposed, the surface Si-H groups are oxidized by H_2O, O_2 or H_2O_2 to Si-OH surface groups (reaction A). The reaction with H_2O_2 is most effective and the rate depends on the H_2O_2 concentration. If the H_2O_2 concentration exceeds a certain limit (3 mmol/L) reaction B takes place: the backbone bonds of the Si to the Si matrix are also oxidized. In effect, the silicon surface is oxidized to a silicon oxide surface. This is what we call a chemical oxide layer. Then, the etching component comes in: reaction C is, compared to reaction D, a fast reaction (note that for reaction C, three Si-Si bonds have to be broken and for reaction D, three times the Si-O bond).

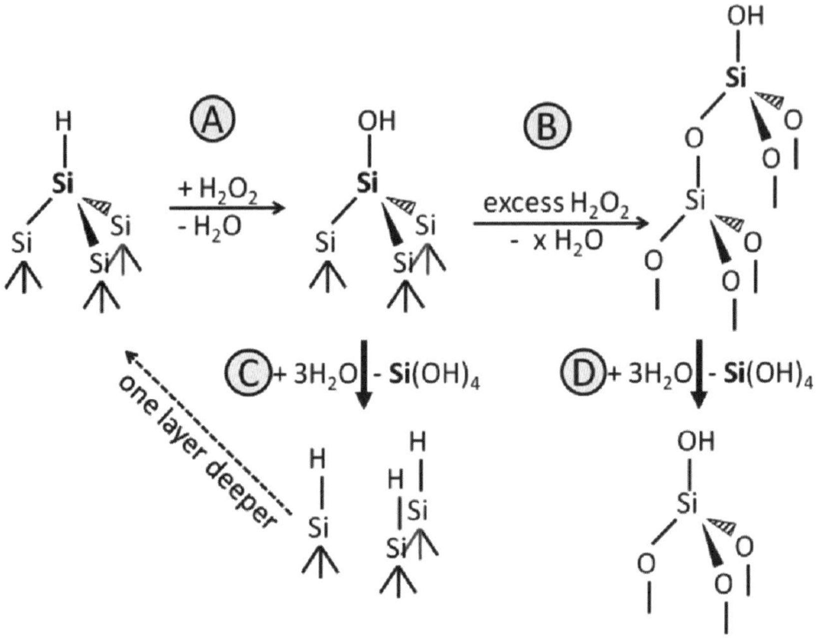

Figure 3.1. *Possible reactions of a silicon surface in ammonia hydroxide solution with hydrogen peroxide resulting in the etching of silicon or silicon dioxide*

The etch rate of silicon in a 1.9 mol/L NH_3 solution at 70°C is 130 nm/min and goes up to 450 nm/min if H_2O_2 is added up to 3 mmol/L [MEE 90]. If the H_2O_2 concentration is further increased, the etch rates seize, because reaction B kicks in. This is the desired reaction, because in cleaning one does not want to etch silicon surfaces and especially not in a much higher rate compared to silicon oxide surfaces. The H_2O_2 concentration of a 1:1:5 APM solution is slightly more than the 1 mol/L that is 300 times higher than the critical limit. Therefore, the APM can safely be used at a 1:1:1,500 ratio. If one wants to reduce the etch rate, the NH_3 should be further reduced to, for example, 0.5:1:200.

The etching of silicon and silica and the subsequent undercutting of particles is the mechanism of removing particles. On the one hand, it is preferable to etch as much as possible, because more etching removes more particles. On the other hand, it is undesirable to lose too much of the substrate, in particular if you want to meet the substrate loss targets set by the ITRS for sub-45 nm technology [ITR 11]. IMEC has shown that 2–3 nm etching is required to remove >90% of the particles (etch) [HEY 99].

Once the particles are detached, redeposition of these particles is prevented by electrostatic double-layer repulsion. Most surfaces immersed in a solution with a pH greater than 10 have a negative surface charge and subsequently a positively charged electrostatic double-layer. As the ion concentration of APM is relatively low, the electrostatic double-layer thickness is significant to prevent a particle coming too close to the surface where it would become entrapped by van der Waals forces.

3.2.1.3. *Problems*

Hydrogen peroxide is a metastable compound and decomposes in water and molecular oxygen (see reaction 3.3).

$$2\ H_2O_2 \rightarrow 2\ H_2O + O_2 \qquad \text{[Reaction 3.3]}$$

The actual decomposition in an APM solution requires a high activation energy of 170 kJ/mol [KNO 99, KNO 11a, b or c] or a catalyst. These catalysts are usually present in the form of minute amounts of metal-ion contamination of <0.2 ppb [SCH 95]. Iron is one of the most common and active contaminants and it can already be present in the chemicals from which the mixture is prepared and/or it is added by the wafers that are processed in these solutions.

Decomposition of H_2O_2 can lead to several process issues:

1) If the concentration of H_2O_2 falls below the critical level of 3 mmol/L, it will etch silicon surfaces. However, in most cases a large safety margin is kept in these processes. Since there is a risk that this failure occurs, a detection method, being a concentration control system, needs to be installed. This can be conductivity (monitoring of H^+, OH^-, NH_4^+ and HO_2^- for instance), pH measurement or near infrared optical technique (monitoring of the HO_2^- species for instance).

2) The gas bubbles that are formed cause collateral damage. They are the agents enabling the catalytic active metal-ion particles to deposit on the wafer surface [KNO 00].

There are several surface roughening mechanisms proposed related with APM [KNO 00]. They are distinguished by the size of the roughness features (see Figure 3.2).

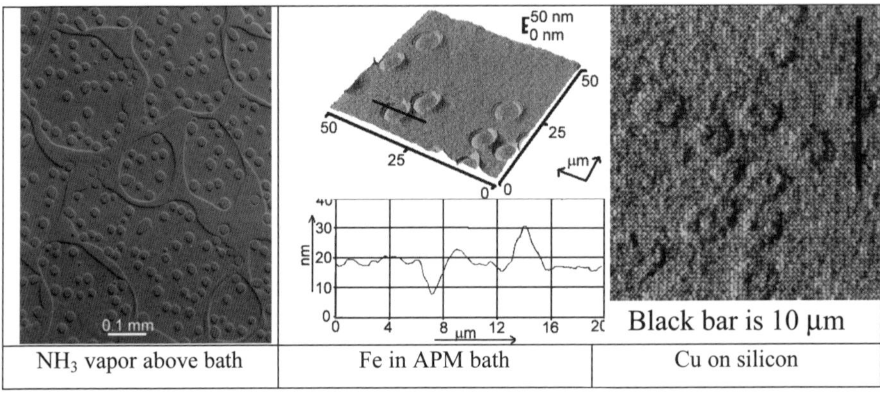

Figure 3.2. *Silicon surface roughness with different root causes. The left part is optical microscope, middle part is AFM and right part is SEM*

NH_3 vapor above bath: NH_3 has a boiling point of –33°C and will evaporate rapidly from a bath operated at 50–70°C. Together with ever present moisture, it condenses on the relatively cold wafer thereby creating an etching environment, that is APM without any H_2O_2. The wafer surface depicted in Figure 3.2 was, prior to this exposure, etched to a hydrophobic state in HF that is a common process step prior to a APM step.

Fe in APM bath: in this second case, the roughening mechanism is a little more complicated, but the essence is the same. The wafer is exposed to H_2O_2 to deplete APM. This depletion is caused by the catalytically active iron particles that have been deposited on the wafer surface during the immersion in the cleaning bath. On the surface, they remain active in the H_2O_2 decomposition creating a H_2O_2-depleted zone around itself with a gradient of H_2O_2 concentration from almost 0 in the center to 3 mmol/L in the ring with the highest silicon etch rate. Outside the zone of <3 mmol/L of H_2O_2, the silicon surface is passivated by the excessive SiO_2 growth. Just outside the ring with highest etch rate, a rim is formed of SiO_2 that relates to the precipitated etch products.

Cu on silicon: in this third case, the origin of the problem is not the APM, but contamination that has been deposited on the silicon surface in a previous step such as Cu deposition from HF immersion steps. The roughness is not caused by lack of H_2O_2, but it is a galvanic corrosion. Cu is present as small pits on the surface. Cu will become a cathode. The cathodic reaction is the reduction of H_2O_2 to OH^- and the two electrons required for this reaction come from the copper pit. This in turn obtains its electrons from the underlying silicon that becomes positively charged. The anodic reaction is the dissolution of Si to Si^{4+} and the four electrons released with this process are flowing to the copper. Si^{4+} will react with H_2O and OH^- to $HSiO_3^-$.

If a silicon surface is cleaned with an APM, most of the particles are removed. The metal-ion contamination present in the APM is easily absorbed on the silica surface due to the relatively high pH. This is because the equilibrium reaction 3.4 is dragged to the right side with the metal ion in the silica (see more about this reaction in the next section).

$$SiOH + M^+ \Leftrightarrow SiOM + H^+ \qquad \text{[Reaction 3.4]}$$

Some silica being etched at the same time as some silica is also grown reduces the amount of deposited metal ions (discussed later in the chapter, see Figure 3.9). However, the metal ion will be throughout the chemical oxide layer, which will make it less easy to remove in a subsequent acidic step, especially if the thickness of the oxide layer is greater than the normal oxide thickness like that in the oxide rims (see Figure 3.2, middle).

3.2.1.4. *Alternatives*

APM is a cleaning solution with a combined oxidation and etching characteristic that enables particle removal. This combined feature is a cause that can result in APM surface roughening. A two-step process is proposed as an alternative for APM where these two characteristics are separated [HEY 99]:

1) SPM followed by HF (hydrophobic IMEC clean).

2) SPM – HF – followed by room temperature APM. APM in this case is only to oxidize the surface. The temperature is too low to have any etching effect.

3) SPM – HF – DI/O_3 (hydrophilic IMEC clean). DI/O_3 is a mixture of ozone dissolved in water and is used to reoxidize the surface.

4) DI/O_3 – HF – DI/O_3.

SPM and DI/O_3 are solutions that oxidize the surface leaving a surface with a fixed SiO_2 thickness. HF is doing the etching that is limited to the amount grown by the oxidation step. Wafers with a surface composed out of hydrophilic and hydrophobic areas tend to have problems in the drying process. Therefore, a reoxidation step following the HF is inserted. This can even be an APM step, which sounds contradictory for something to replace APM.

Diluting APM is a way to save money. It is also a way to reduce the amount of contamination in the mixture that is introduced by the chemicals. Furthermore, the etch rate of the diluted version will be different and can be tuned in such a way that it is minimized. Note always to verify the etch rate if you change the concentration of a running APM process.

To minimize the effects of metal ions in an APM step, complexing agents can be added. These agents will sequester metal ions so they become inactive for the decomposition of metal-ion catalyzed H_2O_2. In addition, the complexing agents also prevent the metal ions from depositing on the silicon wafer. A drawback is that these complexing agents are not always stable in the APM environment and that they can also leave residues on the wafer surface. As some of these complexing agents contain phosphates, doping of the silicon is a potential consequence.

3.2.2. *Hydrochloric acid hydrogen peroxide mixture (HPM)*

3.2.2.1. *Brief history and application*

The second step in the RCA clean is Standard Clean 2 (SC2), also known as the hydrochloric acid hydrogen peroxide mixture (HPM). It is a mixture of two commercially available chemicals in water, that is 35–37% hydrochloric acid in water and 29–31% hydrogen peroxide in water. Similar to APM, its composition is expressed as a ratio x:y:z, where x is the volume of hydrochloric acid instead of the volume of ammonium hydroxide solution. In manufacturing, the mix ratios in use range from 1:1:5 (most concentrated) to 1:1:200 (most diluted version). The temperature for use is from 20ºC to 80ºC with the most common at 50–60ºC.

The original intent of HPM was to remove metal-ion contamination. For this optimization, Kern used as a cleaning challenge a 1 ppm gold contaminated HPM solution. Gold was in the past a common contaminant as it was present in the products. His goal was to minimize the precipitation of gold onto the silicon surface. He came with an optimum of 1:1:6 mix ratio, but for simplification, as with APM, a 1:1:5 mix ratio was also used in both cases at an operating temperature of 80°C. Except for hydrogen peroxide, the choice for these chemicals is less obvious:

– hydrogen peroxide is a strong oxidant that is able to keep 1 ppm of gold at given pH in the oxidized and soluble state. It leaves no residues on the wafer if it is not removed properly;

– hydrochloric acid is a commercially available strong mineral acid, but there could have been more candidates, such as sulfuric acid, nitric acid or even phosphoric acid. A big disadvantage is that hydrogen peroxide is intrinsically unstable in hydrochloric acid (see section 3.2.3). Perhaps it was helpful to use HCl in the removal of gold as chloroauric acid ($HAuCl_4$) is a soluble gold compound. Furthermore, most acids were very impure at that time.

Gold has been banned from most semiconductor manufacturing sites and large amounts of hydrogen peroxide is no longer required for the removal of gold. Today, hydrochloric acid with or even without hydrogen peroxide is used to acidify water to remove the minute amounts of metal-ion contamination precipitated on the wafer in the previous APM step.

Also for HPM, the process condition and composition have been re-evaluated. HPM was used in bath processes with the bath-life well over 12 h. During this time, most of the hydrogen peroxide is decomposed. Since hydrogen peroxide is a costly material, its use could easily be reduced with good understanding that in a large portion of the bath-life successful processing was possible without hydrogen peroxide. Reduction of hydrochloric acid concentration was beneficial for the process performance.

3.2.2.2. *Chemical mechanism*

The goal of HPM is to remove metal-ion contamination mainly left on the surface by the previous APM step. The components in HPM each have a potential removal effect and the combined effect of hydrogen peroxide, acidity and the chlorine content (see Figure 3.3) enables even the removal of gold.

Most metal ions present on a silica surface are already in an oxidized state and are hydrated, for example $M(H_2O)_4^{++}$. Absorbed on a silica surface the complexing water molecules are replaced by the O of deprotonated silanol groups, for example in Figure 3.3 two of the water molecules are replaced and effective $M(H_2O)_4^{++}$ is surrounded by two negatively charged Si-O groups resulting in electroneutrality. Its removal takes place with the following reactions (it can be in any order and not necessarily all are required for the removal of a specific metal ion):

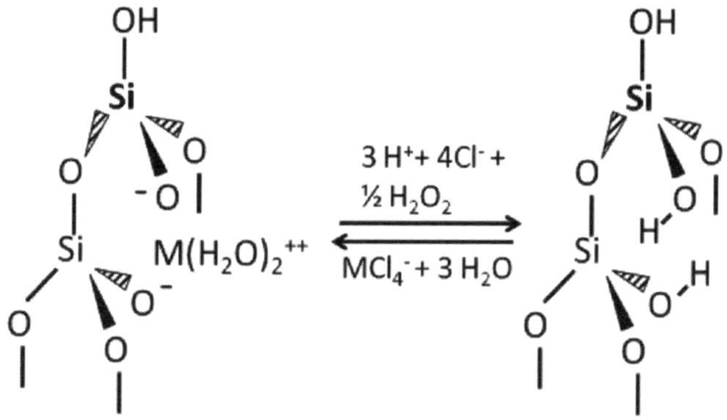

Figure 3.3. *Simplified combined functionality of HPM: oxidation, ion exchange, complexing of metal ion, M*

Oxidation: the metal ion in the given example was in oxidation state II and is oxidized by H_2O_2 to oxidation state III. Each H_2O_2 molecule extracts two electrons. Oxidation changes the solubility properties of the metal ion.

Ion exchange: by exposing the surface to a high concentration of protons, a completion starts with the adsorbed metal ions. Protons take over the coordination sites of the metal ions [LOE 98].

Complexing: chlorine is a hard Lewis acid that likes to combine with other hard Lewis acids such as these oxidized metal ions. The water molecules in the coordination sphere of the metal ion are all replaced by Cl^-; this results in a soluble metal chloride complex.

3.2.2.3. *Problems*

In contrast to APM where H_2O_2 is relatively stable, the H_2O_2 in HPM is intrinsically unstable. Decomposition is fast as within 20 min, 90% of the H_2O_2 was decomposed in a standard HPM solution at 80–85°C [HUR 94]. Chloride is a catalyst for the decomposition reaction of hydrogen peroxide (see reactions 3.5 and 3.6).

$$H_2O_2 + HCl \rightarrow H_2O + HClO \qquad \text{[Reaction 3.5]}$$

$$H_2O_2 + HClO \rightarrow H_2O + HCl + O_2 \qquad \text{[Reaction 3.6]}$$

H_2O_2 will oxidize HCl to hypochlorite, which in turn reacts with a second H_2O_2 molecule to O_2 and HCl. The calculated activation energy for this reaction is 95 kJ/mol [SCH 95], which is lower than the 170 kJ/mol in a stabilized APM solution. This is typical for a catalyzed reaction.

There are a few solutions for this problem among the one-time use of freshly prepared HPM. Lower temperatures mean a lower decomposition rate and a longer bath-life. However, the rate will be unacceptably high for a good process control. An alternative for HPM is diluted hydrochloric acid without H_2O_2 (see the next section).

The APM was a process targeting the removal of particles, but it left metal-ion contamination. As a consequence, an acidic step is required to remove this contamination. Unfortunately, the side effect of HPM is that it adds particles again. Though particles will be present in pre-filtered chemicals in very small amounts, the chance of deposition has significantly increased. At this pH, most surfaces have a positive surface charge and when a positively charged particle approaches a positively charged surface it will be repelled, because the negatively charged electrostatic double-layers start to overlap at close distance. The concentration of Cl^- is that high that the thickness of the electrostatic double-layer is in the order of a few nm, which allows the particle to approach the surface well within the van der Waals interaction sphere.

Besides problems on the wafer, HPM also has problems in the manufacturing environment. HCl fumes will cause the stainless steel parts within the environment of the process to corrode. They can potentially contaminate the carrier and boxes in which the wafers are transported. This can result in cross-contamination to wafers later in the process that might be sensitive for HCl fumes, for example with aluminum surfaces. Furthermore,

HCl will react with NH_3 fumes from, for example, with APM to NH_4Cl particles. These particles can also pop up at unexpected points.

3.2.2.4. *Alternatives*

An acidic step is required, because APM leaves metal-ion contamination on the wafer. Although this can be minimized by using complexing agents, a water-rinse step following this chemical step will show enhanced metal-ion deposition (see later in the chapter).

As already mentioned in the previous section, an alternative for HPM is to leave out the H_2O_2. As gold has been banished from most manufacturing sites, the H_2O_2 is not really necessary. The higher the concentrations of HCl the better the metal-ion removal efficiency, but this adds to the expense of particle addition. It was found that the optimum between the two extremes is at pH 3 [MEU 95]. This means that the 35–37% HCl solution needs to be diluted to 1:10,000 which makes diluted HCl.

3.2.3. *Sulfuric acid hydrogen peroxide mixture (SPM)*

3.2.3.1. *Brief history and application*

Aggressive mixtures of strong acids and oxidants were used in the pre-RCA time for tube cleaning and subsequently for silicon wafer cleaning. They were able to remove great amounts of organic contamination up to the complete removal of photoresistant layers. Examples that were used are sulfuric acid with nitric acid or chromic acid or (fuming) nitric acid. The one that survived safety and environmental requirements and is still in use is the sulfuric acid hydrogen peroxide mixture (SPM), which is also known as Piranha or Caro's acid. It is a mixture of two commercially available chemicals: 96% sulfuric acid in water and 29–31% hydrogen peroxide in water. The composition is expressed as a ratio x:y, where x is the volume of sulfuric acid and y is that of the hydrogen peroxide solution. In manufacturing, the mix ratios in use for recirculated baths range from 4:1 to 8:1. The temperature of use for recirculated baths is from 90°C to 130°C. For point-of-use (POU) mixed single pass implementation, mix ratios range from 2:1 to 8:1. The sulfuric acid preheat temperature can be as high as 180°C resulting in a mixture temperature of about 200°C due to the exothermic mixing reaction of sulfuric acid with hydrogen peroxide and water.

The original intent of SPM was to remove organic contamination. Since the mixture left sulfur contamination on the surface, alternative cleaning mixtures such as APM were proposed to replace it. However, APM was not able to remove most of the target contamination. Optimization in concentration can only be done in the ratio as water addition will reduce the effectiveness of the sulfuric acid. A way to reduce the cost of the SPM bath is by heating sulfuric acid up to 110–130°C and add small amounts of hydrogen peroxide to this bath only if contaminated wafers are immersed. Hydrogen peroxide consumption is reduced and the sulfuric acid is not immediately diluted by the relatively large amount of H_2O/H_2O_2 mixture added to it that would make it less potent. The vigorousness of SPM was enhanced at the beginning of the century by mixing heated sulfuric acid and hydrogen peroxide resulting in operating temperatures greater than 170°C [BUT 06]. Hydrogen peroxide is not preheated before mixing to prevent its decomposition. Even higher SPM temperatures can be achieved with POU mixing by injecting steam into the chamber where the SPM mixture is dispensed on the wafer. Steam has a high internal energy, which is released when it condenses or when it dissolves into the hygroscopic SPM mixture. In order to maximize the effectiveness of steam additions, a SPM mixing ratio of about 10:1 is used. Using SPM with steam in this way has the ability to remove even highly implanted photoresist films [CHR 08, DEK 09].

3.2.3.2. *Chemical mechanism*

Sulfuric acid is a very strong acid. This property makes it a solvent for many organic compounds containing oxygen and nitrogen. The oxygen in the form of alcohol, ketone, carboxylic acid, or ether and the nitrogen in the form of amine or nitro-groups are all protonated making the compound more soluble. To increase the amount of these functionalities in the organic compounds, hydrogen peroxide is added that can oxidize and build in extra oxygen functionality.

The property of concentrated sulfuric acid being such a strong acid is also due to the absence of weaker acids such as water. If too much water is present, the protonation of the organic compound over that of water will become unlikely. The solubility of photoresist sharply decreases if the concentration of sulfuric acid decreases below 80%. Hence, the limitation of the amount of H_2O_2 (that contains also 70% H_2O) to be added to sulfuric acid is related to this critical concentration. In a 4:1 SPM solution, the concentration of H_2SO_4 dropped to 85%.

When photoresist is dissolved in H_2SO_4, the solution turns black. If H_2O_2 is added, the solution starts to decolorize. This is also a method by which SPM is applied: a hot sulfuric acid bath in which wafers are immersed and subsequently an amount of H_2O_2 is added. The decolorization means that the organic photoresist is oxidized. It is unlikely that the photoresist is completely turned into CO_2 and H_2O. Many aromatic compounds will become sulfonated, making relatively stable water-soluble compounds.

The first reaction to take place when adding H_2O_2 to H_2SO_4 is the formation of Caro's acid, H_2SO_5, which is also known as peroxymonosulfuric acid (see reaction 3.7). It is a slightly stronger oxidant than H_2O_2: $E^0 = 1.776$ V versus 1.81 V for H_2SO_5. Since the formation of H_2SO_5 is an equilibrium reaction with water, its concentration will drop if the water concentration increases. This is another reason to have an excess of H_2SO_4 in SPM.

$$H_2SO_4 + H_2O_2 \Leftrightarrow H_2SO_5 + H_2O \qquad \text{[Reaction 3.7]}$$

In some cleaning sequences, SPM is following a HF step. The SPM turns this surface in a controlled way into a surface with approximately 1 nm of SiO_2. The fact that H_2O_2 is turned into H_2SO_5, which is a stronger oxidant, might help, because a HPM solution is not such a strong oxidizing solution for growing a similar oxide layer in the same time frame.

3.2.3.3. *Problems*

If the SPM is used in a recycling mode or in a bath process, dilution of the sulfuric acid lower than the 80% can be a problem. Sulfuric acid is hygroscopic and will attract water from the environment. This is an equilibrium reaction and the equilibrium concentration depends on the temperature. At 110°C, the equilibrium concentration is 85%. It also means that if the concentration of sulfuric acid is lower than 85% and the mixture is heated to 110°C, water evaporates.

One of the most well-known problems of SPM is the formation of particles after the wafers are rinsed and dried. Wafers following the process sequence –HF – water rinse – SPM – water rinse – dry– show, immediately following the drying step, a low particle surface concentration. If the wafers are stored for several hours, they show a significantly higher particle concentration, which becomes even worse over time. These aged particles are also seen after an etch step in hot phosphoric acid. Wafers with a thick thermally grown oxide layer that are exposed to SPM – water rinse – dry do not show these aged particles [KNO 11]. The particles are needle-shaped crystals made up of ammonia and sulfate, either $(NH_4)_2SO_4$ or $(NH_4)HSO_4$.

The mechanism of this phenomenon is that a chemical oxide of approximately 1 nm thick is grown on the initial bare silicon surface. Since the oxide is grown in such a strong acidic environment, many of the oxygen atoms in the newly formed SiO_2 network will be protonated to $SiOH_2^+$ or $Si\text{-}O(H)^+\text{-}Si$ with as a counter ion the hydrogensulfate anion, HSO_4^-. During the rinse step, this bisulfate and proton have to diffuse through the 1 nm thick layer. Apparently, this is a slow process, as even after a 10 min water rinse at 90°C, sulfates are still present in or on the silica surface. If the oxide was not grown by the SPM step, the H_2SO_4 would not be built in with the SiO_2 formation or would not diffuse into the surface layer. Only Si-O groups at the surfaces are protonated and these are relatively easy to remove.

After the wafers are dried, the sulfuric acid diffuses slowly to the surface and reacts with ammonia that is present in most clean room environments (e.g. think of APM). Silica gels are a perfect media to grow crystals [SUI 85]. The slow diffusion in the silica gel environment enables the growth of the needle-shaped crystals. The impact of these crystals is small. If the wafers with these crystals are put in a furnace, the crystals decompose and evaporate to NH_3, H_2O and SO_3. They dissolve in water rapidly. Monitoring particles during SPM processes is a real problem. If a bare silicon wafer is used to determine how many particles are added by the cleaning, the real potential harmful particles are overshadowed by the aged particles. The monitor is not triggered if a real particle problem occurs.

Properties of hot sulfuric acid are quite similar to that of hot phosphoric acid that is used as an etching solution for Si_3N_4 and as a side effect also etches SiO_2 [GEL 67, KNO 05]. In both cases, it is a hot water solution and water etches both Si_3N_4 as well as SiO_2 (see section 4.1). Substrate loss is an issue in the photoresist strip after ultra shallow implantation, where one does not want to lose too much of this highly implanted surface layer and associated dopant.

The waste from SPM is dangerous and a serious problem. Stored in a container for transport, H_2O_2 and H_2SO_5 will continue to decompose to oxygen. The container will come under pressure with a potential explosive situation. Therefore, before the container with this mixture is closed, H_2O_2 or H_2SO_5 needs to be decomposed. This can be done by either storing the container for a long time or by adding nitric acid (HNO_3).

3.3. Wet etching

3.3.1. *Hydrofluoric acid (HF)*

3.3.1.1. *Brief history and application*

Hydrofluoric acid is a chemical that was first synthesized more than 300 years ago. It is used in the glass industry to etch silicate-based glasses. In the semiconductor industry, it is the chemical to etch silicon oxide and is in use since the start of it. Commercial semiconductor grade aqueous HF comes in a concentration from 49% and lower. HF is also used in combination with a 40% aqueous ammonium fluoride (NH_4F) solution to form a so-called buffered HF (BHF) or buffered oxide etch (BOE). The operating temperature is usually slightly higher than room temperature, that is 25°C. The 49% HF solution is not so much in use, because the vapor pressure of HF above such a concentrated solution is very high resulting in a fast decay of the HF concentration in these solutions. BOE is a way to reduce the HF evaporation and still has a solution with a relatively high etch rate for silicon oxide. However, during the development of the semiconductor industry, designed features and used layer thicknesses became smaller and such high etch rates were not necessary. A solution with high etch rates became undesirable, because for the removal of thin oxide layers other functional silicon oxide was exposed at the same time. With high etch rates the control on the removal of this functional silicon oxide becomes very poor.

HF solutions are used for cleaning and etching and sometimes the difference is ambiguous. Therefore, we define cleaning the case where no resist mask (either a hard mask or an organic resist) is present and the etch process stops at bare silicon, that is stripping of silicon oxide (and silicon nitride) layers. In many cases, the SiO_2 layer is contaminated by, for example, ion implantation where SiO_2 is used as a scatter oxide.

3.3.1.2. *Chemical mechanism*

HF is a complicated mixture of partly dissociated fluoride molecules and polymeric fluoride molecules. The main constituents are the dissociated and un-dissociated mono and difluoride species: HF, H^+, F^-, H_2F_2 and HF_2^-. For a long time, it was unknown what the reactive species were in the reaction with SiO_2, but more importantly surface properties of the SiO_2 were completely ignored [OSS 96]. The etch rate of SiO_2 was fitted by many complicated functions, which had no chemical meaning other than that it could predict the etch rate in a limited concentration range.

In order for HF to react with the SiO_2, the surface needs to be in the proper condition. It will react with certain end groups faster than with others. A SiO_2 surface consists of SiOH end groups and these end groups are amphoteric (see Figure 3.4). It means that they (**1**) can either accept a proton (base property to form **2**) or donate a proton (acid property to form **3**). The amount of SiO^-, SiOH and $SiOH_2^+$ groups depends on the pH of the solution in which the surface is immersed. At the point of zero charge, the amount of SiO^- and $SiOH_2^+$ groups is equal and at that pH there charge is zeroed out. For SiO_2, this point is reported to be somewhere between pH 2 and pH 3.

The first step in the reaction is to replace the OH by the fluoride group [KNO 00]. This requires a nucleophilic substitution reaction, either a SN1 or a SN2. For a SN1 reaction to take place, the OH group has to be removed from the Si and on this vacancy an F group can sit. High activation energies are required to break the Si-O bonds. Furthermore, the OH group needs to be a good leaving group. O^{2-} is a very poor leaving group and therefore **3** will not react further. H_2O is a perfect leaving group and reaction from **2** will be relatively fast. OH^- itself is a poor leaving group and it is unlikely that it will be replaced via a SN1 reaction.

The special property of HF and the OH group is that the difluoride can interact with the OH. The proton in HF_2^- or H_2F_2 interacts with a free electron pair of the oxygen. Thereby it brings one fluoride close to the Si. Now the fluoride can start to form a bond with the silicon while the bond between Si and O slowly weakens and finally breaks. It is a kind of SN2 reaction, but with the retention of the Si configuration. Both **1** and **2** react with a difluoride through the coordination complexes **4** and **5** to yield **6**. Etching of silicon nitride which has a similar structure as SiO_2, but all Os are replaced by N, is not going through the difluoride complex, because the protonated NH_2 group (NH_3^+) has no free electron pair for the difluoride to interact with [KNO 11].

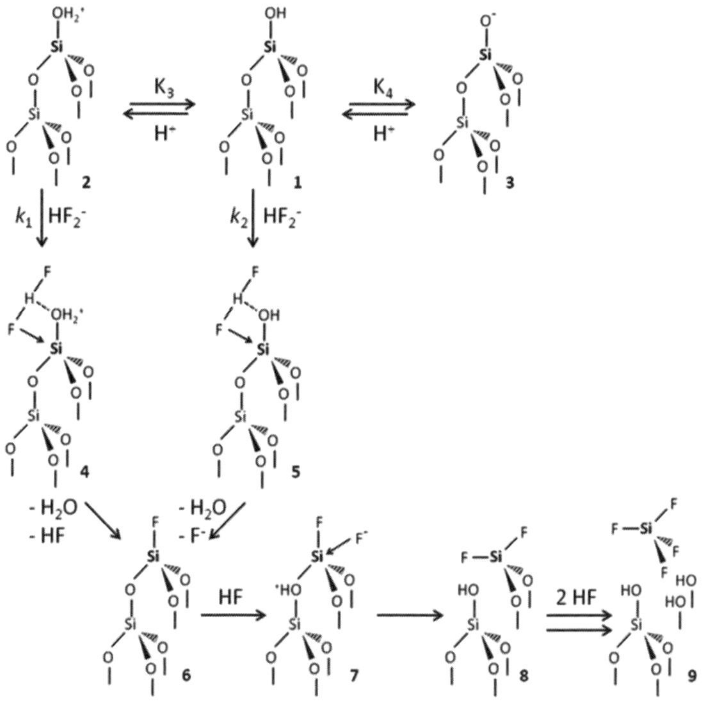

Figure 3.4. *Etch mechanism of SiO_2 etching by HF: a small detail of the SiO_2 surface structure is depicted of which the top unit Si in bold (1, 2 and 3) is etched off in a series of steps to 9. The latter, 9, is the same as 1, but one monomolecular layer deeper in the original surface*

Once the first fluoride is introduced on the silicon, the subsequent steps are much faster. The reaction mechanism of the three following steps, to go to **9**, is pure SN2. One of the oxygen atoms bond to the Si-F group is protonated, which makes it a perfect leaving group, that is Si-OH. From the opposite site F^- approaches the silicon (see intermediate **7**) and while forming the new Si-F bond, the bond with the protonated oxygen breaks to form **8**. In the course of the process, the configuration on the silicon inverts (the so-called Walden inversion). Such an inversion is necessary for a true SN2 reaction and is not possible in the replacement of the first OH group in **1**.

The three SN2 reactions are 18 times faster than the first OH substitution reaction [KNO 00].

$$R = k_0 \frac{K_3[\mathrm{H}^+]}{1 + K_3[\mathrm{H}^+] + \dfrac{1}{K_4[\mathrm{H}^+] + \alpha}}[\mathrm{H_2F_2}] + k_1 \frac{K_3[\mathrm{H}^+]}{1 + K_3[\mathrm{H}^+] + \dfrac{1}{K_4[\mathrm{H}^+] + \alpha}}[\mathrm{HF_2^-}] + k_2 \frac{K_4[\mathrm{H}^+] + \alpha}{1 + \alpha + K_3[\mathrm{H}^+] + K_4[\mathrm{H}^+] + K_3K_4[\mathrm{H}^+]^2}[\mathrm{HF_2^-}] \quad [3.1]$$

Constant	Value	Constant	Value
k_0	0.7 (0.1) $nm.l.mol^{-1}.s^{-1}$	K_3	0.28 (0.06) $l.mol^{-1}$
k_1	1.8 (0.4) 10^3 $nm.l.mol^{-1}.s^{-1}$	K_4	13 (2) 10^3 $l.mol^{-1}$
k_2	1.26 (0.05) $nm.l.mol^{-1}.s^{-1}$	α	0.85 (0.06)

Note: errors are shown between parentheses.

Table 3.1. *Values for reaction rate and equilibrium constants for equation [3.1]*

Based on this reaction mechanism and knowing the rate limiting step, the etch rate of SiO_2 can be derived using standard chemical equations. It results in equation [3.1] and by fitting more than 100 etch rate data having different fluoride concentration and pH, the rate constants and equilibrium constant were determined (see Table 3.1). In the equation, but not explained in Figure 1.4, is the reaction of $SiOH_2^+$ with H_2F_2 that is similar to **4**, but HF_2^- is replaced by H_2F_2 with reaction rate k_0. Furthermore, a correction factor α is included to compensate for the polyacid behavior of the surface: if some SiOH groups on the surface are deprotonated, it will be more difficult for the following SiOH group to lose its proton, because it will make the surface even more negative.

Clearly, the most favorable reaction is HF_2^- with $SiOH_2^+$. Its reaction rate constant (compare k_1 with k_0) is > 2,000 times higher than that of the same reaction with H_2F_2. The reaction HF_2^- with SiOH is < 1,000 times slower (compare k_1 with k_2), but in this case the SiOH surface concentration is much higher than that of the $SiOH_2^+$ groups. Therefore, this reaction will contribute significantly to the total etch rate at pH > 3.

If the bulk of the SiO_2 is removed at a certain moment, the reappearing surface SiOH group will no longer be bonded to the SiO_2 matrix but to silicon. At that moment, the surface will have one monomolecular layer of oxygen left. Removal of this last molecular layer results in F or H passivation (see Figure 3.5).

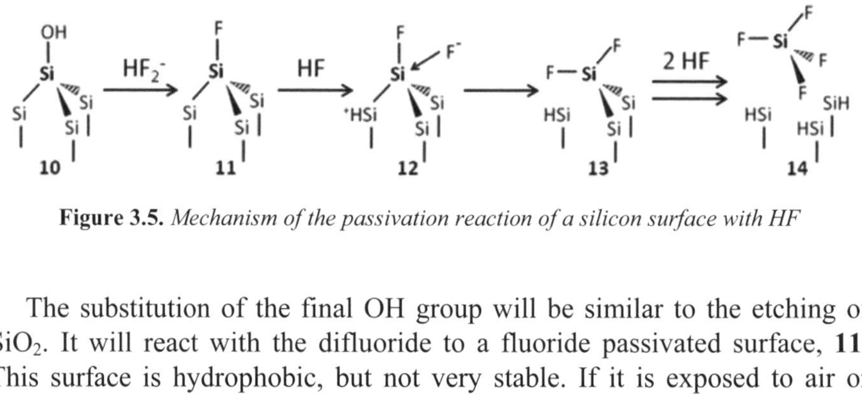

Figure 3.5. *Mechanism of the passivation reaction of a silicon surface with HF*

The substitution of the final OH group will be similar to the etching of SiO_2. It will react with the difluoride to a fluoride passivated surface, **11**. This surface is hydrophobic, but not very stable. If it is exposed to air or water, the surface will rapidly turn into a hydrophilic oxide surface. Removal of the Si-F group is a slow process and is similar to the reaction steps going from **6** to **9**. Owing to three HF addition reactions, the three Si-Si bonds are opened up in going from **11** to the hydrophobic hydrogen passivated surface **14**. If the reaction is complete, the surface is stable and can be stored on air for a couple of weeks without oxide formation.

Two important observations are that the F is not substituted by an H, but the complete Si-F group is etched from the surface [UBA 84]. This would be in agreement with a thermodynamic consideration as Si-F is one of the strongest chemical bonds. The other observation is that this etching of the SiF group is done by a monofluoride species. This will come in handy if selectivity needs to be tuned.

Etching with BOE is not different from what has been described thus far for HF. One of the major differences between HF solutions and BOE solutions is the pH. BOE solutions have a more neutral pH and the silica surface will have more SiO^- and SiOH surface groups. In addition, due to the presence of NH_4F, more of the fluoride species will be dissociated. For this reason, the etch rate given in equation [3.1] will be simplified to equation [3.2] for pH > 4

$$R = k_2 \frac{K_4[\mathrm{H}^+]+\alpha}{1+K_4[\mathrm{H}^+]+\alpha}[\mathrm{HF}_2^-] \qquad [3.2]$$

and equation [3.3] for pH > 6

$$R = k_2 \frac{\alpha}{1+\alpha}[\mathrm{HF}_2^-] = k_2'[\mathrm{HF}_2^-] \qquad [3.3]$$

With etching the surface, several contaminants are targeted to be removed:

– Damaged or contaminated SiO_2 layers are removed.

– Particles and post-ash residues are removed by a lift-off process similar to that of APM. In contrast to APM, the amount of silicon loss is controlled, because it lacks the oxidant. Consequently, HF does not suffer from the surface roughening mechanisms typical for cleaning solutions with combined etching and oxidation properties.

– Metal-ion contamination: HF is an acid and like HCl it extracts metal ions from the surface.

3.3.1.3. *Problems*

Surfaces that need to be cleaned usually consist of areas that need to be etched/cleaned and areas with functional SiO_2, such as isolation oxide (shallow trench isolation). Hydrogen passivation is a slow process and during this reaction the functional oxide is exposed to the HF solution as well. In other words, there is a selectivity requirement.

It is possible to do the hydrogen-passivation reaction selective toward the etching of bulk SiO_2. As the bulk is etched by the difluoride species and the passivation reaction is attained by the monofluoride species, the formation of monofluoride species should be favored over that of the difluoride. The simplest way is to dilute the HF solution and the reaction in reaction 3.8 will shift to the right, that is to the monofluoride species.

$$H_2F_2 \Leftrightarrow 2\ HF \qquad \text{[Reaction 3.8]}$$

By diluting the HF solution, the etch rate will decrease both the bulk SiO_2 etch as well as the H-passivation reaction. The etch rate of SiO_2 decreases much faster than that of H passivation. Unfortunately, the price of this improved selectivity is an increased process time, but to get the removal of a chemical oxide with 0.1% HF, 10 min is sufficient for a bath process. For a single wafer process, this is not an option.

Alternatively, the temperature can be increased, which also results in a shift of the equilibrium to the monofluoride species. In this case, the etch rates of both the bulk and H passivation are increasing, but the rise is faster for the H-passivation reaction. Bulk SiO_2 etch rates double for every 11°C increase in temperature. For H passivation, no data are available for the activation energy.

SiO_2 is a good barrier for metal (M) ions that tend to form silicide compounds (M_xSi_y) or dissolve in bulk silicon. If this layer is removed, the silicon becomes exposed to the contaminants that are present in the HF solution. Metal ions such as Ni, Cu, Ag, Pt and Au all have a tendency to plate on the silicon surface (see reaction 3.9).

$$4/x\ M^{x+}(aq) + Si(s) \rightarrow 4/x\ M(s) + Si^{4+}(aq) \qquad \text{[Reaction 3.9]}$$

On device wafers, this is not a random process and does not occur uniformly over the wafer. Plating has a preference for crystal defects in the silicon substrate, which can be process induced or has a preference for doped silicon. Furthermore, the plating process is a galvanic reaction and plating occurs preferably on silicon that is electrically connected to an anode. Therefore, this failure mode can remain unnoticed as the overall contamination levels on the wafer will not necessarily increase above the detection limits of analytical methods. The failure mode can become visible in an APM step following the HF (see Figure 3.2), but in this case the root cause can be wrongly assigned to the APM.

This problem can be reduced by several methods:

1) use cleaner chemicals: this is not always easy, because the wafer itself can be the source of contamination and will contaminate the HF solution to much higher levels than it was obtained from the supplier;

2) exclude light: in some cases light enables electron transfer from the silicon to the metal ions;

3) add HCl to the HF [TEE 96]: chloride ions can form stable metal-ion complexes that will not plate, for example $CuCl_4^{2-}$. The amount of HCl should be optimized in relation with HF as fluoride ions compete for coordination around the metal ion in which case some metal ions will be less likely to stay in solution.

3.3.2. *Buffered oxide etchant or BOE (HF/NH$_4$F)*

3.3.2.1. *Brief history and application*

A specific HF mixture, known as BOE, standing for buffered oxide etchant (or BHF for buffered HF), is widely used in the semiconductor industry for its etching capability at a controlled pH. BOE is a mixture of NH_4F and HF, usually at 40% (wt%) and 49% (wt%), respectively, in a ratio of X:1, X being the ratio of volumes of the two defined chemicals. As an example, BOE 7:1 is a commercial mixture of seven volumes of 40% NH_4F mixed with one volume of 49% HF. Commercial BOE can be bought from regular chemical suppliers in ratio from 7:1 to 500:1. As supplied, the BOE will be etching regular thermal oxide at a rate between 100 nm/min and 1 nm/min for high X:1 ratio. Typical related pH values of these BOE are from 3.5 to 5.5. BOE 7:1 has been widely used in the technologies with a critical dimension larger than 180 nm but the etching rate is far too high for the nodes at sub-22 nm technologies. In 2011, facilities would rather run BOE 40:1, 100:1 or even 500:1 to reduce that etching rate, thus subsequently having a better etching control.

Because photoresists are usually not attacked by chemicals with a pH greater than 3 (this is true also for the 193-nm current lithography process), buffered oxide etching is preferred for patterned etching of oxides. BOE is therefore widely used in the semiconductor industry to etch oxide layers through a photoresist mask.

While IC makers recognize the benefits of buffered oxide etching, commercial or premixed BOE is more expensive than diluted HF, and is only available in a limited number of blend ratios. In a standard situation, only one or two blend ratios are available at a facility, offering basically two etching rates.

Many works already published on the subject of etching with BOE present the behavior of doped [KUN 95] and undoped [NIE 83, PRO 92, SOM 96] silicon dioxide etching. Many works provide different analyses in regard to the mechanism of SiO_2 etching [PRO 92, VER 94, WAR 71].

3.3.2.2. *Co-injected (blended) BOE*

Making point-of-use BOE *in situ* by co-injection of HF and NH_4OH allows a reduction in cost and the optimization of the etch process. The so called "co-injected BOE" is an HF-based oxide etchant with added NH_4OH. The combination of HF and NH_4OH forms NH_4F in solution, which buffers the fluoride concentration (see reaction 3.10):

$$HF + NH_4OH \Leftrightarrow NH_4F + H_2O \qquad \text{[Reaction 3.10]}$$

There are a large number of species present in BOE, including HF, F^-, H_2F_2, HF_2^-, NH_3, NH_4^+, H_3O^+ and OH^-. The rate of oxide etching depends on the concentrations of HF, HF_2^- and, to a lesser extent, H_2F_2 (see equations [3.2] and [3.3]). Reactions between species are described from reactions 3.11 to 3.15. A system of eight equations (see equations [3.4]–[3.11]) to determine the concentration of each species has to be solved and it is then possible to choose the final experimental settings [LET 00].

HF acidic equilibrium $HF \Leftrightarrow H^+ + F^-$ [Reaction 3.11]

$$Ka_1 = [H^+]\,[F^-]/[HF] = 6.85\text{x}10^{-4}\ \text{mol}\cdot\text{l}^{-1} \qquad [3.4]$$

NH_4OH acidic equilibrium $NH_4^+ \Leftrightarrow H^+ + NH_3$ [Reaction 3.12]

$$Ka_2 = [NH_3]\,[H^+]/[NH_4^+] = 6.31\text{x}10^{-10}\ \text{mol}\cdot\text{l}^{-1} \qquad [3.5]$$

Water dissociation $2\,H_2O \Leftrightarrow H_3O^+ + OH^-$ [Reaction 3.13]

$$Ke = [H_3O^+]\,[OH^-] = 10^{-14}\ \text{mol}\cdot\text{l}^{-1} \qquad [3.6]$$

HF complexation $HF + F^- \Leftrightarrow HF_2^-$ [Reaction 3.14]

$Kd_1 = [HF_2^-]/[HF][F^-] = 3.963\ l\cdot mol^{-1}$ [3.7]

HF dimerization $HF + HF \Leftrightarrow (HF)_2$ [Reaction 3.15]

$Kd_2 = [(HF)_2]/[HF]^2 = 2.7\ l.mol^{-1}$ [3.8]

Electroneutrality equation

$[H_3O^+] + [NH_4^+] = [OH^-] + [F^-] + [HF_2^-]$ [3.9]

Fluoride conservation

$[HF] + [F^-] + 2[(HF)_2] + 2[HF_2^-] = A$ [3.10]

Nitrogen conservation $[NH_3] + [NH_4^+] = B$ [3.11]

$A = (n_1$, number of moles of HF)/total volume [3.12]

$B = (n_2$, number of moles of NH_4OH)/total volume [3.13]

The user chooses A and B values (how much HF and NH_4OH will be mixed in water) from his chemical volumes, which means equations [3.10]–[3.13] are fixed from that choice. The chemical species will accordingly equilibrate themselves through equations [3.4]–[3.11]. During the acidic neutralization of HF by the basic NH_4OH solution, the fluoride element reorganizes itself between its four species. The calculated distribution of the different species is shown in a solution using 1,800 cc of DI water, 100 cc of 49 wt% HF and 0–200 cc of 28 wt% NH_4OH. (see Figure 3.6).

As more NH_4OH is added, more HF reacts to create NH_4F. The fraction of F^- increases, and at the same time the fractions of HF and the dimer, $(HF)_2$, decrease. The HF_2^- behavior is the most interesting. It is well known that HF_2^- is a key etching species (see the previous section). Different works have shown an optimum value of etch rate or the correlation between fluorinated species and pH [PRO 92, VER 94, SOM 96]. Simulation indicates that the optimum $[HF_2^-]$ corresponds to a 1:1 blend of HF:NH_4OH corresponding to a BOE ratio close to 3:1. In such conditions, highest etch rates with no attack of the photoresist is observed [LET 00].

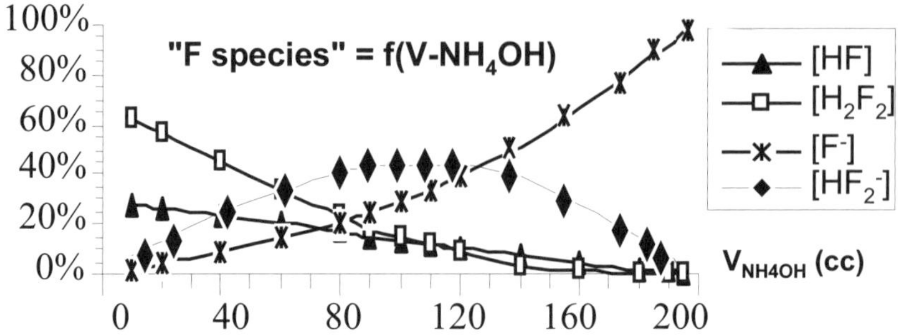

Figure 3.6. *Distribution of the species depending on the mixture from [LET 00]*

It is possible to get relatively low etch rates (less than 0.1 nm/s) or higher values if necessary. The amount of oxide to be etched is the basic criteria for selecting a practical etch rate and, subsequently, the blended mixture (choice of HF, NH_4OH and water volumes). Different blended mixtures can now be prepared to get a similar etch rate. The final co-injection conditions can be selected based on optimum etch uniformity which then refers to the fluid dynamic available from the equipment tool set, from wet bench (usually immersion type) to single wafer platform (usually spray type).

Co-injected BOE can also be part of a full integrated sequence like BOE for oxide etching through a patterned resist, followed by a typical photoresist stripping step followed by a typical RCA clean to clean up the surface (see Figure 3.7). This sequence getting the adequate in-between rinse steps and final dry is being used in production making a "3 steps in 1" full wet solution. Since BOE is a "HF-like" process, careful attention has been paid in the final drying since a BOE-only step is like a "HF-Last step". Integrating subsequent wet cleans to the BOE step without intermediate drying allows more flexibility and less concerns in terms of final defectivity. The yield can then be significantly increased [COZ 00].

More recently, the co-injected BOE has been revisited since the pH value around 3.5 and the optimum of HF_2^- ions at that setup can be seen as an advantage in selective wet etching between SiO_2 and Si_3N_4 for instance. A selectivity of 10:1 between oxide and nitride running co-injected BOE in a 3:1 ratio has been demonstrated compared to a standard 1:1 (very critical) ratio when running a standard diluted HF solution [LET 11a]. Key focus is on offering a HF-Last solution (for pre-Epi, pre-silicide steps, etc.) without consuming Si_3N_4 nitride spacers on the side of the transistors at the same time.

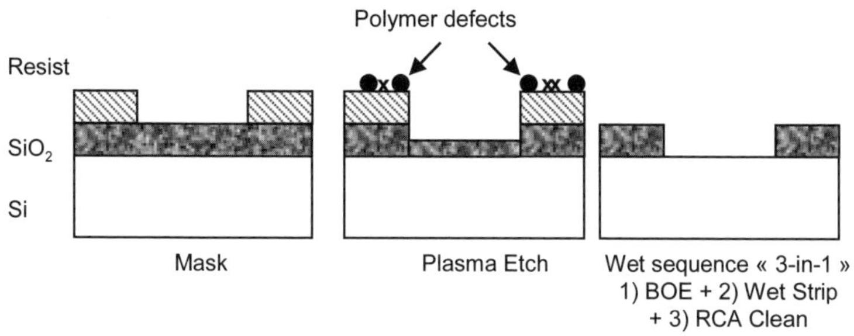

Figure 3.7. *Full wet integration of co-injected BOE from [COZ 00]*

BOE has also been used in recent integration to optimize glass microarrays for protein detection [LIN 11]: an ultrathin SiO_2 coating is used and integrated to allow later on a membrane-based biosensing interface. At that stage, we see the growing interest of sharing methodology and knowledge across different sciences from microelectronics to biology and analytical chemistry.

3.3.2.3. *Problems*

Metal-ion plating is a problem for an HF solution, but is worse for a BOE solution. One reason is the pH is higher, which makes metal ions to plate easier and a second reason is that the chemicals are less pure than HF solutions. BOE are commercially prepared from HF and ammonia. Ammonia is an extra source of contamination. Finally, the concentration of chemicals used to obtain a certain etch rate is higher than for HF that is usually significantly diluted with ultrapure water when it is exposed to the wafers.

Thus far, no attention was paid to the use of BOE over HF. There are a few problems in the use of HF that are reduced by using BOE:

1) If a high concentration of HF is used, a process that uses recycled chemicals could suffer from too much HF evaporation. This would result in an uncontrolled etch rate. With BOE one can have high etch rates without the problem of HF evaporation. This problem does not occur for HF solutions that are used in single shot.

2) If photoresist is still present on the wafer, HF solutions generally have more problems with resist lift-off than BOE solutions. However, in cleaning application, photoresist is not often used. Optimization of the resist application condition can also reduce the problem and enables the use of HF [GAR 09].

3) Selectivity: there is a difference in selectivity between thermally grown SiO_2 and other types of deposited SiO_2 (e.g. plasma TEOS, doped spin-on silica glasses). The rate determining step in the etching of thermally grown SiO_2 is the SN1 step with HF_2^-. In less dense silica, not all Si atoms are bonded to bridging oxygen atoms. Many of the bonded oxygen atoms are terminal OH groups. This means that for the etching of these materials, the slow SN1 reaction step is not required and it can immediately start with the SN2 reaction using the monofluoride. These SN2 steps become rate limiting.

4) Etching of Al: if aluminum is present and a silica surface needs to be cleaned, HF cannot be used, because it etches Al. BOE solutions with a pH over 4.5 are able to etch silica and hardly affect the aluminum surface. The presence of Al is rare in the front end of line.

The waste from fluoride-containing solutions needs special attention. The material can be collected or when presented in wastewater can be reacted with $Ca(OH)_2$ solution to form CaF_2. The latter can be filtered and if one can guarantee the purity it be used for the making of HF. Unfortunately, many waste streams contain more than only fluoride and therefore, this CaF_2 ends up in landfill.

3.4. Rinsing and drying

3.4.1. *Ultrapure water (UPW)*

3.4.1.1. *Brief history and application*

Water is the most used chemical in the semiconductor industry. It is also one of the purest chemicals. The quality used in the semiconductor industry was mainly driven by the suppliers of the UPW installations. In a new installation, contamination levels are expressed in ppq (parts per quadrillion). The application is to remove the chemical from the previous step away from the surface. Ideally, water will evaporate without leaving a trace.

Rinsing is basically the step of removing the active species from the surface. It is generally appointed the “final” rinse, which is the one operated at the end of an etching and/or cleaning sequence prior to the final drying allowing the wafer out of the wet operation before transiting to the next process step in another equipment. But we can also consider “intermediate” rinses that are the ones occurring between two chemicals. The performance can be different in these different cases, the standard goal being to quickly remove the chemicals to avoid cross-contamination of baths or the formation of salt by reaction between two successive mixtures. In the case of wet stripping using hot sulfuric acid (see the previous chapter), a hot rinsing is usually preferred. In the case of an etching mixture, cold rinse is preferred to limit any temperature gradient during the etchant elimination, especially on a single wafer platform. Finally, before the drying step, the rinse is carefully checked to insure minimum residual traces of chemicals and subsequent possible metal deposition on the surface.

On this basis, we still have to consider surfaces and materials: the water, which is the generic solvent used in rinses, will react differently on a hydrophilic or hydrophobic surface. A contact angle measurement will describe the situation. The real surface or product may also offer both types of surfaces, which result in specific process setups (time, flow, temperature, also spin speed and dynamic consideration when running spray tools or single wafer vs. regular wet bench and immersion techniques).

The surface may also offer some pattern structures that also need an optimized rinse setup to insure a sufficient rinse, especially at the bottom of the pattern. The surface may be rinsed, the water analyzed in the drain may show an expected resistivity value (18 MΩ.cm) but the bottom of the structure may still have some metal traces that will impact on the product later on [YAN 09]. The detection and analysis of such a situation have been studied and a new technology called ECRS standing for electrochemical residue sensor has been introduced [YAN 09]. On a single platform, the dynamics of water has also been studied including pattern situation, and reporting centrifugal force optimization during the rinsing steps [MAL 09]. It is also possible to follow and optimize the liquid flows using a (blue) tracer ensuring the rinse of the entire surface during the spinning steps [HAB 11].

3.4.1.2. *Problems*

Making the choice of running through a single wafer platform also requests shorter sequences, from chemical reaction to rinse and dry for the throughput purpose, since only one wafer at the time is treated. High water flows and high rotation speeds have been quickly suggested but charging effects have been suggested to explain arcing and related failing devices lowering the production yield. Surface potential difference (SPD) measurements have shown to be effective in in-line characterizing charging non-uniformity and non-visual residues for both process and tool development (see Figure 3.8).

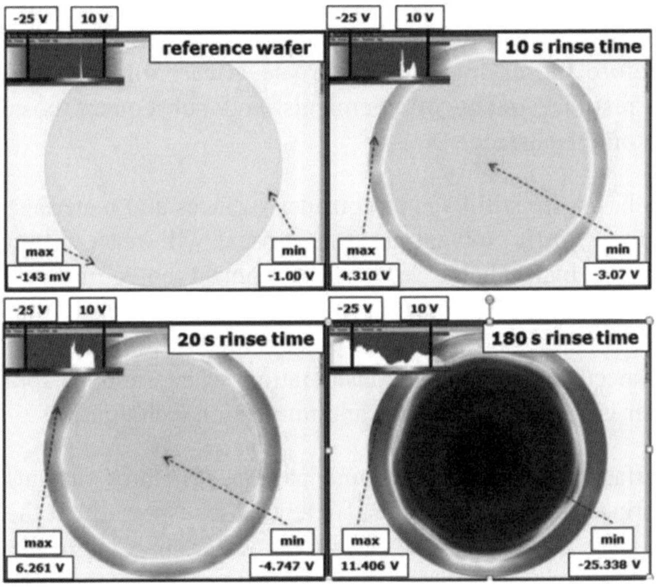

Figure 3.8. *A charge map of thermal oxide wafers after DI water rinse (single wafer processing) by [HOP 09]*

One implemented solution is to add CO_2 inside the water: it limits electrostatic discharge (ESD) and also particle redeposition [GOT 10, YAM 10]. ESD may result in oxide punch through junction damage or resistor/conductor fusing. $DiCO_2$ is particularly described in Chapter 2 of this book by Gottschalk and her co-authors. This charging effect is a detrimental effect that occurs on single wafer equipments.

If wafers are transferred from a chemical bath to a water-rinse step, the pH will change from high or low to neutral. The effect is that the deposition efficiency of metal ions in water will increase. As an example, the amount of deposited Cs as a function of pH is plotted (see Figure 3.9). At high pH, the surface concentration is low, because the silica is etched and if Cs is absorbed on a surface SiO^- group, this group is etched off together with the Cs ion. At low pH, the metal ion is in competition with the protons. Clean silica on the wafer surface is a very effective ion exchange filter at neutral pH, that is the reverse reaction as depicted in Figure 3.3.

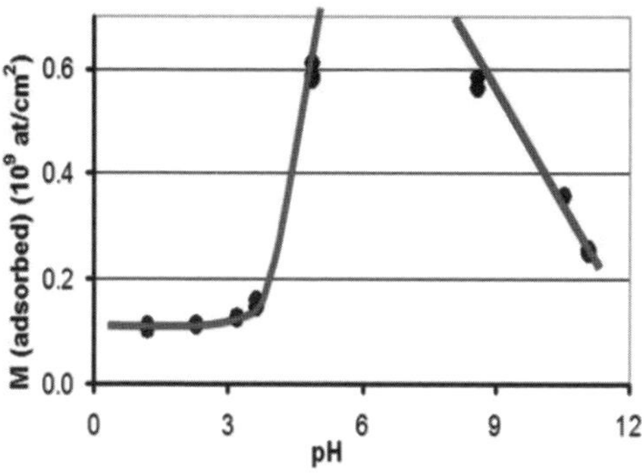

Figure 3.9. *Adsorbed Cs on silica deposited from a 200 ppt Cs solution as a function of pH [KNO 07]. Data between pH 6 and pH 8 are not available, because pH is uncontrolled if no buffer can be used*

The adsorption of metal ions is described by the Langmuir adsorption isotherm (see equation 3.14):

$$\sigma_{SiOM} = S\frac{[M]-\sigma_{SiOM}/B}{K[H]+[M]-\sigma_{SiOM}/B} \qquad [3.14]$$

As the dissolved metal-ion concentration close to the surface becomes depleted, it needs to be replenished from the bulk. The diffusion of metal ions to the silica surface is described by Fick's diffusion equation (see equation [3.15]):

$$\frac{d\sigma_{SiOM}}{dt} = D\frac{d[M]}{dx} \qquad [3.15]$$

Here, σ_{SiOM} is the surface metal concentration of metal *M*, *D* is the diffusion coefficient of *M* in water, *x* is the distance in the liquid from the surface, S is the number of sites to which M can absorb, K is the adsorption coefficient of M and B is the thickness of the diffusion layer. K values have been determined by ion chromatography using a pure silica column [KNO 07]. For Cs^+, K is 2,700. A simulation of Cs^+ adsorption on a silica surface is presented (see Figure 3.10). Other metal ions such as earth alkali metal ions absorb much better; one can anticipate on the basis of Figure 3.7 that Ca^{2+}, for example, ends up in a much higher surface concentration after a water rinse. Note that equation [3.14] becomes more complicated as Ca^{2+} replaces two protons from the surface, which is also the reason for a stronger adsorption.

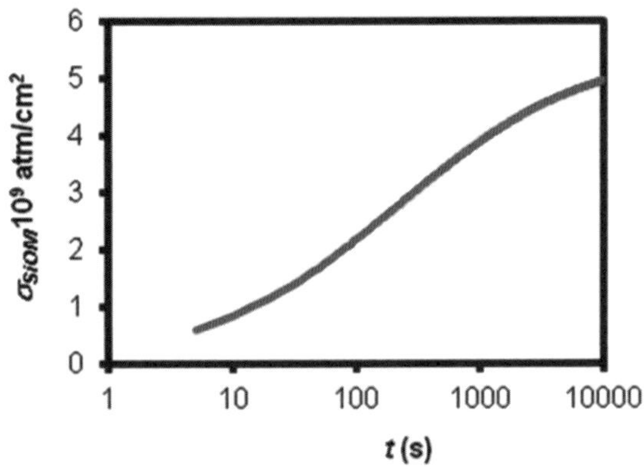

Figure 3.10. *Cs^+ adsorption on silica in water at pH 6 contaminated with 10 ppt Cs^+*

In a wet bench, water is considered a safe position for wafers to be stored in before drying. In these process tools, the waiting time in the final rinse position after a chemical step is uncontrolled. They run from 10–30 min. The problem with uncontrolled rinse times is evidenced (see Figure 3.10). Metal surface concentrations are four times higher in going from 10 to 1,000 s rinse. The great benefit of a single wafer process is that the rinse times are controlled and short (in the order of 10 to 30 s). This gives little time for metal ions to be deposited. Furthermore, the protons from the previous acid rinse will remain, in the first few seconds of the rinse, on the surface layer creating a local lower pH with less metal ion deposition. An alternative for this metal ion deposition problem is to acidify the final rinse.

Particles from water can also deposit on wafer surfaces and have an effect on yield [WAL 09]. This occurs mainly in the drying stage. After the rinse step the water that stays on the surface, the so-called carry-over layer, has the same particle concentration as the bulk. These particles will not leave from the surface anymore and will get attached to the surface if the carry-over layer is dried in. During this drying, the particles have a preference to deposit on specific places on the surface [WAL 09]. This can be positive for yield if these particles would deposit on non-critical areas, but it could also be detrimental for the yield if this preferential deposition is on critical areas. In the latter case, the effect of the particles is enhanced.

The rinse process suffers from extreme particle addition in the case where hydrophobic wafers are rinsed using a process where droplets can be formed on the wafer. For example, a spin step where the amount of dispensed water is not enough to wet the complete wafer or a rinse step where wafers are placed vertically and water is sprayed of the wafer. Due to surface energy, particles have a tendency to be located at the gas/water interface. If a water droplet with surface particles rolls over a dry (hydrophobic) surface, the particles will be printed on the wafer surface [MOU 94].

Particles do not have to be killed for the device. Pure silica particles have shown not to increase the defect level in gate oxide structures [RIN 09]. It is expected that some or most particles in UPW are colloidal silica. Similar to the previous discussion on the adsorption of metal ions on silica surfaces, metal-ion contamination in UPW will adsorb on the colloidal silica causing a non-uniform distribution of the metal-ion contamination in water. A colloidal silica particle will be a carrier of metal-ion contamination and it is the local higher metal-ion concentration that results in electrical defects on the semiconductor device. That water etches silica is known in the geological science for a long time [LEN 17]. Although the etch rates are very low, it can become significant for the current challenges with regard to the amount of silica loss per cleaning step. For products with a gate length of 20 nm, it is expected that a clean process is allowed to remove no more than 0.2 Å [ITR 11]. The solubility of silica at room temperature is 0.012% or 2 mmol/L [ALE 54]. During a rinse step, approximately 10 mL of water per cm^2 of silicon surface is used. Potentially this can etch off 1.2×10^{19} atoms Si/cm^2 that is equal to 6 µm of silica! It does not happen, because the reaction is kinetically controlled. It is very slow: the geologist would speak in terms of years [LAS 98].

The etch rate of silica depends on the temperature and the amount of dissolved silica [HUR 72]. Lower dissolved SiO_2 levels mean a higher etch rate. This would mean that the forecast for required UPW quality with regard to silica content needs to be changed. Older ITRS roadmaps predicted an ever decaying concentration, but these have been recently removed out [ITR 11]. However, silica could still form colloidal particles and have an effect on the particle requirement of UPW.

When silicon is exposed, the etch rate can increase significantly. With regard to etching of SiO_2, etching of silicon is faster because Si-Si bonds need to be broken instead of the much stronger Si-O bonds (see the section on mechanism of HF etching). The etch rate depends on the Si-termination step in the previous step: if the process was stopped at F passivation (see Figure 3.11, **11**), the surface is more prone to silicon etching than if the surface is H passivated (see Figure 3.11, **14**). An additional requirement for the etching of silicon is oxygen. Oxygen oxidizes the Si-H group that immerses the surface after the Si-OH group is etched off (see Figure 3.11).

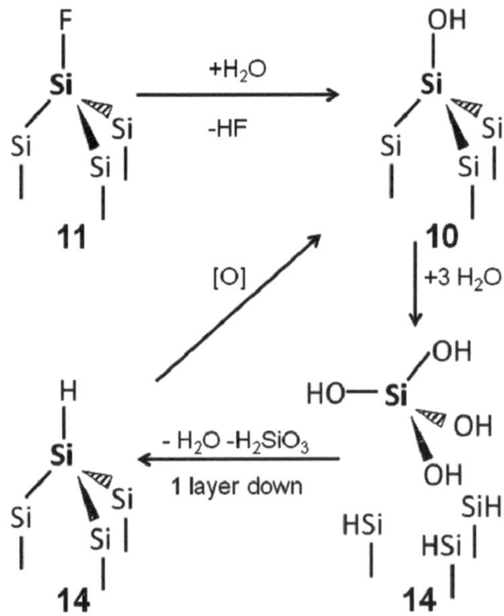

Figure 3.11. *Oxidation and dissolution of silicon in UPW*

SiO_2 and Si etching itself causes a secondary problem. Diffusion of the *m*-silicate (H_2SiO_3) away from the surface is slow and particularly if the drying process is already started it will remain at the surface. After drying, H_2SiO_3 will fall apart in SiO_2 and H_2O. The water close to the surface that is only removed by drying will be enriched with silica and the silica will be concentrated during the further drying process. If the last droplets are dried, drying stains or drying marks will be visible on the wafer [REI 11].

Recent studies regarding the key role of oxygen dissolved in water and chemicals (like HF) investigate the impact on wafer processing [GOT 11, GOT 12] and this is much more detailed in Chapter 4.

The aspect ratio (ratio of the width with regard to the depth of a given structure) is also investigated since it becomes more difficult to address efficient rinses for deep and tight patterns. Same concerns will apply to the drying, which is to insure a small and deep structure can be properly rinsed and dried.

3.4.2. *Drying*

Drying is the final step of the wet sequence (etching or cleaning, or both) and this is very important and unique since it leaves the surface in a given situation that has to be the most stable before the next process step occurs. Many cases have to be distinguished: from drying a hydrophilic surface or a hydrophobic surface or a mix of both situations, drying a flat or a patterned surface including small or high aspect ratio (relation between the width and the depth of the structure to dry), drying dynamics from a classical wet bench immersion system to an advanced single wafer platform offering spray, liquid, or even immersion as well.

Wafers might be in a vertical position or in a horizontal position. The key point is to remove the chemical or the rinsing liquid away from the surface with limited visual or non-visual traces left behind (like charges as discussed in section 4.1). The traces being left can be in very little concentration (typically 5E+9 at/cm^2) but could kill the device: we might see some corrosion if oxygen is not in very limited content [LAU 09], we might see some watermarks during HF-Last steps with inherent bad material regrowth during a following epitaxial growth step [HIR 05], we can imagine bad wafer bonding if the drying is not carefully handled [LET 09, LET 11b].

The first concept is to investigate the ways to eliminate the water out from the surface. Depending on the surface property, the water droplet will show a different contact angle in terms of evaporation as observed by Kulinich (see Figure 3.12).

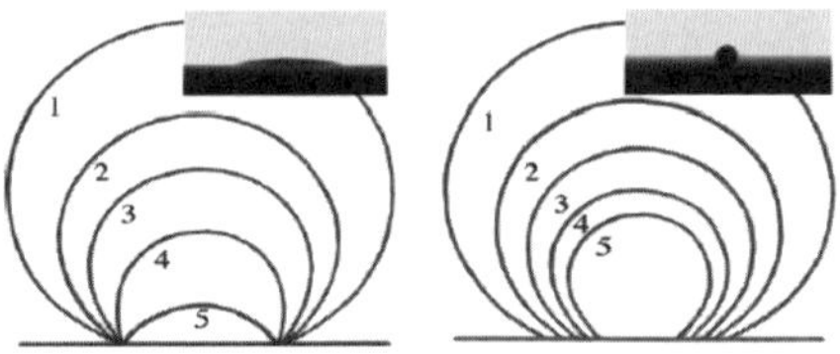

Figure 3.12. *Schematic shape evolution of water droplets during evaporation from surfaces. Droplet profiles are shown for 1 (1), 20 (2), 27 (3), 34 (4) and 37 min (5). Insets are photographs of droplets just a few seconds before disappearance [KUL 09]. Left part demonstrates a hydrophilic character of the surface whereas the right part shows a hydrophobic property of the surface*

Focus on droplet evaporation has been studied to fully understand the drying rate, the flow patterns observed inside drying drops and the residual deposits [BEL 11]. Basically, a water droplet removal requires specific forces because the contact angle hysteresis generates a force in opposition with the motion. A liquid and surface react leading to residues formation as described in chemical reactions 3.16–3.18.

$Si + O_2 \Rightarrow SiO_2$ [Reaction 3.16]

$SiO_2 + H_2O \Rightarrow H_2SiO_3$ [Reaction 3.17]

$H_2SiO_3 + 6\ HF \Rightarrow H_2SiF_6 + 3\ H_2O$ [Reaction 3.18]

Crystals have been observed [HIR 05]; they correspond to the formation of H_2SiO_3 salt. It is also suggested that some HF residues may also react with H_2SiO_3 to form H_2SiF_6. After drying, it would decompose to SiF_4 in a gas phase. From these reactions, we can see that we can limit reaction 3.16, and subsequently reactions 3.17 and 3.18, by reducing the O_2 concentration, which can be done by a specific treatment of the water supply but inside the equipment during the drying step. In another way, the more the HF is concentrated, the more the reaction produces H_2SiF_6 salt.

For more than decades, the standard drying solution was about the "Spin Rinse Dry" technique: spinning wafers under nitrogen gas flow. To improve the drying efficiency, a new technique appeared based on a surface tension gradient effect using isopropyl alcohol (IPA) called the Marangoni effect [MAR 91, THE 99]. The alcohol vapor is delivered near the wafer–water interface, the absorbed alcohol reduces the surface tension of water, the alcohol concentration in water is higher, close to the wafer surface and the water flows out of the wafer surface. The residual liquid film on the dry wafer is very thin, resulting in minimized non-volatile residues. From there, it is possible to characterize the IPA adsorption on surfaces using wafer-thermal desorption gas chromatography mass spectroscopy (WTD-GC-MS).

The relationship between the IPA process time and the amount of IPA residue on the surface has been reported [HAG 10]: optimizing the wafer temperature and the IPA supply rate helps to limit water mark formation so that an ultraclean and IPA-saving drying process with a watermark-free performance can be achieved. The drying performance is definitively related to the materials and the pattern: in hydrophilic or hydrophobic cases, the remaining water that has to be removed will need different drying process adjustments (see Figure 3.13).

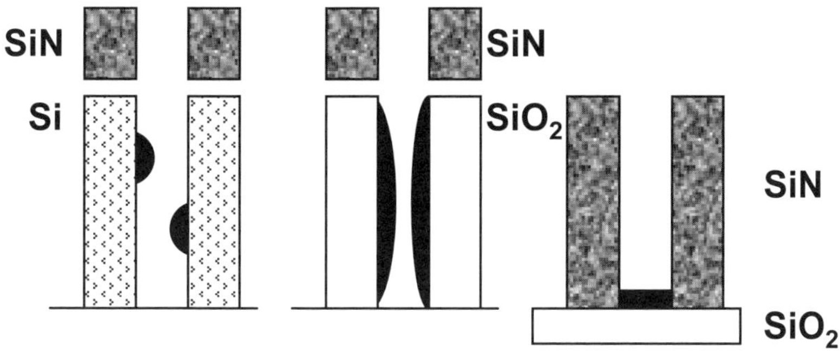

Figure 3.13. *Remaining DI water on various patterns shape [EOM 11]*

It is even more challenging to address the drying challenge when the pattern structure offers a high pattern ratio (ratio between the height and the distance between two material structures called the pitch). Dedicated patterns of different sizes have been studied ("pillars" shape) and for that particular shape, a critical aspect ratio of 18 has been found for one given cleaning-drying technique [VOS 11].

Depending on the cleaning equipment, going through a single wafer platform or a wet bench running 50 or 100 wafers at the time, drying step will have different options. An IPA vapor dryer can prevent a water mark on an hydrophobic surface (HF treated Si for example) but watermarks are easily generated on even hydrophilic surfaces, such as thermally grown oxide SiO_2 and nitride Si_3N_4 [EOM 11].

Finally, the replacement of IPA alcohol in the drying sequence by new solvents, called "low gamma solvents" (LGS), allows for the pushing of the surface tension gradient effect to further limits: it is then possible to operate the drying in specific conditions of pressure and temperature that would not be possible with IPA because of its flammable properties [LET 09]. Other considerations like cost or specific hardware needs in cleaning equipments could limit the implementation of such new solvents, but the technical background shows that for specific applications, it could be the way to proceed since the drying specification can request extreme optimization [LET 12].

From the chemistry prospective, the main focus of the drying optimization through engineering studies is more about chemical replacement (water being replaced by another fluid, in a liquid or gas phase) than chemical reaction. The key focus is on the chemical bindings left after such drying steps and the strengths of such final surface preparation with regard to the following steps. In the semiconductor industry, the key point is to control the surface between two steps/equipments; therefore, the stability of the surface between two steps is a very important parameter: users usually implement a specific queue time between the most critical steps, for example a clean before an epitaxial regrowth.

3.5. Conclusion

In the semiconductor industry, wet chemistries have been widely used for cleaning and etching for decades. Basic mixtures, turn around, using acidic solutions to avoid metallic contamination (which is detrimental to the device performance), HF or HF mixtures like BOE to address wet etching and integration, and a basic solution for particle removal purpose. All these basic chemical sequences are currently revisited to take into account the material loss challenges, especially since every Ångström has to be controlled and possibly saved where new materials are also simultaneously integrated such as Germanium, Silicon Germanium, III-V materials among others. The main chemical reaction is still about oxidation and reduction or acidic and basic. The theoretical approach of chemical basics like solving multiple equation systems has still to be applied but with a even more careful attention to address the future transistor technology nodes and challenges. Material losses, selectivity between exposed materials are key focuses.

3.6. List of Abbreviations

APM:	Ammonium (hydroxyde) (hydrogen) Peroxide Mixture
SC1:	Standard Cleaning 1
HPM:	Hydrochloric (acid) (hydrogen) Peroxide Mixture
SC2:	Standard Cleaning 2
SPM:	Sulfuric (acid) (hydrogen) Peroxide Mixture
BHF:	Buffered HydroFluoric acid
BOE:	Buffered Oxide Etch
SN1:	Substitution Nucleophilic 1
SN2:	Substitution Nucleophilic 2
ECRS:	ElectroChemical Residue Sensor
SPD:	Surface Potential Difference
ESD:	ElectroStatic Discharge
WTDGCMS:	Wafer Thermal Desorption Gas Chromatography Mass Spectroscopy
LGS:	Low Gamma Solvent

3.7. Bibliography

[ALE 54] ALEXANDER G.B., HESTON W.M., ILER R.K., "The solubility of amorphous silica in water", *Journal of Physical Chemistry*, vol. 58, no. 6, pp. 453–455, 1954.

[BEL 11] BELMILOUD N., TAMADDON A.H., MERTENS P.W., *et al.*, "Investigating on the drying dynamics of millimetric water droplets: source of watermarks on silicon wafers", in HATTORI T., RUZYLLO J., MERTENS P., NOVAK R. (eds), *ECS Transactions, Semiconductor Cleaning Science and Technology 12*, vol. 41, no. 5, pp. 205–212, 2011.

[BUT 06] BUTTERBAUGH J.W., CHRISTENSON K.K., LEE N.P., "A high-temperature batch spray process for implanted resist stripping", *Solid State Technology*, vol. 49, pp. 83–85, 2006.

[CHR 08] CHRISTENSON K.K., BUTTERBAUGH J.W., WAGENER T.J., *et al.*, "All wet stripping of implanted photoresist", *Solid State Phenomena*, vol. 134, pp. 109–112, 2008.

[COZ 00] COZE A., PHILIT G., LE TIEC Y., *et al.*, "An integrated wet chemical etch-strip-clean sequence", *Solid State Technology*, vol. 43, pp. 146–149, 2000.

[DEK 09] DEKRAKER D., PASKER B., BUTTERBAUGH J.W., *et al.*, "Steam-injected SPM process for all-wet stripping of implanted photoresist", *Solid State Phenomena*, vol. 145–146, pp. 277–280, 2009.

[EOM 11] EOM D., KIM K., SHIN Y., "Drying performance of single IPA dryer to prevent pattern collapse and watermark", in HATTORI T., RUZYLLO J., MERTENS P., NOVAK R. (eds), *ECS Transactions, Semiconductor Cleaning Science and Technology 12*, vol. 41, no. 5, pp. 197–204, 2011.

[GAR 09] GARNIER P., PERNET B., GOMEZ Y., *et al.*, "Photoresist adhesion during wet etch on single wafer tool", *Solid State Phenomena*, vol. 145–146, pp. 219–222, 2009.

[GEL 67] GELDER W., HAUSER V.E., "The etching of silicon nitride in phosphoric acid with silicon dioxide as a mask", *Journal of Electrochemical Society*, vol. 114, no. 8, pp. 869–872, 1967.

[GOT 10] GOTTSCHALK C., SCHMIDT T., WOLKE K., "Controlled DI water CO_2 gasification to eliminate ESD effects and corrosion", *Proceedings of the Surface Preparation and Cleaning, Conference,* Austin, TX, March 2010.

[GOT 11] GOTTSCHALK C., BRAMMER U., LE TIEC Y., "Deoxification of liquids for advanced semiconductors processing", in HATTORI T., RUZYLLO J., MERTENS P., NOVAK R. (eds), *ECS Transactions, Semiconductor Cleaning Science and Technology*, vol. 41, no. 5, pp. 287–294, 2011.

[GOT 12] GOTTSCHALK C., "The challenge to reduce dissolved oxygen for advanced processing", *Future Fab*, vol. 41, pp. 60–63, 2012.

[HAB 11] HABUKA H., OHASHI S., TSUCHIMOCHI T.-A., *et al.*, "Water motion over a wafer surface rotating in a single wafer wet cleaner", in HATTORI T., RUZYLLO J., MERTENS P., NOVAK R. (eds), *ECS Transactions, Semiconductor Cleaning Science and Technology*, vol. 41, no. 5, pp. 279–286, 2011.

[HAG 10] HAGIMOTO Y., TETSUKA T., IWAMOTO H., *et al.*, "Elucidation of an isopropyl alcohol (IPA) adsorption phenomenon on a wafer surface for achieving an ultra-clean and IPA-saving drying process in the batch cleaning system", *Ultra Clean Processing of Silicon Surfaces (UCPSS) Conference Proceedings, Session 11*, Ostend, Belgium, September 2010.

[HEY 99] HEYNS M., BEARDA T., CORNELISSEN I., *et al.*, "Advanced cleaning strategies for ultraclean silicon surfaces", in NOVAK R., RUZYLLO J. HATTORI T. (eds), *6th International Symposium on Cleaning Technology in Semiconductor Device Manufacturing*, The Electrochemical Society Proceedings Series, Pennington, NJ, vol. 99–36, pp. 3–16, 2000.

[HIR 05] HIROMITSU N., TAKEHIKO O., HIROKI O., *et al.*, "Insights into water mark formation and control", *Solid State Phenomena*, vol. 103–104, pp. 83–86, 2005.

[HOP 09] HOPPNER K., MANUWALD R., FAHR T., *et al.*, "Novel in-line inspection method for non-visual defects and charging", *20th Advanced Semiconductor Manufacturing Conference Proceedings*, Berlin, 2009.

[HUR 72] HURD D.C., "Factors affecting solution rate of biogenic opal in seawater", *Earth and Planetary Science Letters*, vol. 15, no. 4, pp. 411–417, 1972.

[HUR 94] HURD T.Q., MERTENS P.W., SCHMIDT H.F., *et al.*, "The role of hydrogen peroxide in the SC2 clean", *Proceedings of the Institute of Environmental Science 40th Annual Technical Meeting*, IES, Chicago, IL, p. 218, 1–6 May 1994, available at https://lirias.kuleuven.be/handle/123456789/64035.

[ITR 11] I.T.R.S., International Technology Roadmap for Semiconductor, available at http://www.itrs.net.

[KER 70] KERN W., POUTINEN D.A., "Cleaning solution based on hydrogen peroxide for use in silicon semiconductor technology", *RCA Review*, vol. 31, pp. 187–205, 1970.

[KNO 99] KNOTTER D.M., DE GENDT S., BAEYENS M., *et al.*, "Hydrogen peroxide decomposition in ammonia solutions", *Journal of the Electrochemical Society*, vol. 146, no. 9, pp. 3476–3481, 1999.

[KNO 00a] KNOTTER D.M., "Etching mechanism of vitreous silicon dioxide in HF-based solutions", *Journal of the American Chemical Society*, vol. 122, no. 18, pp. 4345–4351, 2000.

[KNO 00b] KNOTTER D.M., DE GENDT S., MERTENS P.W., *et al.*, "Silicon surface roughening mechanisms in ammonia hydrogen peroxide mixtures", *Journal of the Electrochemical Society*, vol. 147, no. 2, pp. 736–740, 2000.

[KNO 05] KNOTTER D.M., STEWART N., SHARP I., *et al.*, "Performing selective etch of Si_3N_4 and SiO_2 using a single-wafer wet-etch technology", *Micro*, vol. 23, p. 47–53, 2005.

[KNO 07] KNOTTER D.M., RUSMAN I.E., NOIJ G.P.A., "Equilibriums between metal ions in aqueous solutions and on silica", *Proceedings o the 7th Surface Preparation and Cleaning Conference (SPCC)*, Austin, TX, 2007.

[KNO 11a] KNOTTER D.M., DENTENEER T.J.J., "Etching mechanism of silicon nitride in HF-based solutions", *Journal of the Electrochemical Society*, vol. 148, pp. F43–F46, 2011.

[KNO 11b] KNOTTER D.M., "The chemistry of wet cleaning", in REINHARDT K.A., REIDY R.F. (eds), *Handbook of Cleaning for Semiconductor Manufacturing: Fundamentals and Applications*, Scrivener Publishing, Salem, ME, pp. 39–94, 2011.

[KNO 11c] KNOTTER D.M., "The chemistry of wet etching", in REINHARDT K.A. REIDY R.F. (eds), *Handbook of Cleaning for Semiconductor Manufacturing: Fundamentals and Applications*, Scrivener Publishing, Salem, ME, pp. 95–141, 2011.

[KUL 09] KULINICH S.A., FARZANEH M., "Effect of contact angle hysteresis on water droplet evaporation from super-hydrophobic surfaces", *Applied Surface Science*, vol. 255, no. 7, pp. 4056–4060, 2009.

[KUN 95] KUNII Y., NAKAYAMA S., MAEDA M., "Wet etching of doped and nondoped silicon oxide films using buffered hydrogen fluoride solutions", *Journal of the Electrochemical Society*, vol. 142, no. 10, pp. 3510–3513, 1995.

[LAS 98] LASAGA A.C., *Kinetic Theory in the Earth Science*, Princeton University Press, Princeton, NJ, 1998.

[LAU 09] LAUERHAAS J.M., "Suppression of galvanic corrosion in advanced BEOL integration", *Solid State Technology*, vol. 52, no. 9, pp. 12–15, 2009.

[LEN 17] LENHER V., MERVILLE H.B., "The solubility of silica", *Journal of the American Chemical Society*, vol. 39, pp. 2630–2638, 1917.

[LET 00] LE TIEC Y., RIGAUDIERE J., PIZZETTI C., "The chemistry of co-injected BOE", in NOVAK R., RUZYLLO J., HATTORI T. (eds), *6th International Symposium on Cleaning Technology in Semiconductor Device Manufacturing*, The Electrochemical Society Proceedings Series, Pennington, NJ, vol. 99–36, pp. 377–384, 2000.

[LET 09] LE TIEC Y., FOURNEL F., ROCHAT N., *et al.*, "Drying impact on semiconductor surfaces after innovative solvent exposure", in HATTORI T., RUZYLLO J., NOVAK R., MERTENS P., BESSON P. (eds), *ECS Transactions, Semiconductor Cleaning Science and Technology 11*, vol. 25, no. 5, pp. 131–138, 2009.

[LET 11a] LE TIEC Y., GRENOUILLET L., POSSEME N., *et al.*, "Selective wet etching for sub22nm node", *11th Surface Preparation and Cleaning, Conference Proceedings*, Austin, TX, March 2011.

[LET 11b] LE TIEC Y., VENTOSA C., ROCHAT N., *et al.*, “Water management on semiconductor surfaces”, *Microelectronic Engineering*, vol. 88, no. 12, pp. 3432–3436, 2011.

[LET 12] LE TIEC Y., VINET M., GRENOUILLET L., *et al.*, “FDSOI technology for sub22nm nodes: a surface preparation view”, *12th Surface Preparation and Cleaning Conference Proceedings*, Austin, TX, March 2012.

[LIN 11] LINMAN M., ABBAS A., ROBERTS C., *et al.*, “Etched glass microarrays with differential resonance for enhanced contrast and sensitivity of surface plasmon resonance imaging analysis”, *Analytical Chemistry*, vol. 83, no. 15, pp. 5936–5943, 2011.

[LOE 98] LOEWENSTEIN L.M., MERTENS P.W., “Adsorption of metal ions onto hydrophilic silicon surfaces from aqueous solution: effect of pH”, *Journal of the Electrochemical Society*, vol. 145, no. 8, pp. 2841–2847, 1998.

[MAL 09] MALLET A., PRAT M., BROUSSOUS L., *et al.*, “Cleaning process in single wafer tool: impact of dispersion phenomena on rinsing time”, in HATTORI T., RUZYLLO J., NOVAK R., MERTENS P., BESSON P. (eds), *ECS Transactions, Semiconductor Cleaning Science and Technology 11*, vol. 25, no. 5, pp. 147–154, 2009.

[MAR 91] MARRA J., HUETHORST J., “Physical principles of Marangoni drying”, *Langmuir,* vol. 7, pp. 2748–2755, 1991.

[MEE 90] VAN DEN MEERAKKER J. E. A. M., van den STRAATEN M. H. M., “A mechanistic study of silicon etching in NH_3/H_2O_2 cleaning solutions”, *Journal of the Electrochemical Society* Vol. 137, no. 4, pp. 1239–1243, 1990.

[MEU 95] MEURIS M., MERTENS P.W., HOPEDEBEEK A., *et al.*, “The IMEC clean: a new concept for particle and metal removal on Si surfaces”, *Solid State Technology*, vol. 38, no. 7, pp. 109–114, 1995.

[MOU 94] MOUCHE L., Etude des mécanismes de contamination particulaire et métallique des substrats de silicium en solution, PhD Thesis, Faculty of Luminy, University of Aix-Marseille II, 1994.

[NIE 83] NIELSEN H., HACKLEMAN D., “Some illumination of the mechanism of SiO_2 etching in HF solutions”, *Journal of the Electrochemical Society*, vol. 130, no. 3, pp. 708–712, 1983.

[OHM 90] OHMI T., “Foresightedness in RCA cleaning concept and their present problem”, *8th Workshop in Ultra Clean Technology*, Tokyo, Japan, pp. 5–15, 1990.

[OSS 96] OSSEO-ASARE K., "Etching kinetics of silicon dioxide in aqueous fluoride solutions: a surface complexation model", *Journal of the Electrochemical Society*, vol. 143, no. 4, pp. 1339–1347, 1996.

[PRO 92] PROKSCHE H., NAGORSEN G., "The influence of NH_4F on the etch rates of undoped SiO_2 in buffered oxide etch", *Journal of the Electrochemical Society*, vol. 139, no. 2, pp. 521–524, 1992.

[REI 11] REINHARDT K.A., REIDY R.F., MARSELLA J.A., "Surface phenomena: rinsing and drying", in REINHARDT K.A. REIDY R.F. (eds), *Handbook of Cleaning for Semiconductor Manufacturing: Fundamentals and Applications*, Scrivener Publishing, Salem, ME, pp. 143–168, 2011.

[RIN 09] RINK I., WALI F., KNOTTER D.M., "Impact of metal-ion contaminated particles on gate oxide integrity", *Solid State Phenomena*, vol. 145–146, pp. 131–134, 2009.

[SCH 95] SCHMIDT H.F., MEURIS M., MERTENS P.W., *et al.*, "H_2O_2 decomposition and its impact on silicon surface roughening and gate oxide integrity", *Japanese Journal of Applied Physics*, vol. 34, pp. 727–731, 1995.

[SOM 96] SOMASHEKHAR A., O'BRIEN S., "Etching SiO_2 films in aqueous 0.49% HF", *Journal of the Electrochemical Society*, vol. 143, no. 9, pp. 2885–2891, 1996.

[STO 99a] STORM W., GERBER H.A., HOHL G.F., *et al.*, "Determination of SC1 etch rates at low temperatures with microscope interferometry", in HEYNS M.M., MERTENS P.W. (eds), *Proceedings of the 4th International Symposium on Ultra Clean Processing of Silicon Surfaces (UCPSS) Conference*, Antwerp, Belgium, Acco Leuven/Amersfoort, p. 275, 1998.

[STO 99b] STORM W., GERBER H.A., HOHL G.-F., *et al.*, "Determination of SC1 etch rates at low temperatures with microscope interferometry", *Solid State Phenomena*, vols. 65–66, pp. 275–278, 1999.

[SUI 85] SUIB S.L., "Crystal growth in gels", *Journal of Chemical Education*, vol. 62, no. 1, pp. 81–82, 1985.

[TEE 96a] TEERLINCK I., *et al.*, in HEYNS M.M., MEURIS M., MERTENS P. (eds), *3rd International Symposium on Ultra Clean Processing of Silicon Surfaces*, [s.n.], Atnwerp, Belgium, p. 21, 1996.

[TEE 96b] TEERLINCK I., MERTENS P.W., MEURIS M., *et al.*, "Optimized HF solutions for ultra-clean silicon surfaces", *Symposium on VLSI Technology Digest of Technical Papers*, abstract 21.1, pp. 206–207, 1996.

[THE 99] THESS A., BOOS W., "A model for Marangoni drying", *Physics of Fluids*, vol. 11, no. 12, pp. 3852–3855, 1999.

[UBA 84] UBARA H., IMURA T., HIRAKI A., "Formation of SiH bonds on the surface of microcrystalline silicon covered with SiO_x by HF treatment", *Solid State Communications*, vol. 50, no. 7, pp. 673–675, 1984.

[VAN 90] VAN DEN MEERAKKER J.E.A.M., van den STRAATEN M.H.M., "A mechanistic study of silicon etching in NH_3/H_2O_2 cleaning solutions", *Journal of the Electrochemical Society*, vol. 137, no. 4, pp. 1239–1243, 1990.

[VER 94] VERHAVERBEKE S., TEERLINCK I., VINCKIER C., *et al.*, "The etching mechanisms of SiO_2 in hydrofluoric acid", *Journal of the Electrochemical Society*, vol. 141, no. 10, pp. 2852–2857, 1994.

[VER 96] VERHAVERBEKE S., PARKER J.W. and MCCONNELL C.F. "A quantitaive model for SC1 processing", in HEYNS M., MEURIS M., MERTENS P. (eds), *Proceedings of the 3rd International Symposium on Ultra Clean Processing of Silicon Surfaces (UCPSS)*, Acco Leven/Amersfoort, Antwerp, Belgium, p. 153, 1996.

[VOS 11] VOS I., HELLIN D., VERTOMMEN J., *et al.*, "Silicon nano-pillar test structures for quantitative evaluation of wafer drying induced pattern collapse", in HATTORI T., RUZYLLO J., MERTENS P., NOVAK R. (eds), *ECS Transactions, Semiconductor Cleaning Science and Technology 12*, vol. 41, no. 5, pp. 189–196, 2011.

[WAL 09] WALI F., KNOTTER D.M., MUD A., *et al.*, "Impact of particles in ultrapure water on random yield loss in IC production", *Microelectronic Engineering*, vol. 86, no. 2, pp. 140–144, 2000.

[WAR 71] WARREN L.J., "The measurement of pH in acid fluoride solutions and evidence for the existence of $(HF)_2$", *Analytica Chemica Acta*, vol. 16, pp. 53199–53202, 1971.

[YAM 10] YAMADA Y., EDA H., KODERA M., *et al.*, "Charged-induced attraction of particles in surface preparation and clean processes", *10th Surface Preparation and Cleaning Conference Proceedings*, Austin, TX, March 2010.

[YAN 09] YAN J., DHANE K., VERMEIRE B., *et al.*, "*In situ* and real-time metrology during rinsing of micro and nano-structures", *Microelectronic Engineering*, vol. 86, no. 2, pp. 199–205, 2009.

Chapter 4

The Use and Management of Chemical Fluids in Microelectronics

4.1. Ultrapure water

Semiconductor device manufacturing consumes large amounts of water, particularly ultrapure water (UPW). UPW is widely employed in the many cleaning sequences that represent more than 20% of the total process steps in on semiconductor manufacturing line. Its use is critical for the removal of contaminants including organics, particles and metals that can lead to electrical failure in the finished devices. As device line widths continue to shrink, the requirements for ever greater water purities in semiconductor applications continue to increase. Indeed, today's semiconductor standards for ionic contaminants in quality analysis are so stringent that some analytical specifications are beyond current analytical detection limits. UPW is also used in other wet processes, including etch, solvent processes and chemical mechanical planarization (CMP); the latter is one of the most UPW-consuming and polluting processes with its use of a high volume of water and slurries (particles in aqueous solution containing additives). The broad application of UPW in semiconductor device manufacturing thus necessitates the use of large amounts of energy and chemicals for its production.

Chapter written by Christiane GOTTSCHALK, Kevin MCLAUGHLIN, Julie CREN, Catherine PEYNE and Patrick VALENTI.

The 2011 edition of the *International Technology Roadmap for Semiconductors* established a consumption goal of 7 L/cm^2 per wafer and a long-term goal of 4.5 L/cm^2 for 2020. This means that processing a 200 mm wafer needs 2.8 m^3 of UPW in an up-to-date fab [ITR 11a]. As a point of reference, a typical 200 mm wafer fab processes 40,000 wafers per month and can use up to 3,000 m^3 of UPW per day. This is similar to the water requirements for a city of 20,000 people.

The critical parameters for specifying UPW for semiconductor processing are developed below; typical design elements and functionalities will be reviewed.

4.1.1. *Parameters of UPW*

The following parameters define and characterize the quality of UPW and also of high-purity chemicals and formulated blends. They must be regularly tested and monitored to ensure that the liquid being produced matches the specifications established for the manufacturing process. No particular method is specified for each parameter measurement, but certain methods have become de facto norms as detection limits have decreased and specifications tightened [ITR 11b].

Resistivity (or its inverse, conductivity) is measured online and is a fundamental metric for UPW. The resistivity of water should be 18.2 MΩ-cm when corrected to 25°C (18.2 MΩ-cm is the theoretical maximum resistivity for pure water at 25°C). Since the current state-of-the-art UPW system is stable enough, the resistivity readings are generally only valuable for detecting large process shifts.

Dissolved oxide level (DO), measured as ppb, is primarily process step dependent. The DO content may need to be controlled for rinsing hydrogen-passivated silicon to prevent etching by oxygenated water and to control gate oxide thickness. On the other hand, low DO levels can cause corrosion in some metal processing steps. Tool operation and configurations influence DO levels at the wafer, and N_2 environments may be required to control DO at the point-of-use. DO is frequently considered to be a process variable and is measured for stability.

Total oxidizable carbon (TOC), measured as ppb, is an important indicator in the final filter UPW, as the TOC indicates the presence of halogenic compounds, organic nitrogen and organic acids, all of which may affect process yields. The TOC that survives the treatment steps in a UPW system is thought to be either "light" molecules or small fragments of larger molecules such as acetates, formats and organic nitrogen. Organics found in the UPW included chloroform, urea and trimethylamine (TMA).

Particles that adhere to wafer surfaces during IC device manufacture may impair the application of thin films and photolithographic substances. The presence of particles can cause physical imperfections, which, in turn, may lead to the failure of discrete and/or integrated transistors. Particles are reported as particles/L. Particles as small as 0.04 μm can be detected when measured using laser light scattering techniques. Equipment is available that can measure particles in a liquid in both offline (static) and online (dynamic) configurations. Static measurement is most often used in the production of high-purity chemicals while dynamic measurement is used for fabrication process control in chemical feed lines.

Bacteria that survive the water treatment pose both a biological and particulate threat to integrated devices. Bacterial adhesion occurs naturally in water as pipe walls attract minute quantities of organic nutrients, which attach to the wall and initiate the biofilm process. While regular sanitization programs are employed by some and provide safeguards against microbial activity, biofilm can prove resistant and may permanently coat the inaccessible surfaces of valves and dead legs. Proper design and adequate flow velocity are more important for the cleanliness of the system than periodic sanitization. The current test for bacteria and other organisms in UPW is a culture method test for viable bacteria. However, the standard methods lack sensitivity in that only viable bacteria are recovered and large volumes of water may need to be sampled to provide adequate reliability. A new technology called scan RDI may offer a solution for the testing of total viable organisms. The method is able to detect a single cell based on direct measurements of cell activity and includes bacteria and other live organisms that may be present in a biofilm. Another method for bacteria detection is epifluorescence, in which a microscope is used to visually identify both viable and non-viable bacteria that have been stained with dyes that cause the biological materials to fluoresce under ultraviolet (UV) light.

Silica exists primarily in two forms: as silicates and as polymeric silica. Silicates, which are also referred to as reactive silica or dissolved silica, are reactive to molybdate and can be measured online. The polymeric forms are called colloidal silica or particulate silica; these are not measured by online analyzers. Colloidal silica is typically determined by measuring total silica (by ICP/MS) and subtracting the reactive silica measurement.

Anion and cation measurement at the point-of-use (POU) (e.g. in the wafer rinse bath) may be important to identify and prevent certain sources of contamination. Certain types of haze formation on wafer surfaces are thought to be caused by concentrations of anions and cations (e.g. ammonium chloride or ammonium sulfate residues) on the silicon surface at concentrations in excess of 10^{14}–10^{15} ions/cm^2. Since there are equivalent amounts of anions and cations present, the haze formed on the silicon wafer surface is stable. The haze is caused by poor rinsing of the wafers after chemical cleans and inadequate exhaust at wet benches. Ionic (anions and cations) removal in UPW systems can be quantified at very sensitive levels using ion chromatography (IC). Anions and cations measurable by IC include fluoride (F^-), chloride (Cl^-), nitrite (NO_2^-), bromide (Br^-), nitrate (NO_3^-), sulfate (SO_4^{2-}), phosphate (PO_4^{2-}), lithium (Li^+), sodium (Na^+), potassium (K^+), magnesium (Mg^{2+}) and calcium (Ca^{2+}).

Trace metals have long been identified as a critical source of defects on the wafer. Certain processing steps can expose a wafer to contaminants in UPW in a unique way. For example, in the case of a wafer cleaning step using HF, the HF strips the native oxide and leaves bare silicon exposed. Certain elements such as copper, gold, silver and mercury have a high electro-potential and will plate out onto the wafer if they are present in the UPW. With expanded use of copper materials in some manufacturing environments, the chance for cross-contamination is increased. For this reason, it is useful to include copper in the elements to be tested in the POU rinse baths.

Table 4.1 shows the general test methodologies for monitoring contaminants in UPW and gives the required range of performance according to the SEMI standards F63-0211 [SEM-11-F63].

Parameter	Measured	Test method	Range of performance
Resistivity	Online	Electric cell	> 18.18 MΩ-cm
Dissolved O_2	Online	Electric cell	< 10 ppb
TOC	Online	Conductivity/CO_2	< 2 ppb
Particles (micron range) > 0.05	Online	Light scattering	Particules/L < 500
Bacteria	Lab	Incubation (1 L sample)	< 1 CFU/L
Total Silica Dissolved Silica	Lab Online or lab	ICP-MS or GFAAS Colorimetric	< 0.5 ppb < 0.5 ppb
Cations, anions, metals	Lab	Ion Chromatography, ICP-MS	1 to 50 ppt

Table 4.1. *General test methodology for UPW and range of performance*

4.1.2. *UPW system unit operations*

There are a number of unit operations required to generate UPW from typical water sources. A certain number of these, including reverse osmosis (RO) and on exchange bed technologies, are used in nearly every fab; others are dependent on the feed water characteristics; some are optional, depending on user preferences. UPW systems can generally be divided into three stages:

– pretreatment;

– primary treatment;

– polishing and distribution.

Figure 4.1 shows a typical setup.

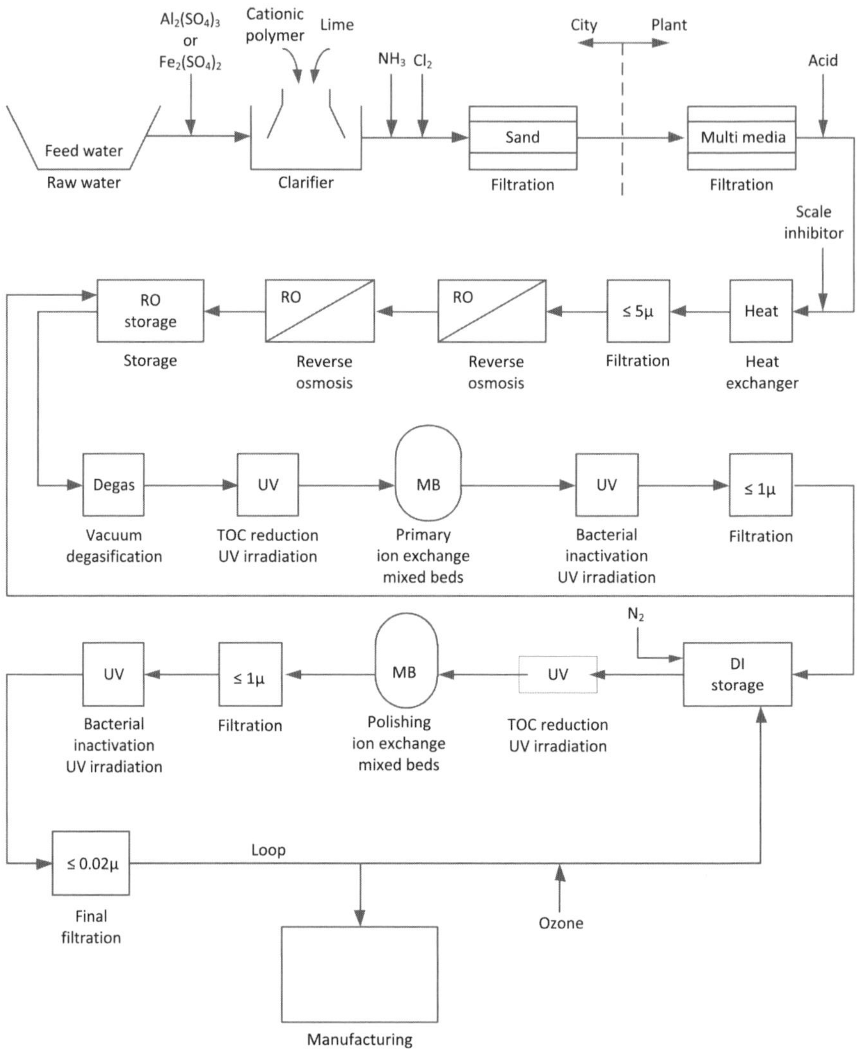

Figure 4.1. *Schematic of a typical ultrapure water system starting with raw water [SEM 09]*

4.1.2.1. *Pretreatment*

Untreated water, whether from a surface source such as a lake or a river, or from a groundwater source, is the raw material from which UPW is made. The most frequent raw water source used as a feed in UPW systems is treated municipal drinking water.

Most natural water must be treated before it can be used for human consumption. Drinking water has upper limits for inorganic contaminants (e.g. arsenic, copper and lead), pesticides, volatile organic chemicals (e.g. benzene, trichloroethylene and toluene), turbidity and microbiological contaminants. For certain groundwater sources, municipal water treatments may only utilize chlorination or filtration or a combination of both. For many surface water sources, coagulation (injection of aluminum or iron salts), flocculation (injection of a long-chain polymer), sedimentation, lime or lime/soda ash softening, filtration, chlorination or chloramination (chlorine plus ammonia and other steps) may be required.

Pretreatment includes all water treatments prior to the RO membrane treatment. These steps protect the membrane units from scaling due to sparingly soluble salts; fouling due to living or non-living suspended particles; and chemical attack by pH, oxidizing agents or other dissolved contaminants. Pretreatment equipment may include:

– *Surface filtration*: microfiltration, or ultrafiltration that removes bulk suspended solids; 1–5 µm cartridge filtration that is a polishing step for the removal of suspended solids.

– *Ion exchange*: sodium-cycle cation exchange (commonly called softening, removes scale-forming cations like Ag^{2+} and Ca^{2+}).

– *Activated carbon filtration*: removes oxidizing agents and certain organic molecules.

– *Chemical addition*: acid injection that minimizes cellulose acetate membrane damage and/or to control carbonate scales; scale inhibitor injection that controls scaling and sulfite ion injection that removes the oxidizing agent.

4.1.2.2. *Primary treatment*

The goal of primary treatment in UPW systems is to produce purified water with the following characteristics: a low amount of dissolved gases, less than 1 mg/L of organic and silica contaminants and a measured resistivity between 0.5 and 4 MΩ cm.

Common unit operations in primary treatment include:

– two-pass reverse osmosis;

– degasification.

Reverse osmosis membrane treatment: this unit is the workhorse for water purification in most semiconductor UPW operations. This technology will be discussed in more detail later in this section. It provides an extremely high rejection of dissolved ions (charged atoms and molecules), organic (carbon-containing) compounds, silica (silicon-containing) compounds and near complete rejection of suspended contaminants. RO processing will not, however, reject dissolved gases and volatile organic compounds and an additional unit operation is therefore required.

Degasification: volatile organic compounds and dissolved oxygen, carbon dioxide and other gases must be removed in the pretreatment step in a process known as degasification. This process removes volatiles that can form ions and the volatile organic compounds, reducing the loading on downstream polishing equipment. The volatile contaminants are removed in vacuum degasifiers and/or in membrane degasification units. Vacuum towers were previously the standard but now, in almost all plants, membrane degasification technology is used. Typically, a two-membrane degasification unit is installed for the removal of oxygen (O_2) and volatile organics.

4.1.2.3. *Polishing and distribution*

The relatively low levels of contaminants that remain after RO membrane treatment and degasification steps are reduced further in the polishing steps. UPW is a powerful solvent and it can dissolve materials from the surfaces it contacts to form ionic and organic contaminants. These are removed in continual polishing steps that maintain the UPW within specifications.

Polishing unit operations differ from fab to fab, depending on operating and investment choices. They may include:

– UV sterilization;

– ion exchange;

– electrodeionization;

– surface filtration – ultrafiltration;

– distribution loops.

UV sterilization: this process is commonly used to address the TOC and microorganism contamination found downstream of the RO membrane. The water passes through channels irradiated by UV light at 185 nm. This generates ozone (O_3), an extremely reactive oxidant, in the water. Ozone kills microorganisms and converts low levels of TOC to ionized compounds that can be removed by the primary ion-exchange units. Further irradiation by UV light at 254 nm catalyzes the destruction of dissolved ozone that could damage equipment, particularly filtration media.

Ion exchange: organic and silica contaminants are removed in this step, typically in a two-stage ion-exchange unit with primary and polishing beds. This technology will be discussed further later in this section.

Electrodeionization: this is a relatively new technology that can replace the primary mixed bed ion-exchange units. This technology will be discussed further later in this section.

Surface/ultrafiltration: filters with a pore size less than or equal to 0.45 µm are used after UV irradiation to remove inactivated bacterial bodies. The final filter prior to distribution typically has a pore size of less than 0.2 µm.

Distribution loops: distribution of UPW employs one or more loops. Within each loop, UPW continuously circulates from the final filters of the polishing step to those manufacturing areas requiring UPW (end users) and back to a storage tank, where it re-enters the polishing process. The loop, with a significant velocity and without dead ends, is designed to prevent bacterial contamination of the water. The specifications for polymer materials and components used in UPW and liquid chemical distribution systems are described in the standard SEMI F57-0312 [SEM 12b].

UPW supplied to the end-user area is commonly called the *supply*. UPW moving from the end-user area back to the water treatment area is commonly referred to as the *return*.

4.1.3. *Fundamentals*

4.1.3.1. *Reverse osmosis*

Principle: RO is a filtration technique using a selective membrane that allows the removal of ions and large molecules from solutions on one side when a pressure is applied on the solution.

Osmosis is a natural phenomenon that can be illustrated by the following experiment: if two water volumes with different concentrations of a common contaminant are separated by a semi-permeable membrane; water molecules move across the membrane from lower to higher contaminant concentrations until equilibrium is achieved in the chemical potential of the contaminant. Once equilibrium is achieved, the height of the water corresponding to the volume having crossed the membrane is equal to the osmotic pressure and this is proportional to the concentration of the dissolved salts. If a pressure greater than the osmotic pressure is applied to the high concentration side of an osmotic cell, water will cross the membrane in the opposite direction; this is known as *reverse osmosis*. The Van't Hoff equation links osmotic pressure and concentration:

$$\Pi = \Delta C \times R \times T, \text{ where} \quad [4.1]$$

Π = osmotic pressure in Pa

ΔC = concentrations difference in moles/m^3

R = ideal gas constant (8.314 J/mol·K)

T = temperature in K

Osmotic pressure thus depends on temperature and concentration differences between the two solutions. Figure 4.2 shows a schematic representation of the osmotic processes.

RO membrane elements: the RO membranes in industrial units are typically made from polyamide or cellulose acetate, formed as either spiral-wound or hollow fiber membranes. Spiral wound membranes are the most common form in UPW systems. Typically, cross-flow spiral wound membranes are cast as a film onto a flat sheet. These are sandwiched between a feed spacer having a thickness of 7–25 mm and a permeate carrier. The sandwich is sealed at each edge and wound around a perforated tube. The feed solution circulates in the spacer and the permeate carrier allows drainage of the permeate to the axial collector (see Figure 4.3).

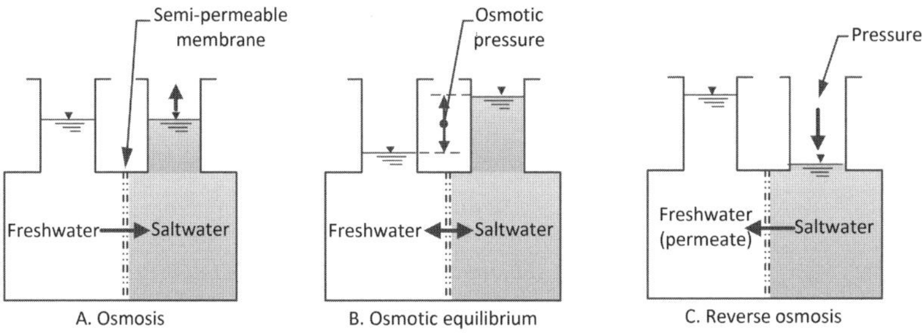

Figure 4.2. *Simplified concept of osmosis, osmotic pressure and reverse osmosis*

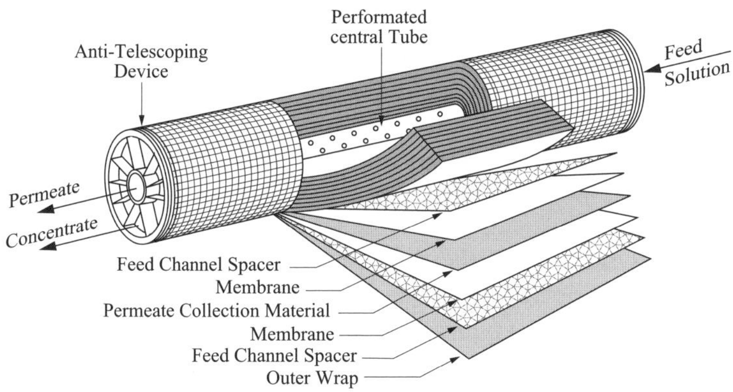

Figure 4.3. *Typical spiral-wound RO membrane element and pressure vessel*

Implementation: the simplest RO system involves the following elements (see Figure 4.4):

– pretreatment stage;

– high pressure pump to supply energy to the system;

– RO vessels;

– control valve on the reject piping that maintains the system pressure.

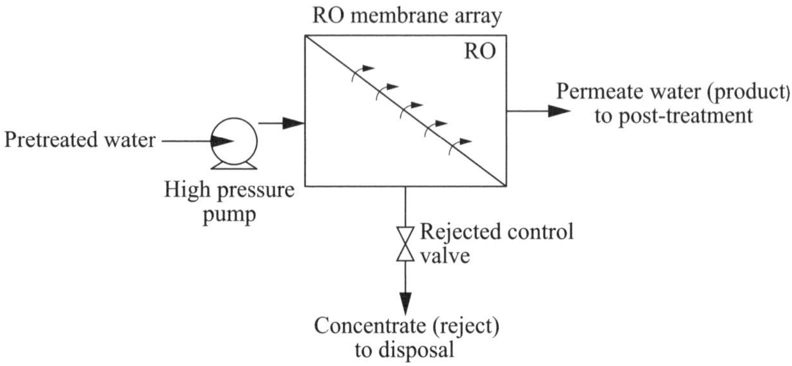

Figure 4.4. *A flow schematic for a simplified RO unit*

Feed water, pretreated to meet any established RO membrane feed water quality guidelines, is pressurized before entering the RO membranes. Two flows exit the membranes: the combined product (permeate) and the combined concentrate (reject). The fraction of feed water in permeate is called the recovery and this is normally expressed as a percentage. The maximum allowable ratio of permeate to reject depends on the water's scaling potential, which is a function of the feed water quality. This ratio is maintained by the use of a control valve on the reject piping, which controls the flow rate of the reject, forcing the permeate flow rate to the desired value.

The pump system raises the pressure of the pretreated feed water to the level required for operation of the desalting system. For RO, the pump system discharge pressure typically is 8.5–27 bar (125–400 psi) for low-TDS and brackish water systems and 55–82 bar (800–1,200 psi) for seawater systems.

Performance: the two main parameters for the design of an RO stage are as follows:

– salt rejection or selectivity;

– conversion or recovery.

Salt rejection: the value of this parameter increases with the molecular weight and the ion valence of the salt. It is an indicator of the quality of the performance of the RO and is calculated using the following equation:

% Rejection = ((Feed TDS–Product TDS)/ Feed TDS) × 100 [4.2]

feed TDS: total dissolved solids content of water supplied to module;

product TDS: total dissolved solids content of the purified water.

For monovalent ions (Cl^-, NO_3^-, HCO_3^-, Na^+, K^+ ...) the selectivity rate is around 85–95%, for multivalent ions (Ca^{2+}, Mg^{2+}, SO_4^{2-} ...) it is 98%. Selectivity rates are almost 100% for bacteria, virus and colloids.

Recovery: the % recovery or conversion of the RO unit is defined as the amount of water recovered for use as a percentage of the water fed into the RO unit. This parameter defines the productivity of the unit. The % recovery is calculated using the following equation:

% Recovery = (Permeate Flow Rate / Feed Water Flow Rate) × 100 [4.3]

The rate of purified water production depends strongly on the desired quality of permeate and on the solubility limits of the salts in the concentrate. Also, the temperature of the water is an important factor in determining the productive flow: near 15°C the permeate flow increases by 2.5–3% for every additional degree Celsius of the feed water temperature. Usually, the pretreatment stage includes a heat exchange step to raise the feed water temperature up to 20°C. Besides temperature, other design parameters for RO units include pressure, characteristics of the feed water, fouling index, conductivity and pH value.

Arrangement in UPW system: depending on the required performance of an RO system, one or more pressure vessels containing RO membranes can be arranged in parallel, in series, or both. This membrane-pressure-vessel arrangement is often called a membrane array or a pressure-vessel array. The UPW specifications for semiconductor applications are so stringent that RO units usually have a two-stage permeate arrangement, called *two-pass reverse osmosis*, in which the permeate from the first-pass RO is fed to a second-pass RO system for further purification. For most two-pass RO systems, the second pass concentrate is recycled as the feed to the first-pass RO to increase water use efficiency.

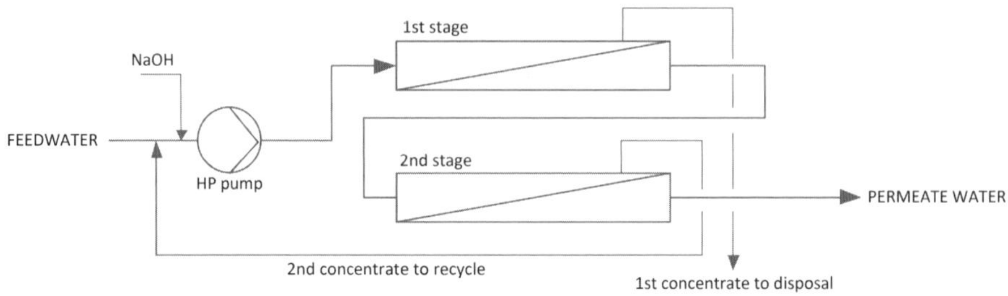

Figure 4.5. *Schematic of a two-stage permeate arrangement*

For more details on RO technology, see the references [KUC 10, AIM 10].

4.1.3.2. *Ion exchange and demineralization*

Ion exchange is a powerful, well-proven technology developed in the 1950s for the demineralization of water to an extremely high quality. It remains the best technology for the production of UPW. Ion exchange is also used in a variety of industrial and domestic applications that include the removal of color from cane sugar syrups to make white sugar, purification of antibiotics and other pharmaceuticals and extraction of uranium from ores.

Ion-exchange resins are insoluble solids containing mobile cations or anions that are capable of being exchanged, in a reversible and stoichiometric manner, with ions of a similar charge present in an electrolyte solution. Ion-exchange resins can be non-selective or they may exhibit binding preferences for certain ions or classes of ions, depending on their chemical structure. This preference may depend on the size of the ions, their charge or their structure.

Polymer resins with polystyrenic or polyacrylic skeletons are most commonly used for demineralization. They become ion-exchange resins following the grafting of functional groups that determine their class. There are two classes of ion-exchange resins:

Cation exchangers with the general formula $R\text{-}H^+$ exchange protons for the positively charged ions that are present in the electrolyte solution (e.g. Ca^{2+}, Mg^{2+}, Na^+, K^+). Strong acid cation exchangers have a sulfonic group SO_3H, and are capable of fixing absolutely all cations. Weak acid cation exchangers have a carboxylic group COOH and can fix only cations associated with weak anions (HCO_3^-).

Cation exchange selectivity: in general, a resin's affinity for ions increases with the valence of the ions and, for similar valences, with an atomic number.

Strong acid cation resin:

Monovalent cations: $Ag^+ > Cs^+ > Rb^+ > K^+ > NH_4^+ > Na^+ > H^+ > Li^+$

Divalent cations: $Ba^{2+} > Sr^{2+} > Ca^{2+} > Cu^{2+} > Mg^{2+} > Be^{2+}$

Anion exchangers with the general formula $R\text{-}OH^-$ exchange hydroxyl groups for the negatively charged ions that are present in the electrolyte solution (e.g. Cl^-, SO_4^{2-}, NO_3^-, HCO_3^-). Strong base anion exchangers have a quaternary ammonium group and are capable of fixing absolutely all anions. Weak base anion exchangers have an amine group and can fix only anions associated with strong acids (Cl^-, SO_4^2, NO_3^-).

Anion-exchange selectivity:

Strong base anion resin: $I^- > NO_3^- > Br^- > Cl^- >$ Acetate $> OH^- > F^-$

Total demineralization is accomplished by using, in series, both kinds of ion exchangers to replace all electrolyte cations by H^+ and all anions by OH^-. Thus, ions in the electrolyte solution are replaced with the components of water (H^+ and OH^-) that are released by the ion-exchange resins.

The cation-exchange reactions are as follows, with R the cation-exchange resin:

$$2\,R'H + Ca^{++} \rightarrow R'_2Ca + 2\,H^+ \qquad [4.4]$$
$$R'H + Na^+ \rightarrow R'Na + H^+$$

The anion-exchange reactions are as follows:

$$R''OH + Cl^- \rightarrow R''Cl + OH^- \qquad [4.5]$$

$$2\,R''OH + SO_4^{=} \rightarrow R''_2SO_4 + 2\,OH^- \qquad [4.6]$$

The recombination reaction of water is as follows:

$$2H^{+} + OH^{-} \rightarrow H_2O \qquad [4.7]$$

Implementation: the process takes place in one or more columns containing resins through which the solution passes. Deionizers may be categorized as "multiple-bed", in which each resin type is in a separate vessel; and "mixed-bed" (Figure 4.6), in which the cation-exchange and anion-exchange resins are intimately mixed and contained in a single pressure vessel. The thorough mixture of cation exchangers and anion exchangers in a single column makes a mixed-bed deionizer equivalent to a lengthy series of two-bed plants. As a result, the water quality obtained from a mixed-bed deionizer is appreciably better than that produced by a two-bed plant, and mixed-bed deionizers are commonly used in UPW manufacturing. A single polishing treatment after an RO unit produces water with a resistivity of 15 MΩ cm. To achieve the maximum theoretical resistivity (18.2 MΩ cm at 25°C), two mixed-bed deionizers in series are necessary.

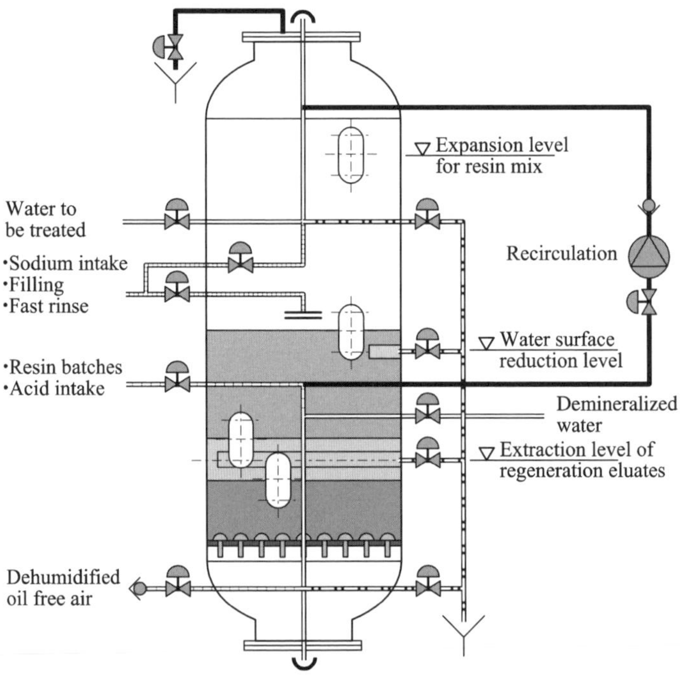

Figure 4.6. *Schematic of a mixed-bed deionizer [DEG 89]*

Regeneration process: when the majority of ions in the resin have been replaced by ions from the solution, the resin is considered exhausted. When this happens, the resistivity of the water begins to decrease and it is necessary to initiate a regeneration cycle. A regenerant solution is percolated through the resin bed, and regenerant ions are fixed by the resins to be exchanged again in the following production cycle:

– the hydrogen ion form is produced by the use of a strong acid regenerant such as hydrochloric acid (HCl) or sulfuric acid (H_2SO_4);

– the hydroxide ion form is produced by the use of a strong base regenerant such as sodium hydroxide (NaOH).

There are two types of regeneration processes:

– *Co-current regeneration*, in which both feed and regenerant solutions make contact with the resin in a downflow mode. These units are less expensive in terms of initial equipment cost, but the efficiency of use of the regenerant chemicals is lower than in a counter-current regeneration since the lower resins are never fully regenerated.

– *Counter-current regeneration* takes place using a flow that is counter to the production mode. The regenerant solution first meets weakly saturated resin layers. The fixed ions are driven to the upper layers already saturated and cannot be fixed again. This regeneration mode yields improved performance and lowers reagent consumption, producing a higher quality of treated water due to the improved regeneration of the last layers.

For mixed-bed deionizers, the regeneration takes place after separation of the two resins. The bed is subjected to an upflow of water. Anionic resin beads are less dense than the cationic ones, and they rise to the top so that the bed is separated into two layers of resin. Each is regenerated *in situ* with the appropriate regenerant and then rinsed with clean water. The internal piping of the vessel is arranged so that regenerants and washes enter at the point separating the two resins and flow either up or down as required. An upflow of compressed air then mixes the resins again.

The wastewater from the regeneration process contains all the minerals removed from the water plus acid and basic components from the spent regenerants, and must be neutralized.

There are also non-regenerable resins used as a polishing step. When the resin is exhausted, it is removed for disposal (outside regeneration or destruction) and replaced by a new resin.

For further reading, please refer to references [OWE 95, ARD 98, DAR 12].

4.1.3.3. *Fundamentals of electrodeionization*

Electrodeionization (EDI) is a trend in UPW processing that has been gaining a market share for the primary treatment and polishing stages. EDI is a mixed process combining ion exchange and electrodialysis (see Figure 4.7). Under the influence of an electric field, ions are continuously removed, attracted to one of two electrical charges. Ion-selective membranes separate the positive ions in the water that move toward the negative electrode and the negative ions that move toward the positive electrode.

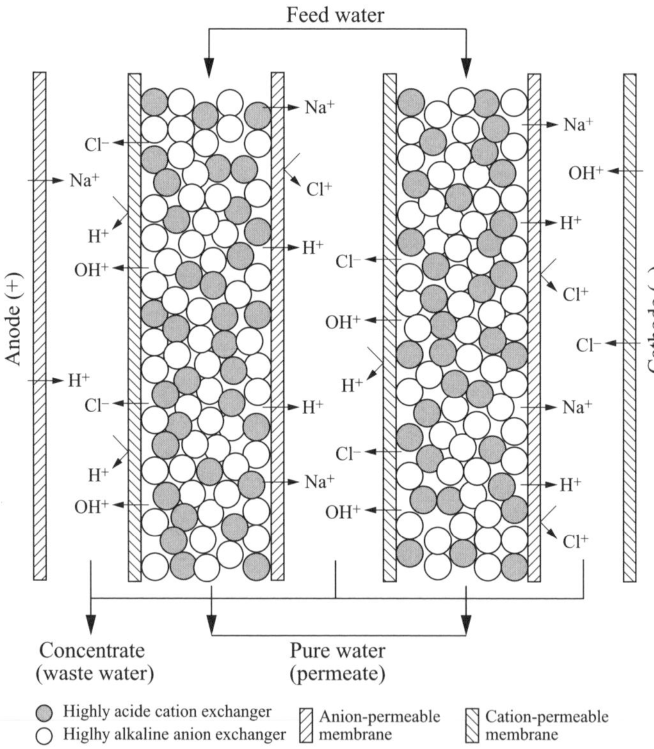

Figure 4.7. *Principle of EDI*

Ions are removed from the process system by the standard ion-exchange reactions in a conventional resin bed. However, in traditional ion-exchange units, after the contaminants are trapped onto the resin sites, the resin continues to be exhausted and lose capacity. In EDI, the contaminants are continuously removed as they are attracted to one of the two electrical charges, and then migrate through the resin bed, through ion-exchange membranes and into the concentrate stream where they are removed from the device. The applied current continuously regenerates the ion-exchange resin since it induces the dissociation of the water in hydrogen ions (H^+) and hydroxyl ions (OH^-):

$$H_2O \rightarrow H^+ + OH^- \qquad [4.8]$$

Water splitting replaces the chemical regeneration process.

EDI is a polishing technology and requires RO as a pretreatment. The ionic quality of the water produced is in the range of 15 to 18 MΩ-cm. EDI units typically require higher capital and ongoing electrical costs than conventional ion-exchange units. However, the combination of RO-EDI provides the customer with a continuous, chemical-free system. It allows reduction of over 95% of chemical usage compared to conventional exchange processes. EDI thus replaces on-site regenerated mixed-bed resin where end users do not wish to store, handle and neutralize chemicals, or off-site regenerated mixed beds.

4.1.4. *Future trends*

Due to the high UPW usage, semiconductor industries are facing issues dealing with incoming water availability, wastewater treatment discharge requirements and costs of manufacturing.

Today, industries are taking steps to significantly lower the use of UPW through equipment, process optimization and to develop solutions of reuse such as reclaimed water for technical installations. This will be developed in section 4.5.

4.2. Gases for semiconductors

Gases are ubiquitous throughout the processes employed in semiconductor device fabrication, from the fabrication of the single crystal silicon from which silicon wafers ("wafers") are made, to their final passivation, when components and circuits are laid out. Stages involving gases are called "dry", as opposed to "wet stages" involving liquids. Since the properties of these gases are well-described in the literature, our discussion focuses on safety in their use. Similarly, techniques and processes used for fabrication in microelectronics are only briefly mentioned to explain when gases are used during the wafer fabrication process. For more details on fabrication stages, refer to S. Franssilla's book, *Introduction to Microfabrication* [FRA 10].

This section discusses the main gases used in wafer and device fabrication, their use and the means required for their implementation.

4.2.1. *Main gases used in the semiconductor fabrication process*

A table listing the gases used in the different fabrication stages is available in the SEMI C3-0812 standard [SEM 12a]. The table includes the most common gases: nitrogen, oxygen, argon, ammonia, hydrogen, silane, nitrogen trifluoride, dichlorosilane, nitrous oxide, phosphine, arsine, boron trifluoride, diborane, etc. We present in the next section the main stages in question.

4.2.1.1. *Substrate fabrication*

Single crystal silicon is manufactured from polycrystalline silicon melted in an oven under an argon blanket. The single crystal silicon boule is slowly grown from the melt by the Czochralski process [ZUL 83] under inert conditions maintained by this blanket. The argon used must be contaminant-free to produce a pure single crystal.

4.2.1.2. *Wafer preparation*

This stage consists of creating a layer of silicon dioxide (SiO_2) on the wafer surface. It is carried out at a very high temperature (700–1,300°C) in a diffusion furnace, in which ultrapure oxygen is injected to oxidize surface silicon (dry thermal oxidation) according to the reaction [FRA 10]:

$$\mathrm{Si\ (solid)} + \mathrm{O_2\ (gas)} \rightarrow \mathrm{SiO_2\ (solid)} \qquad [4.9]$$

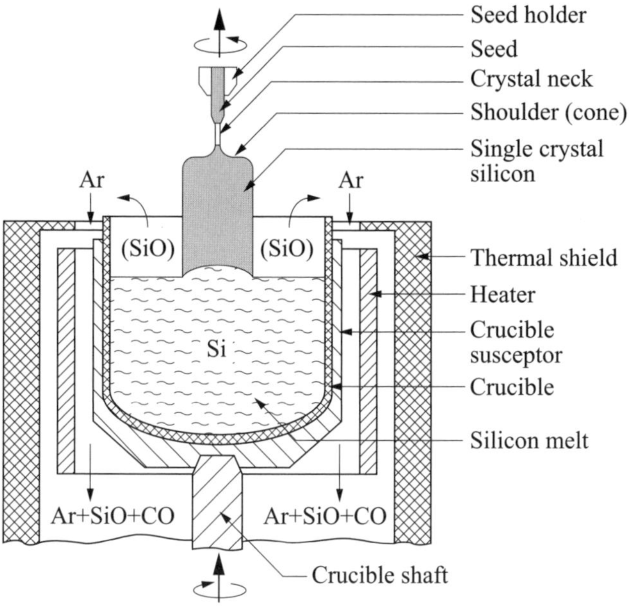

Figure 4.8. *Beginning of the growth of single crystal silicon (Source: http://www.tf.uni-kiel.de/matwis/amat/elmat_en/kap_6)*

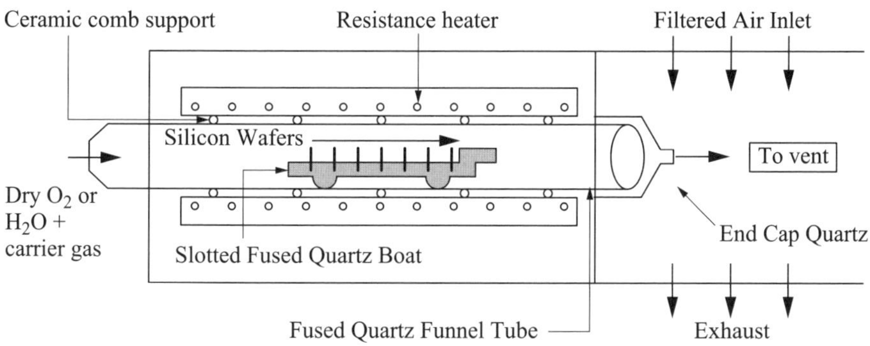

Figure 4.9. *Diagram of a thermal oxidation furnace (Source: http://cnx.org/content/m36703/1.2)*

4.2.1.3. *Doping*

This operation introduces dopants into the silicon to create n- and p-zones [IAE 12]. The two most frequently used techniques are diffusion and ion implantation [WAL 06]. The usual gas sources of dopants are [SHE 76]:

– n-doped zones: arsine (AsH_3) and phosphine (PH_3) are the gaseous reagent sources for arsenic and phosphorus;

– p-doped zones: boron trifluoride (BF_3) and diborane (B_2H_6) are the gaseous boron sources.

In the case of ion implantation, atoms supplied by the gas source are first ionized and then move through a particle accelerator to reach an energy from a few keV to a few hundreds of keV before being implanted in the material.

4.2.1.4. *Deposition*

The most common dry processes are based on thin film deposition from vapor phase precursors, commonly called chemical vapor deposition or CVD (LPCVD, APCVD, PECVD, VPE). A new variant has been introduced over the last few years: atomic layer deposition (ALD). These processes produce ultrapure thin films on the substrate surface from a "precursor" gas containing the element(s) to be deposited. Often a carrier gas such as nitrogen (N_2) or hydrogen (H_2) and, if necessary, a dopant gas may be present in the vapor phase as well [LEN 00]. For instance, silane (SiH_4) and dichlorosilane (SiH_2Cl_2) are common gaseous sources of silicon [LEV 89], while ammonia (NH_3) and nitrogen protoxide (N_2O) can be used as sources of nitrogen [BUC 03].

In the case of silicon deposition, dichlorosilane or silane undergo pyrolytic decomposition according to the following reactions:

$$SiH_2Cl_2 \text{ (gas)} \Leftrightarrow Si \text{ (solid)} + 2\ HCl \text{ (gas)} \quad [4.10]$$

$$SiH_4 \text{ (gas)} \Leftrightarrow Si \text{ (solid)} + 2\ H2 \text{ (gas)} \quad [4.11]$$

In the case of silicon nitride (Si_3N_4) deposition, a dichlorosilane flow (LPCVD) or a silane flow (PECVD) is mixed with a flow of ammonia in a reducing environment. For example, silicon nitride is deposited according to the following equation [DAH 84]:

$$3\ SiH_2Cl_2 + 10\ NH_3 \Leftrightarrow Si_3N_4 + 6NH_4Cl + 6\ H_2 \quad [4.12]$$

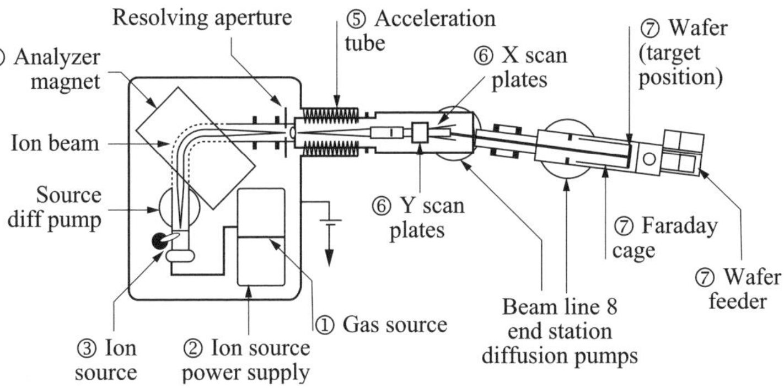

Figure 4.10. *Diagram of an ion implanter*
(Source: http://www.ece.gatech.edu/research/labs/vc/theory/ionimplant.html)

Arsine, phosphine and diborane are the most common gaseous dopant sources for processes such as PSG, BPSG and *in situ* doped polysilicon.

In the case of PECVD deposition processes, plasma is produced by radiofrequency or microwave excitation of gases injected into the process chamber. The plasma creates highly chemically reactive ions and radicals that react to produce thin films on the substrate surface at significantly reduced temperatures as compared with purely thermal processes. In addition, creative manipulation of the electric fields within the deposition chamber permits physical modification of the growing thin films through actions such as field-induced ion bombardment by molecular ions created from the gaseous reagents present in the plasma.

4.2.1.5. *Dry etching*

Dry etching employs gases that react physically and/or chemically with the unprotected substrate. In etching processes, material is removed from the substrate as volatile products of either a physical or chemical interaction, as described for physical sputtering or ion beam milling, vapor phase etching, plasma etching or reactive ion etching [JAN 96].

Argon is sometimes used in physical etching processes. Argon ions are created and accelerated to the substrate surface where the physical impact produces etching.

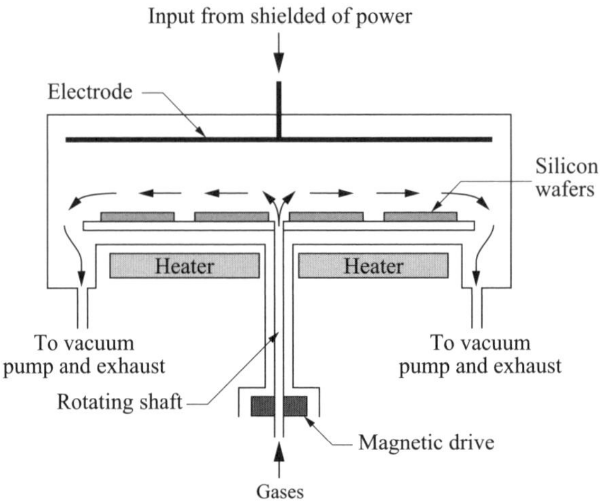

Figure 4.11. *Schematic representation of a radial flow PECVD reactor (Source: http://cnx.org/content/m25495/latest/)*

Chemical etching processes most often employ fluoride-based gases, including:

– tetrafluoromethane (CF_4) and other fluoromethanes (CHF_3, CH_2F_2, CH_3F), fluorobutanes (C_4F_8, C_4F_6) or perfluorocyclopentene (C_5F_8) to etch silicon, polysilicon, silicon dioxide and, with the addition of argon, oxygen or hydrogen, silicon nitride [SAK 09];

– sulfur hexafluoride (SF_6), silicon tetrafluoride (SiF_4) and nitrogen trifluoride (NF_3) to etch silicon dioxide [YR 98].

– Other sulfur-containing additives such as sulfur dioxide (SO_2) or carbonyl sulfide (COS) are sometimes used to favor the verticality of etching trenches [NAL 02, NAG 95].

4.2.1.6. *Cleaning*

Cleaning is important for surface preparation prior to any new operation, as well as for routine equipment maintenance to eliminate any deposits accumulated during the processing of a batch.

In the case of wafers, this preparation can be carried out using plasma. Oxygen plasma is used to remove organic compounds by oxidation [CHO 03] or while hydrogen removes oxides by reduction. In the latter case, a hydrogen–argon mixture is sometimes used to combine the chemical action of hydrogen and a physical action by argon [ZHA 97].

Nitrogen trifluoride is primarily used for cleaning reaction chambers [CHE 03]; sometimes fluorine (F_2) is used or, in the case of plasma-assisted cleaning, hexafluoroethane (C_2F_6) or octafluoropentane (C_3F_8).

4.2.2. *Gas quality requirements for semiconductor fabrication*

The integration rate of functions in a microelectronic component requires levels of etching such that the smallest impurity or particle deposited on the component during its manufacturing makes it unsuitable for use. Thus, the materials used must be ultrapure and the fabrication environment, storage devices, as well as distribution lines toward the point of use must be contaminant-free. For example, most current fabrication units require a class 1 clean room according to the US FS209E standard, i.e. at maximum: there is one particle of a diameter smaller than 0.5 μm per cubic foot in the fabrication ambient air. This standard has been used for a long time to characterize semiconductor production environments. It has been replaced by the ISO14644-1 standard [LIE 03] in which the reference particle diameter is no longer 0.5 μm but is 0.1 μm, this being defined for a maximum concentration of 10 particles (of 0.1 μm in diameter) per cubic meter.

Members of the semiconductor industry sector founded SEMI to standardize the quality of the gases used in the fabrication of semiconductor devices, along with safety, uses, etc. Gas quality standards are available on the SEMI website at www.semi.org. The quality level is expressed in both gas purity and particle concentration in the gas. Semiconductor manufacturers can require percentages up to 99.9999% in terms of purity, notably for gas sources in ion implantation.

4.2.3. *Implementation of gases used in the manufacturing of semiconductors*

4.2.3.1. *Source of the main gases*

4.2.3.1.1. Air gases

The designation “harmful gases” is used for most of the gases present in the atmosphere: nitrogen, oxygen and argon; this is perhaps surprising, since these gases make up 99.97% of the air that we breathe. However, the quantities consumed in semiconductor manufacturing units require that these gases be stored in liquid form and there is the potential for a leak from this liquid storage. That makes these gases dangerous within the context of the semiconductor environment:

– Nitrogen and argon are chemically neutral gases and therefore they would seem to be relatively innocuous from a general safety point of view. However, they present a risk of asphyxiation if they are present in an area in sufficient concentration to lower the ambient concentration of oxygen to less than 17%. This risk is very high in the case of a liquid leak, because of volume expansion during vaporization. It is especially hazardous in poorly ventilated areas.

– Oxygen is a highly oxidizing agent. Excessive concentrations of oxygen in an area will therefore promote ignition of any flammable materials that may be present. Obviously, then, a leak from a liquid oxygen storage vessel constitutes an extreme fire hazard [EIG 00].

Table 4.2 shows the essential physical properties of these gases [YAW 98].

	Nitrogen	Oxygen	Argon
Molecular weight (g/mol)	28.013	31.999	39.948
Freezing point temperature (K)	63.15	54.36	83.80
Boiling point temperature (K)	77.35	90.17	87.28

Table 4.2. *Essential physical properties of nitrogen, oxygen and argon*

These three gases are typically produced using an air separation unit (ASU) that carries out a cryogenic fractional distillation of air (see Figure 4.12) that separates the different gases according to their boiling point [EIG 08]. This technique allows for purities of 99.9999%. The global process is illustrated in Figure 4.12.

Air is first compressed and then cleaned by adsorption, in order to dry it and erase any trace of carbon dioxide (CO_2). It is then expanded and cooled to liquefaction in a cold box. The latter is made up of heat exchangers enabling the temperature to be lowered, and of a distillation column for the separation. It is obtained via the exchange between the descending liquid (which is undergoing oxygen enrichment) and the ascending gas (enriching in nitrogen), because of its higher volatility.

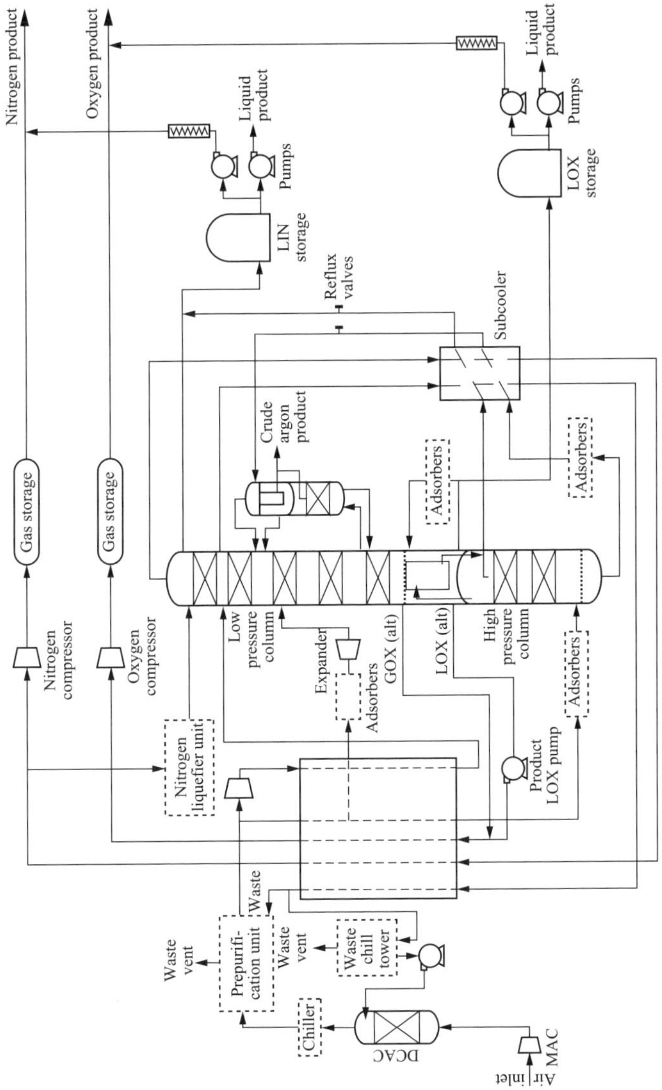

Figure 4.12. *Representative air separation plant flow diagram [EIG 00]*

4.2.3.1.2. Hydrogen

Hydrogen is the lightest gas; it is found in the atmosphere in trace amounts. It is a powerful reducer and highly flammable. Its main physical properties are summed up in Table 4.3 [YAW 98].

Molecular weight (g/mol)	2.016
Freezing point temperature (K)	13.95
Boiling point temperature (K)	20.39
Lower explosive limit (%)	4.0
Upper explosive limit (%)	75.0
Autoignition temperature (°C)	400

Table 4.3. *Main physical properties of hydrogen*

Steam methane reforming (SMR) is currently the main industrial fabrication method. In the SMR process, sulfur is first removed from natural gas by hydrodesulfurization. The resulting purified natural gas (mostly methane, CH_4) is then steam reformed at high temperature (840–920°C) in the presence of a nickel catalyst via the reaction [EIG 09]:

$$CH_4 + H_2O \Leftrightarrow CO + 3\ H_2 \quad [4.13]$$

The CO product is further reacted with water vapor according to the water–gas shift (WGS) conversion:

$$CO + H_2O \Leftrightarrow CO_2 + H_2 \quad [4.14]$$

This combination gives the following global reaction:

$$CH_4 + H_2O \Leftrightarrow CO_2 + 4\ H_2 \quad [4.15]$$

The hydrogen obtained contains, on average, 16–25% of CO_2 that must be eliminated by pressure swing adsorption purification.

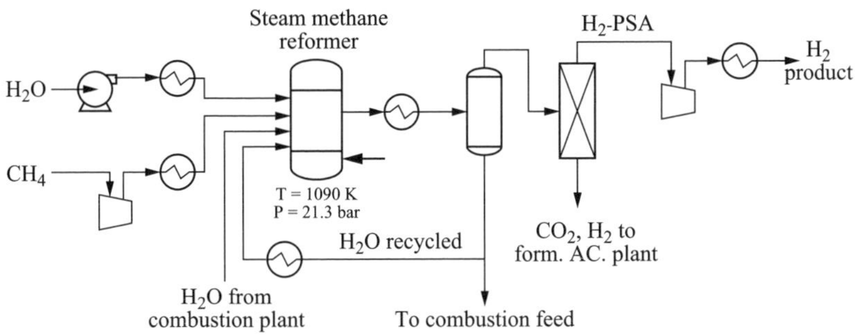

Figure 4.13. *Steam methane reforming process and hydrogen PSA flow sheet [PEN 11]*

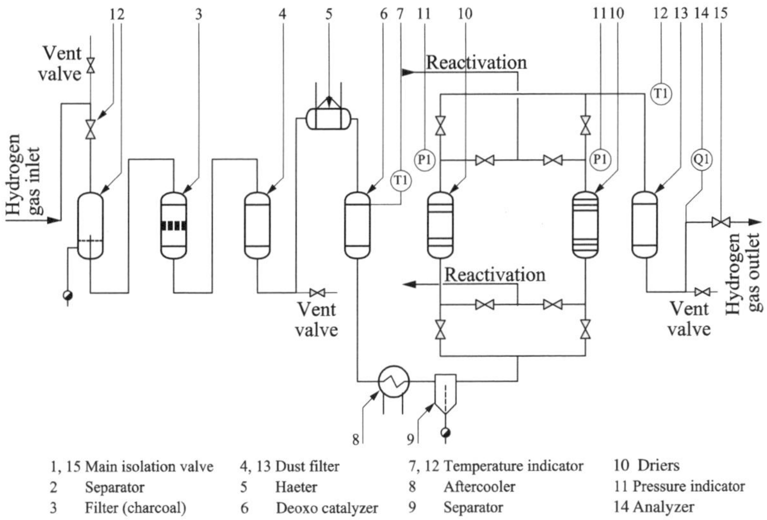

Figure 4.14. *Flow sheet of a typical hydrogen purification system [EIG 06]*

4.2.3.1.3. Ammonia

Natural ammonia comes from biodegradation of the nitrogenous matter found in organic waste and in the ground. It can form flammable or even explosive mixtures with air, and it is toxic in high concentrations. Its main physical properties are summed up in Table 4.4 [YAW 98]:

Molecular weight (g/mol)	17.031
Freezing point temperature (K)	195.41
Boiling point temperature (K)	239.72
Lower explosive limit (%)	16.0
Upper explosive limit (%)	25.0
Autoignition temperature (°C)	650
Threshold limit value (mg/m^3)	17

Table 4.4. *Main physical properties of ammonia*

On an industrial scale, this gas is mainly produced according to the Haber–Bosch process, which consists of hydrogenating nitrogen in the presence of catalysts [APP 97]. Hydrogen, which is generally produced by SMR, and nitrogen, which comes from an ASU, are injected into a high pressure (200 bar) reactor where nitrogen hydrogenation occurs through the action of a catalyst according to the following equation:

$$N_2 + 3\ H_2 \Leftrightarrow 2\ NH_3 \qquad [4.16]$$

The ammonia obtained is then purified by liquefaction (cryogenics).

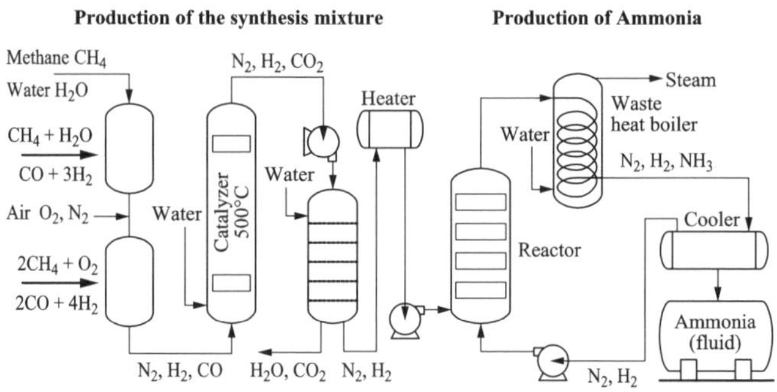

Figure 4.15. *Haber-Bosch process. (Source: http://need-media.smugmug.com)*

4.2.3.1.4. Nitrogen trifluoride

Although fluorine is the most reactive halogen because of its high electronegativity, it does not react directly with nitrogen to form NF_3 [MAY 64]. Therefore, this gas does not exist naturally. It is only formed by industrial production. It is highly oxidizing, even if the reaction is slow and, at high concentration, it is toxic. NF_3 is more stable, easier to store and use than fluorine and less harmful for the environment than C_2F_6 and sulfur hexafluoride (SF_6); it has replaced them for chamber cleaning since the beginning of the 2000s. Its main properties are as follows [YAW 98]:

Molecular weight (g/mol)	71.002
Freezing point temperature (K)	66.36
Boiling point temperature (K)	144.09
Threshold limit value (mg/m^3)	29

Table 4.5. *Main physical properties of ammonia trifluoride*

At industrial scales, it is formed by the direct reaction of fluorine with ammonia with the help of a catalyst depending on the global reaction scheme [HIR 06]:

$$NH_3 + 3\ F_2 \Leftrightarrow NF_3 + 3\ HF \qquad [4.17]$$

$$2\ NH_3 + 3\ F_2 \Leftrightarrow N_2 + 6\ HF \qquad [4.18]$$

Reaction 4.17 is predominant when the temperature is lower than 50°C. The latter is highly exothermic; therefore, a continuous cooling is required, as well as a dilution of the reactants in inert gases, in order to avoid any risk of explosion. The nitrogen trifluoride formed is then purified by distillation.

4.2.3.1.5. Silane and dichlorosilane

Silane and dichlorosilane are available exclusively from industrial production. Silane is part of the hydride family. It is highly unstable and ignites spontaneously in air [NAK 98]. Dichlorosilane has a flammability range wider than that of hydrogen and it is corrosive and reacts violently with water to form hydrogen chloride [NAK 98]. The main physical properties of these two gases are summed up in Table 4.6 [YAW 98, EIG 10].

	Silane	Dichlorosilane
Molecular weight (g/mol)	32	101.007
Freezing point temperature (K)	88.15	151.15
Boiling point temperature (K)	161	281.45
Lower explosive limit (%)	1.37	4
Upper explosive limit (%)	96	96.0
Autoignition temperature (°C)	58	55
Threshold limit value (mg/m^3)	5	–

Table 4.6. *Main physical properties of silane and dichlorosilane*

The most common industrial process for the production of dichlorosilane uses the disproportionation of trichlorosilane ($SiHCl_3$) on an anion-exchange resin at controlled temperature, in order to separate the products from the reaction by evaporation [BAK 76] according to:

$$2\ SiHCl_3 \Leftrightarrow SiH_2Cl_2 + SiCl_4 \qquad [4.19]$$

Silane is obtained according to the same principle by temperature control [BAK 76]:

– disproportionation of dichlorosilane into trichlorosilane and chlorosilane by lowering of the bed temperature below trichlorosilane evaporating temperature:

$$2\ SiH_2Cl_2 \Leftrightarrow SiHCl_3 + SiH_3Cl \qquad [4.20]$$

– and then further disproportionation of chlorosilane into silane and dichlorosilane:

$$2\ SiH_3Cl \Leftrightarrow SiH_4 + SiH_2Cl_2 \qquad [4.21]$$

The advantage of this process lies in the recycling of trichlorosilane from the manufacturing of metallurgical silicon. This minimizes costs.

4.2.3.2. *Storage and distribution*

Because of their purity levels, these gases are ultra-concentrated and extremely reactive. This reactivity presents risks for explosion and/or fire and for toxic releases. Perfectly leak-tight and robust storage installations are therefore essential to prevent accidents.

Equipment design forces the classification of gases according to the nature of the risk that they pose so as to separate incompatible gases:

– toxic gases: arsine, phosphine, diborane, ammonia, silicon tetrafluoride, dichlorosilane, boron trifluoride;

– flammable gases: silane, dichlorosilane, hydrogen, arsine, phosphine, diborane;

– powerful oxidizers: oxygen, nitrogen trifluoride;

– generators of anoxic atmospheres: nitrogen and argon.

Generally:

– flammable gases are isolated to avoid propagation in the case of fire;

– toxic gases are contained to avoid external contamination;

– oxidizers are separated from flammable gases to avoid accelerating the flames in case of fire.

There are three levels of safety methodologies associated with gases in semiconductor practice:

– leak prevention and containment, if necessary;

– detection and alarm;

– treatment of the event.

The most dangerous gases are the dopant reagents since they have very high toxicity levels (e.g. TLV-TWA arsine = 0.05 ppm) and because they are highly flammable (e.g. LFL-UFL arsine = 4.5–78% vol.). Therefore, they are subjected to the most restrictive safety measures. Because of its pyrophoric nature, silane also requires special attention.

4.2.3.2.1. Leak prevention

Leak prevention consists of ensuring the leak integrity of containers. Leak integrity is ensured through the use of materials that are resistant to pressure and any molecule reactivity, and through the use of suitable connections for the assembly of distribution lines.

The component material most widely used is austenitic stainless steel of grade X2CrNiMo18-10 1.4404 (EN10027), commonly called 316L (AISI). Piping is manufactured by cold drawing in order to obtain elements of very specific dimensions. Piping and all ancillary components (valves, fittings, etc.) are treated by electrolytic polishing, to ensure a high-quality internal surface with an average roughness not exceeding 0.2 μm. This surface quality limits adsorption/desorption phenomena and thereby reduces the contamination risks.

Piping assembly is carried out using orbital welding. This technique produces high-quality welds and minimizes the risk of leakage. Connections to system components such as valves are made using metallic gaskets that are also electrolytically polished. Leak integrity is assured by pressurizing the elements of the system under test using a mixture of nitrogen and helium at 4% and checking for leaks using a mass spectrometer.

Safety for highly toxic gases must be enhanced by containment since the consequences of leakage are so severe. Barriers to propagation are found on three levels:

– packaging: gases are put in bottles containing adsorbents, which trap the gas molecule on the adsorbent surface. The gas can only be extracted from its container by pumping;

– storage: gas bottles are stored in gas rooms that are kept under slight vacuum pressure via extraction of air; air is treated before being discharged.

Distribution:

– bottles are connected with distribution pipes in gas cabinets that are kept under a slight vacuum; the air is treated before being discharged and the cabinets are set up in gas rooms;

– pipes are orbital welded and contained in a double tube in which nitrogen flows through the outer tube, in order to drive any blow-by gases toward detection and treatment prior to discharge;

– valves are installed in a sealed box, referred to as a valves manifold box or VMB. This enables containment in the case of leaks.

Pyrophoric gases, such as silane (LFL = 1.4%; UFL = 100%), or highly flammable gases, such as hydrogen (LFL = 4%; UFL = 77%), are preferably stored outside to prevent risks of explosion caused by a leak. Propagation risk, in case of fire, can be limited: either by respecting safety distances or by implementing partitioning between storage sites.

4.2.3.2.2. Detection and alarm

Abnormality detection refers to loss of containment, gas leaks, as well as fire.

Containment is monitored using a pressure difference measurement between the area to be contained and the outside ambient. Typically, in the case of a loss of containment, a safety programmable logic controller (PLC) will close gas distribution valves and actuate alarm devices.

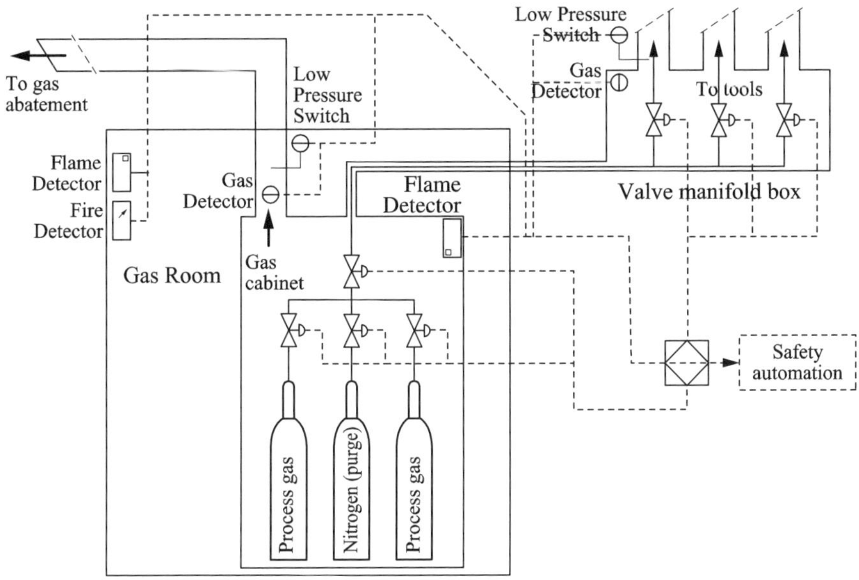

Figure 4.16. *Safe storage and distribution of harmful gases*

Gas leaks are typically detected using catalysis sensors for flammable gases and electrochemical sensors for toxic gases and oxygen. It is called "ambient" detection when it concerns geographical areas (gas room, maintenance corridor, manufacturing area, etc.) and "local" detection when it relates to a piece of equipment, a valve manifold box or a gas cabinet. In the latter case, samples extracted at the "local" site are transported to a detector. If gas is detected, the safety PLC will close the distribution valves for the detected gas and actuate alarm devices. Alarm thresholds are generally adjusted to be 20% of TLV or LEL for maintenance operations and 50% of TLV or LEL for the evacuation of the area.

Fire detection is most often carried out using smoke detection (generally opacity or ionization) or flame detection (UV and/or IR radiation). Smoke detection is sensitive to convection; therefore, it is typically used within premises, flame detection is normally used outside, as well as in valve manifold boxes and gas cabinets.

Alarms are typically given in a visual (bright flash) and audible manner (siren sounds).

4.2.4. *Conclusion*

Gases are used in the fabrication of semiconductor components, from the manufacture of the silicon wafers used as substrates through to the passivation of the protection of implanted components and circuits. The number of stages and the diversity of techniques and processes used involves the implementation of a wide range of gases and purity levels that make these gases ultra-concentrated and highly reactive products.

Dangers resulting from this reactivity require the deployment of safety measures at all levels of use, from packaging up to the point of use. They concern facilities, as well as distribution equipment and fabrication machines.

The nature and rapid evolution of the risks generated by these gases in modern fabrication processes will require a broad industry collaboration on safety and applications as the technology goes forward. For gas suppliers, this will involve, among other things, the development and testing of alternative gaseous reagents that can reduce the economic impact of the risk prevention measures on the production costs of the components.

4.3. Dissolved gases

Wet chemical treatments employing dissolved gases play an important role in semiconductor manufacturing; these treatments are especially critical in surface preparation applications. Front end of line (FEOL) and back end of line (BEOL) processes (as described in Chapters 1 and 2 of this book) use gases such as N_2, H_2, O_2, CO_2 and O_3 in enhanced cleaning and/or process control protocols. Dissolved ozone, with its high redox potential (see Table 4.3), is used in processes requiring an oxidizing agent for cleaning and hydrophilization [GOT 04]. Carbon dioxide, which ionizes in water, is used to adjust the pH value and the resistivity of solvents, highly sensitive parameters that must be controlled in metallization processes [GOT 10]. In addition, oxygen-free processing, a longstanding requirement in certain semiconductor manufacturing protocols, has become more critical as stricter control over surface oxidation has become important [VER 00]. By decreasing the dissolved oxygen in the UPW used for rinsing, native oxide formation on a silicon surface can be reduced [LI 00].

In most semiconductor applications, the process window for dissolved gas levels is very tight and the concentrations of dissolved gases must be monitored in real time. The optimal yield management thus requires a highly accurate control of dissolved gas concentrations, in particular for O_3, CO_2 and O_2 dissolved in UPW (DI-O_3, DI-CO_2 and DO). A variety of control issues related to this requirement can be resolved through the use of appropriate dissolved gas delivery systems. These systems are designed with precision concentration control that can cope with the demands of varying process recipes and the duty cycles inherent in single wafer processing equipment. In the following discussion, we describe the chemistry related to the formation of relevant dissolved gas solutions and various practical issues of relevance to the generation and control of these solutions.

4.3.1. *DI-O3*

4.3.1.1. *Application*

Ozone provides an alternative to traditional sulfuric acid–peroxide and RCA wet cleans that improves cleaning efficiency while simultaneously lowering chemical consumption and disposal costs. As a consequence, its use in wafer cleaning and photoresist stripping applications has been explored by semiconductor researchers for more than 20 years. A number of different process cleaning protocols employing dissolved ozone have been developed. The details of these protocols depend on the primary purpose of the cleaning step: i.e. light and heavy organic cleans or metal and particle removal.

Wafer cleaning using dissolved ozone always involves chemical oxidation. Surface oxides are created in DI-O_3 cleaning processes and these are self-limiting at the nm scale. Oxidation growth rates are directly proportional to dissolved ozone concentrations and process temperatures. Since the solubility of O_3 in water decreases at higher temperatures, optimal process windows for producing a dense surface oxide and the most effective cleaning protocols are normally achieved at ambient temperature [NEL 97]. Particle and metal removal employ a mixture of DI-O_3 and HF/HCl, and accurate and repeatable chemical concentrations are paramount in these protocols. Control of the dissolved ozone content represents the most challenging aspect of system design in this case. The effectiveness and reproducibility of these cleaning processes can only be ensured when the point-of-use concentrations of the dissolved ozone remain above some application-specific minimum value throughout the cleaning process or inside a process window.

4.3.1.2. *Chemistry*

Ozone (O_3) is an allotrope of oxygen. It is a highly reactive, gaseous oxidizing agent and it is widely used for the treatment of drinking water and wastewater [GOT 10]. Much of the effectiveness of dissolved ozone in water as a cleaning and disinfecting agent is due to its high oxidation potential (Table 4.7). Ozone dissolved in UPW can react directly with a compound or it can react indirectly through the production of hydroxyl radicals, OH°, that have a higher oxidation potential than ozone.

Figure 4.17 shows reaction mechanisms for oxidation by ozone and by hydroxyl radicals with different reaction pathways that lead to different oxidation products. The direct ozone reaction pathway is selective, with normally slow reaction-rate constants. The indirect OH reaction is fast and non-selective, requiring activation by initiators such as a high pH, hydrogen peroxide or UV radiation. Although a fast reaction is desirable, reactions that depend solely on radicals generated in the bulk solution are usually avoided since, in many cases, the active species must act directly on the surface. Reactive radicals generated in the bulk of the cleaning solution are relatively far away from the surface and can react out during diffusion to the surface. A comprehensive review of reaction pathways and important process parameters is available [GOT 00], as are kinetic parameters for a variety of reactions employing DI-O_3 [HOI 83a, HOI 83b, HOI 83c].

Element	Symbol	Electrochemical potential (eV)
Fluorine	F_2	3.06
Hydroxyl radical	OH°	2.80
Ozone	O_3	2.08
Hydrogen peroxide	H_2O_2	1.78
Chlorine	Cl_2	1.36
Oxygen	O_2	1.23

Table 4.7. *Oxidation potentials*

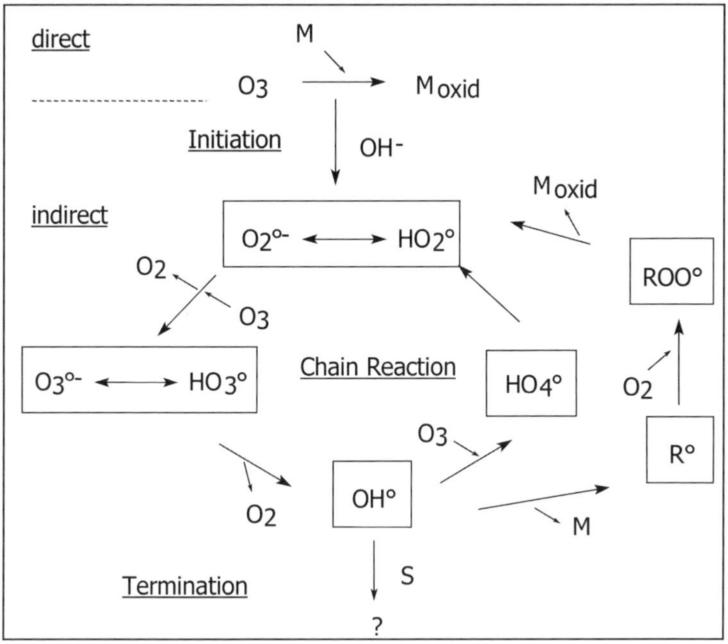

Figure 4.17. *Model of the indirect and direct ozonation, S: Scavenger, R: Reaction product, M: compound (modified from [STA 83])*

4.3.1.3. *Generation and control of DI-O$_3$*

Ozone is an unstable molecule and it must be generated onsite through the reaction:

$$3\ O_2 \rightarrow 2\ O_3 \qquad [4.22]$$

Methods of ozone production differ in their working principle and ozone sources. Werner von Siemens developed the first ozone generator in Germany in 1857. Ozone was produced by an electric discharge through an oxygen-containing gas. For semiconductor applications, clean ozone gas is produced in ozone generators that employ pure oxygen and a silent discharge (also known as a dielectric barrier discharge or DBD) in which oxygen is passed between two electrodes separated by a solid dielectric. An alternating current high voltage field is applied to the electrodes and this produces a microfilamentary electrical discharge in the gap. Impact events between high energy electrons and oxygen molecules in the discharge gap dissociate molecular oxygen into O atoms; ozone is produced through the recombination of these oxygen atoms with molecular oxygen.

Ozone generators based on the silent discharge principle require small quantities of a dopant with the oxygen feed. Typically, these dopants are N_2 and CO_2. The use of noble gases as dopants, such as helium (He), neon (Ne) and argon (Ar), has also been proposed [MAN 00]. The presence of the dopant helps to produce long-term stability in the ozone-generation process. The dopant concentration in the feed may range from 10 parts per million (ppm) to 10%, dependent upon the ozone generator design and the quality of the feed gas. Some ozone generators can operate without the addition of a dopant gas, providing that the oxygen feed is not "too clean" (i.e. the oxygen already contains small amounts of impurities such as nitrogen). Semiconductor applications require a minimum oxygen purity of 99.995% with low levels of total organic carbon and humidity.

Ozone production is an endothermic process (ΔH = +143 kJ/mol) [MAN 01]. In addition, the process is relatively inefficient, with only 4–12% of the electrical energy supplied going to the DBD and the formation of ozone; the rest of the input energy is transformed into heat [OZO 94]. Since the ozone molecule decays quickly at elevated temperatures, efficient cooling is required for the production of high concentrations of ozone.

Ozone can also be produced either through the electrolysis of water or by UV irradiation of an oxygen containing gas at wavelengths below 200 nanometers (nm). The latter route is similar to the natural process occurring in the stratosphere. These methods have capacity limitations, and the cost and size of the generation equipment are high relative to the silent discharge method.

Silent discharge (DBD) is the most common method employed for ozone production in electronics fabrication since these applications require ultrapure ozone gas to avoid particle and metal contamination. An ozone production of 100–500 g/h at concentrations between 100 and 400 g/Nm^3 is typical for today's practice (N defining "normal condition": 0°C, 1,013.25 mbar).

Wet (aqueous) applications require a process step that transfers the ozone gas into the water solvent. This can be done by injection into the water within an application tool, or using a separate contacting system to produce the necessary concentrations of dissolved ozone upstream of the application tool. Intimate contact between the gaseous ozone gas and the water can be achieved through various configurations of gas diffusers that must be constructed of materials resistant to ozone.

The injection of ozone into a solution can be accomplished using fine porous plates immersed in fluid baths, membranes, injector nozzles, static mixers or bubble columns. PTFE membrane mixers direct the liquid phase through a bundle of microporous tubing that is housed in an appropriate shell. The gas is passed through the outer shell counter-current to the liquid and diffuses through the porous tube into the liquid. To avoid bubbles, the pressure on the gas side must always be lower than on the fluid side. Static mixers consist of several mixing elements arranged in a series within a pipe. These systems are small and easy to handle and the hydrodynamic conditions provide for a uniform distribution of small bubbles of dissolved ozone over the entire cross-section in which the fluid is contained. The ozone gas does not completely dissolve in the fluid in static mixers and the surplus gas appears as bubbles in the outlet of the mixer. If gas bubbles are undesirable in a particular process, a debubbler must be installed downstream of a static mixer. Bubble columns are very simple contactors that have a relatively low transfer efficiency. Modifications that increase gas/fluid contacting surfaces are known to enhance the efficiency of ozone dissolution, but are seldom reported. Each type of contactor has unique hydrodynamic behavior and mass transfer characteristics. It is important to consider these factors when evaluating experimental results and optimizing processes that require dissolved ozone. Stand-alone systems for dissolved ozone in UPW are commercially available. Closed loop control of the dissolved ozone concentration in these systems guarantees effective process control even in single wafer applications where flow rates fluctuate between 0.5 and 60 L/min. Current developments in contacting technology can produce ozone concentrations in ambient temperature DI water between 25 ppm at flow rates of 140 L/min to over 115 ppm at flow rates of 5 L/min.

4.3.1.4. *Some practical considerations*

Ozone is a very strong oxidant, and therefore all "wetted" materials within a system must be highly corrosion resistant. Table 4.8 provides an overview of specific material compatibilities for use with ozone. It should be noted that most, if not all, compatibility tables found on the Internet relate to the relative low dissolved ozone concentrations normally used for drinking water applications.

Material	O_3 Gas	O_3 dissolved	Comment
Metals			Metals can suffer severe corrosion
Stainless steel	✓	✗	
Silver	✗	✗	Silver and other metals can destroy ozone catalytically
Inorganic oxides			
Glass, quartz (SiO_2)	✓	✓	
Aluminum oxide (Al_2O_3)	✓	✗	
Fe-, Cu, Mn-oxide	✗	✗	Efficient catalyst
Organics	Most organics severely attacked		
PTFE, PFA	✓	✓	Good for process lines
PVDF, PVC	✗	(✓)	Useable in wet drain lines, but not process lines
PP, PE	✗	✗	Will be destroyed at fast rate
Kalrez, Chemraz	✓	✓	Sealing

Table 4.8. *Material compatibility with high ozone concentration*
✓ = *material compatible,* ✗ = *material not compatible*

Some purity and process requirements preclude the use of certain materials with DI-O_3, despite the fact that they are ozone resistant. For example, high quality stainless steel, which is resistant to gaseous ozone, produces unacceptable metal contamination when used with ozonated water. Great care must therefore be taken in screening all wetted components for material compatibility with respect to ozone and DI-O_3. For instance, silver, especially as its oxide, cannot be used in ozone systems, since it will catalytically decompose ozone to produce particle and metal contamination. If conventional VCR fittings are used in the equipment configured for ozone generation and transport, then problems will arise, since standard VCR gaskets are silver-plated. Thus, in ozone systems, non-plated stainless steel gaskets must be specified to avoid this effect. Similarly, other metal oxides (e.g. manganese [Mn] or palladium [Pd]) will also lead to the catalytic decay of ozone and, in fact, may be used for the efficient destruction of ozone. In addition, while it may be desirable to replace polytetrafluoroethylene (PTFE)/perfluoroalkoxy (PFA) system components with polyvinyl chloride (PVC) and polyvinylidene fluoride (PVDF) in order to reduce costs, these latter materials are less stable to ozone exposure over time. Thus, PVC can be used for drain lines, but it is not recommended for process lines. Polypropylene (PP) and polyethylene (PE) must be avoided in all cases since they will be immediately destroyed, producing dangerous ozone leaks.

Because ozone is a toxic gas (TLV: 0.2 mg/Nm^3), ozone generation systems must convert excess ozone to oxygen using an ozone destruct unit. These are usually based on catalytic or thermal ozone decomposition. As noted, manganese or palladium can be used as the catalyst material. An ambient ozone safety monitor for continuous monitoring of any ozone leakage should be provided in areas where equipment operators could face possible exposure. These monitors should be coupled to an automatic ozone generator shut-off. Typically, UV-photometers are preferred for the measurement of both low and high ambient ozone concentrations. Different concentration ranges (low ambient levels vs. high process concentrations) require different detector configurations (different cuvette lengths) for accurate measurement. Other ozone concentration measurement techniques based on electrochemical or semiconductor principles are less reliable.

Tubing connecting the ozone generator to a wet bench or to single wafer equipment should be doubly contained to protect against exposure to an ozone leak. For similar reasons, the dissolved ozone generation system and those process areas employing ozone or DI-O_3 must be connected to a fume exhaust system for operator protection.

4.3.2. *DI-CO2*

4.3.2.1. *Applications*

Shrinking feature sizes make semiconductor devices more fragile and more vulnerable to electrostatic discharge (ESD) effects. Conductive materials can be employed in a single wafer chamber as one means of preventing ESD damage. Proper grounding and the use of ionization to neutralize static charges on electrically isolative materials can provide further protection against ESD damage [AVE 05].

Elimination of ionic charges is important for both FEOL and BEOL processing. In FEOL semiconductor processing, ionic residues from preceding chemical steps (e.g. from steps employing H_2SO_4, HF and HCl), while they still reside on the wafer, can prevent charge build-up. However, the final wafer rinsing step is typically performed using non-conductive UPW, and this can initiate charge build-up and damage. During the UPW rinse step in a single wafer tool, a wafer can develop a charge of up to 25 kV when placed on isolated chucks. Many devices having tiny structures can be severely damaged by electrostatic stress well below this value. At worst, even a complete lift-off can occur. A possible solution to this problem can be found in the use of controlled levels of dissolved ionizing gases in the rinse step. In particular, carbonated (CO_2) rinsing provides a solution for damage-free cleaning and charge reduction in single wafer cleaning tools. The addition of CO_2 as an ion source in the rinse reduces resistivity to a level that prevents surface charging and provides yield improvement [HAL 08, SPI 09].

The addition of DI-CO_2 in BEOL rinses simultaneously controls both pH (> 4) and conductivity of the UPW, preventing corrosion of Al and Cu [MAT 07, NEL 97]. The controlled addition of DI-CO_2 to the rinse allows the establishment of real-time control of the conductivity and acidity over the entire flow range in the process window. Various patents have been issued on the use of carbonized water for prevention of Cu corrosion in wet cleaning steps. The most recent ones focus on single wafer equipment [AOK 05, CHO 06].

Another potential area of DI-CO_2 lies in megasonic cleaning. The cavitation created in these processes cleans, but it can also lead to damage. Therefore, it is necessary to control cavitation in the process. The light emitted from cavitation bubbles is called sonoluminescence (SL) – light from sound – and the level of sonoluminescence is a sensitive indicator of the presence of a cavitation event. Research has shown that the addition of DI-CO_2 inhibits SL generation, thus reducing damages [KUM 11].

4.3.2.2. *Chemistry*

CO_2 is a colorless, odorless gas. It is the anhydride of carbonic acid (H_2CO_3). By dissolving CO_2 in water, carbonic acid is generated. In an aqueous solution, H_2CO_3 dissociates, primarily to bicarbonate (HCO_3^-) and protons (H^+). At higher pH values (pH > 10.3), the carbonate is the dominant species in aqueous solution (see also TOC above).

$$CO_2 + H_2O \rightarrow H_2CO_3 \qquad pK = 2.6 \qquad [4.23]$$

$$H_2CO_3 + H_2O \rightarrow HCO_3^- + H_3O^+ \qquad pK = 3.8 \qquad [4.24]$$

$$HCO_3^- + H_2O \rightarrow CO_3^{2-} + H_3O^+ \qquad pK = 10.3 \qquad [4.25]$$

The concentration of CO_2 in solution depends, as with all gases, on the partial pressure and the temperature. The dissolved CO_2 concentration in water increases with increasing partial pressure and reducing temperature. The concentration of dissolved CO_2 influences the conductivity and resistivity, respectively, of UPW. Since the conductivity/resistivity itself is influenced by the temperature in the opposite way, a maximum is formed in the conductivity curve, as shown in Figure 4.18. For a given temperature and conductivity, the pH value can be calculated. If it is desirable to stay above a pH of 4, the conductivity should never exceed 40 μS/cm, as shown in Figure 4.19.

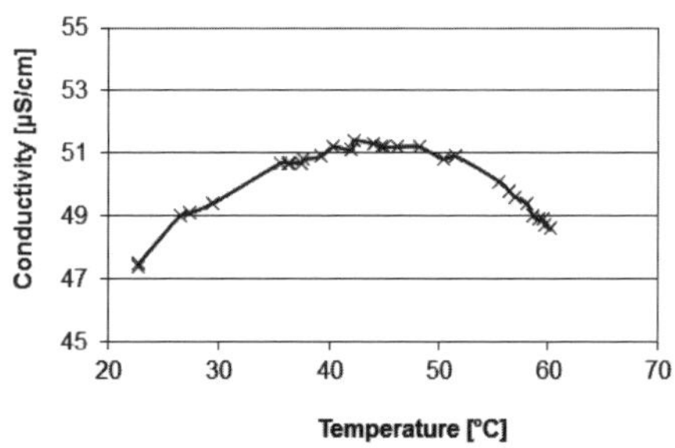

Figure 4.18. *Measured conductivity at saturation point by varying the UPW temperature, 1 bar (absolute)*

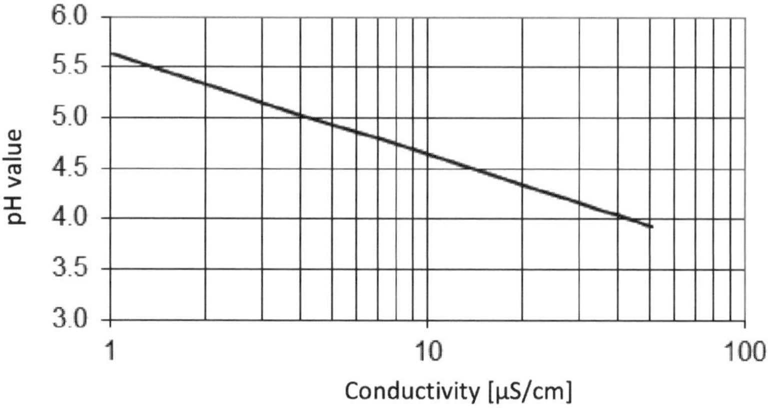

Figure 4.19. *Calculated pH-value as a function of DI-CO_2 conductivity at 25°C*

4.3.2.3. *Generation and control*

The main application of DI-CO_2 systems is in single wafer tools employing multiple chambers to achieve high throughput. In these wafer processors, a CO_2 dosage system that can serve multiple chambers over a fairly wide range of UPW flows while maintaining tight control of the conductivity/resistivity is essential. The following are the basic requirements necessary for such a DI-CO_2 system:

– feed-forward and immediate conductivity control;

– stability of conductivity with fluctuating UPW flow rate;

– automatic supply pressure control;

– low cost of ownership → environmental aspects;

– compact design.

Conductive water can be achieved with simple setups such as CO_2 bubbling or contactors (e.g. membranes or inline mixers). Similar contacting approaches to those used in the DI-O_3 applications are common for DI-CO_2, and the addition of CO_2 into rinse water is widely implemented, especially in single wafer equipment. Where it is necessary to fulfill the conductivity control requirements described above in relatively tight process windows, the systems must employ more elaborate closed-loop control.

Figure 4.20 shows a typical example of a single wafer processor's UPW flow pattern and the UPW conductivity maintained by a commercially available CO_2 dosage system. The conductivity range is seen to maintain well within 5% of the set point while the UPW flow rates change every 20 s, to simulate a multi-chamber cleaning tool. The outgoing pressure is not influenced at all. Since the CO_2 dosage concentration is a quadratic equation of the conductivity, extreme control of the ingoing gas flow of CO_2 for the required accuracy is necessary.

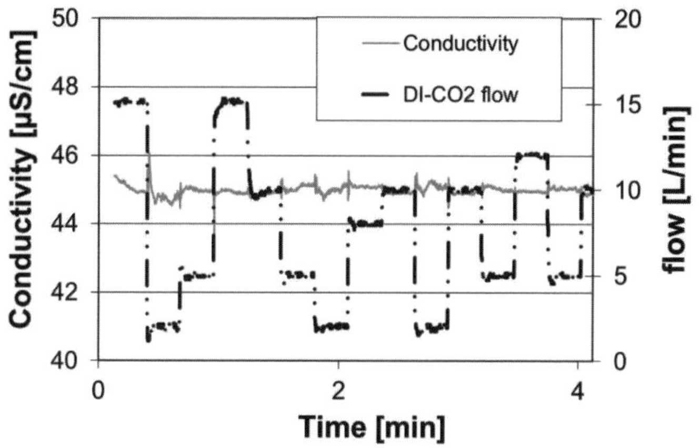

Figure 4.20. *UPW flow change at constant conductivity set-point of 45 μS/cm*

4.3.2.4. *Practical aspects*

Colorless and odorless CO_2 gas gives little warning of exposure. The main health hazard associated with CO_2 gas is asphyxiation by displacement of oxygen. Exposure to CO_2 concentrations of 10% in the air can produce rapid unconsciousness and death from oxygen deficiency, while concentrations as low as 3% may produce shortness of breath and headaches. Should a leak occur, CO_2 may accumulate in the working area and it is strongly recommended that an ambient CO_2 concentration monitor with safety interlock be installed in the CO_2 dosage system. The TLV for CO_2 is 5,000 ppm [OSH 89]. For further information, refer to the material safety data sheets (MSDS) for CO_2.

4.3.3. *DO*

4.3.3.1. *Application*

As semiconductor processing changes owing to decreased dimensions and new architectures (i.e. structures now from typical planar manufacturing to new finfet structures, on bulk or SOI substrate), new fabrication strategies are emerging. These include novel approaches such as gate-last versus standard gate-first integration and new materials such as High K dielectrics and metal gates. To avoid Complementary Metal Oxide Semiconductor (CMOS) variability issues, these new fabrication strategies demand careful attention to the way that layers are deposited and etched, and to related interfacial concerns [KUH 11]. Recently, the integration community has focused on oxygen control in FEOL at the gate stack level, specifically during epitaxial growth steps and during the contact cleaning step for improved performance. In diluted hydrogen fluoride (dHF), for example, control of the oxygen content is mandatory [BIL 09], for example, for stable hydrogen-terminated surfaces [PAG 03]. Also, oxygen control for BEOL applications has been widely reported [PER 07, LAU 09]. Control of the oxygen concentration in UPW thus appears as a new critical process requirement.

4.3.3.2. *Chemistry*

The preparation of hydrogen terminated silicon (Si-Hx) surfaces by HF acid etching/UPW rinse is a common step in wafer surface preparation. The surface terminates with a monolayer of hydrogen, which passivates the chemically cleaned reactive surface and suppresses native oxide growth. Native oxide will start to regrow on the passivated surface, depending on the available concentration of dissolved oxygen in the UPW:

$$SiH_{2x} + x\,O_2 \rightarrow SiO_x + 2\,H_2O \qquad [4.26]$$

Metals (Me), for example copper [Cu] and cobalt [Co], will also be oxidized by dissolved oxygen in the UPW, producing species that are soluble in water:

$$Me + O_2 + 4\,H^+ + 2\,e^- \rightarrow Me^{2+} + H_2O \qquad [4.27]$$

Also, a galvanic cell is formed in cases where two metals are present. The less noble material is corroded in the presence of oxygen. The amount of galvanic corrosion can thus be decreased by decreasing the content of dissolved oxygen in the UPW [PAG 03, GOV 11].

4.3.3.3. *Generation and control*

UPW in new semiconductor fabs is typically de-oxygenated to less than 10 ppb. Details about the technique are described elsewhere in this section. From the point of delivery (POD) to the point of process (POP), oxygen diffuses back into the water. As a result, point of use de-oxygenation of UPW and chemical mixtures of reducing agents become necessary for critical processes.

One possible approach to the drastic reduction of oxygen in liquids is the use of one or more contactors wherein an inert purge gas – in counter-current flow – contacts the oxygen-containing liquid. The contacting technologies used are as described for DI-O_3 and DI-CO_2. The outlet oxygen concentration that can be achieved in the liquid depends strongly on the oxygen content of the purge gas: the higher the grade, the cleaner the purging gas and the lower the outgoing oxygen concentration. The saturation concentration as a function of the oxygen content in N_2 marks the physical/chemical limit for this method (Table 4.9).

Grade	O_2 gas content	O_2 saturation concentration
4.6	5 ppm	0.22 ppb
5	3 ppm	0.13 ppb
6	0.5 ppm	0.02 ppb
7	0.03 ppm	0.001 ppb

Table 4.9. *Purging gas N_2 at different grades, its maximum oxygen content and the resulting saturation concentration of oxygen in water at 20°C and ambient pressure*

4.3.3.4. *Practical aspects*

POU de-oxygenation is only one of the measures necessary to achieve controlled, low oxygen concentrations in process fluids. Also, measures must be taken to avoid oxygen contamination of process water. After de-oxygenation, tubing length to the process should be as short as possible to avoid any air permeation. Any oxygen contamination from this source will depend on the tubing material, tubing diameter and residence time. The data in Figure 4.21 demonstrates that when using PFA tubing having a length of 10 m, flow rates above 20 L/min are necessary to minimize the addition of oxygen in the UPW. Flow rates below 20 L/min can lead to oxygen contamination of the UPW in the double-digit region.

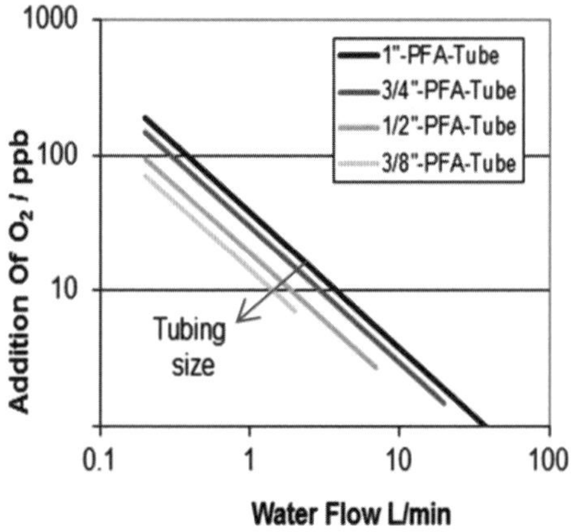

Figure 4.21. *Permeation of oxygen through PFA tubing as a function of flow rate and tubing diameter, length 10 m, 20°C, ambient environment*

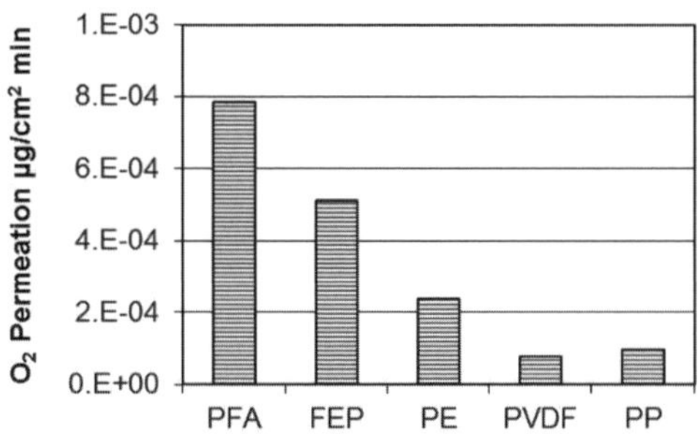

Figure 4.22. *Oxygen permeation rates depending on tubing type (after 30)*

By changing the material, the permeation rate can be drastically reduced (Figure 4.22). Carr [CAR 00] conducted tests that evaluated many types of tubing by measuring the inlet and outlet oxygen concentration from a known tubing length and controlled UPW flow rates. Based on these data, permeation constants through the tubing wall were calculated. PFA turned out to be the "worst" material while PVDF showed the best results. In the case of de-oxygenating mixtures with HF, PVDF is unsuitable due to the high permeation rate of HF [VAN 09]. Ideally, the final de-oxygenation step for chemicals and UPW is integrated or in close proximity to the tool and the process chamber itself should have a nitrogen atmosphere.

Monitoring the DO content requires reliable measurement techniques with detection limits suitable for low concentration. Electrochemical sensors exist that have the potential for *in situ*, continuous and automatic measurements of DO in a liquid. In these sensors, oxygen in the liquid being tested diffuses through a membrane in the sensor into a reaction chamber. The amount of oxygen diffusing depends on the concentration of oxygen in the liquid being tested. The oxygen in the reaction chamber reacts at a cathode to produce an electrical current that is measured. This current is proportional to the oxygen concentration in the liquid being sampled. The application range and accuracy of these sensors differ, depending on the kind of electrode. In case of combination with HF, the electrode in these sensors will be immediately destroyed. Therefore, further development of DO sensors is necessary.

4.4. High-purity chemicals

Microelectronics involves the manufacture of ever-smaller electronic devices in dense arrays that have increasingly complex interconnect structures. Manufacturing at this level requires extremely high-purity chemicals. Indeed, as device dimensions have shrunk, the allowable contaminant level in process chemicals has gone from parts per million through parts per billion to parts per trillion in the most demanding advanced applications. This, in turn, has placed increasingly stringent purity restrictions on the manufacturer, transport and storage of these chemicals and on the metrology used to test and confirm their purity.

High-purity chemicals are used in all phases of semiconductor device manufacturing. Processes requiring such chemicals include wet etching of dielectric and conductive films, removal of residues following dry (plasma) etching of these films, and cleaning of film surfaces before and after other process steps in device manufacturing. The classes of chemicals employed include aqueous solutions (acids, bases and oxidizers), solvents and semi-aqueous (solvent/aqueous mixtures) solutions. Aqueous solutions are normally used in FEOL applications either alone or in mixtures. Individual chemicals are distributed to process tools from central supply systems and are either used as is or mixed on-tool to form the final process solution. Solvents and semi-aqueous solutions are typically used for BEOL applications. They are fed directly to process tools from central distribution systems and used as is.

Chemical purity requirements are driven by a multitude of factors. Surface contamination on wafer substrates falls into several categories: particles, metals, mobile ions and surface carbon. Particles are potentially killer defects in both FEOL and BEOL processes. Shrinking device dimensions at each node reduce both the size and number of particles that can be tolerated on the wafer surface at each process step. The tolerance to metal contamination is dictated by the metal's effect on device performance. The worst forms of metal contamination include highly mobile metals such as Na and K; silicon-compatible metals such as Ni and Cr that either dissolve in Si or easily form silicides and gate oxide integrity (GOI) killers such as Ca and Sr. Mobile ions (such as Na^+ and K^+) have similar severe effects on transistor performance. Surface carbon due to residual organic contamination must be minimized to avoid inadvertent masking or passivation of semiconductor surfaces.

On-wafer cleanliness requirements are enumerated by the *International Technology Roadmap for Semiconductors* (ITRS) and form the basis for the equipment, processes, chemical and UPW purity requirements. The relevant portions of the ITRS FEP and Interconnect surface prep tables can be found on the Internet (http://www. itrs.net/Links/2011ITRS/Home2011.htm). Also, SEMI has developed specifications for many, if not all, of the high-purity chemicals used in semiconductor manufacturing. SEMI specifications define Grades 1–4 in order of increasing purity; these grades correlate with the informal purity designations of VLSI, ULSI, SLSI and XLSI, respectively. The latter designations, in turn, are defined by trace metals contaminant levels of <100 ppb, <10 ppb, <1 ppb and <0.1 ppb, respectively. (Refer to Yield Table YE3 at http://www.itrs.net/Links/2011ITRS/2011Tables/Yield_2011Tables.xlsx.)

These specifications define the baseline requirements for the formulated aqueous, solvent and semi-aqueous blends used in semiconductor processing.

4.4.1. *Techniques of purification for high-purity chemicals*

As ppm impurity levels in high-purity chemicals have given way to ppb and ppt levels, it has become increasingly important to require high purity ever further upstream in the chemical supply chain. Manufacturing products using high-purity feedstock and production methods are preferable to purifying chemicals later in the process. Despite the relatively small volumes used in the semiconductor industry (relative to the chemical industry as a whole), chemical suppliers are recognizing the inherent value and profit potential in supplying high-purity products to this manufacturing sector. To this end, manufacturers of high-purity chemicals have adopted many of the practices of the semiconductor industry, using vessels, piping and other fluid-handling components that are highly resistant to leaching, corrosion and other modes of contamination.

Despite the improved purity of the raw materials, further on-site purification steps are almost always required to make high-purity chemicals meet the stringent specifications of advanced semiconductor manufacturing. Several of the most common methods are discussed in this section.

4.4.1.1. *Filtration*

Filtration for particulate removal is perhaps the most common purification technique for high-purity semiconductor process fluids. Particulate removal is critical in semiconductor manufacturing processes since particles are the primary killer defect in device manufacturing. The ITRS Roadmap defines the "critical particle size" as half the design rule for a given node, typically the DRAM half-pitch. This implies that, at a DRAM half-pitch of 40 nm, the critical particle size is 20 nm or 0.02 µm. Particle contamination can damage devices through multiple mechanisms. Particles can form bridges between device elements such as interconnect lines, producing short circuits. Particles can obstruct BEOL trenches and vias, creating voids during the Cu fill process, resulting in open circuits. Also, particles can act as unwanted photolithography masks, changing the shape of the device structure and potentially altering or destroying device functions.

Particle filtration is employed at all stages of the supply chain including the manufacturing of chemical raw materials and finished goods, the filling of chemical distribution systems and the device manufacturing process.

The filters used for the production of high-purity chemicals must combine an ability to remove extremely small particles with high flow rates and compatibility with a wide range of acidic, basic and solvent chemicals. Filter media are distinguished by their nominal pore size; in general, smaller pore filters are able to capture smaller particulates. However, filters can capture particles below the nominal pore size due to certain mechanisms of particle capture inherent in the media. Membrane filters typically provide a tortuous path for liquid flow, enabling particles to be captured at multiple points in the flow path regardless of particle size. This flow pattern enables fluid flow to remain high despite the increasing loading of particles on the filter media over the filter life. In addition, particles can be adsorbed on to the filter surface, providing a secondary means of particle capture unrelated to the particle size of the contaminant.

Microporous polymer membranes supplied by manufacturers such as W. L. Gore® and Pall® are the most common filter materials. Table 4.10 lists typical filter media.

PTFE (hydrophobic/hydrophilic)
PES/PAS (polyethersulfone/polyarylsulfone)
UPE (ultra-high molecular weight polyethylene)
PP (polypropylene)
Nylon

Table 4.10. *Filter media*

4.4.1.2. *Ion exchange*

An ion exchange is commonly used for the production of UPW as a final step to remove trace metals (for details see section 4.1.3.2); it can also be used in the production of high-purity chemicals, especially those chemicals employed in FEOL processing that demand ppt-level purity [ITR 2011b, Table YE3; http://www.itrs.net/Links/2011ITRS/2011Tables/Yield_2011Tables.xlsx]. Trace metals such as Na, Mg, Ca and Fe can be effectively removed from liquid chemicals using ion exchange. This type of purification and indeed all purification of high-purity chemicals other than filtration is normally performed at the point of chemical manufacture. On leaving the manufacturer, great care must be taken in the packaging, transportation, delivery and distribution of the chemicals to and within the fabrication facility to insure that no metallic or ionic contamination is re-introduced by local chemical handling.

4.4.1.3. *Distillation*

Distillation, most often associated with the production of deposition precursors, is also a common method for purification of the solvents used in semiconductor manufacturing. As with other liquid purification methods discussed in this section, distillation is mostly used in the production of the pure component raw materials rather than in the creation of final products such as formulated chemical blends. Semiconductor-grade solvents, manufactured to ppb- and ppt-standards of purity, are generally cleaner than their industrial-grade counterparts by an order of magnitude or more. Multi-stage distillation is required to achieve these purity requirements.

4.4.1.4. *Other purification techniques*

The manufacture of both organic and inorganic high-purity chemicals can involve multiple purification steps, including recrystallization, sublimation, absorption, extraction and other techniques. The particular combination of purification methods for a specific chemical is often proprietary to the chemical manufacturer.

The main goal of blending, handling and delivering the high-purity chemical to the fabrication is to insure that management of the chemical following its manufacture does not allow any impurities to re-contaminate the chemical prior to use.

4.4.2. *Handling systems for high-purity chemicals*

As noted, the goal of the systems designed to transport and deliver high-purity chemicals throughout the fabrication is to insure that the chemical purity is not degraded prior to use. To ensure this goal, chemical handling within semiconductor fabs has transitioned over time from a largely manual operation to a highly automated one. This automation has provided the added benefits of maintaining much higher purity levels in greater volumes in fabrication chemicals while achieving greater safety in chemical handling.

Chemical distribution systems (CDS), also called bulk chemical distribution (BCD) systems [REI 11]), are designed to achieve a variety of goals:

– to provide a central point for storage of high-purity chemicals destined for delivery throughout the fab;

– to maintain chemical purity;

– to ensure operator safety;

– to avoid exposure of the process equipment and semiconductor products to harmful vapors from the chemicals.

The design of a successful BCD system requires consideration of the leakage characteristics and materials compatibility of a variety of interfaces and wetted surfaces. Interfaces include those between the supply container and the delivery system, those at which the delivery system is manifolded to multiple tools, and those between the BCD and the process tool itself. Each of these interfaces constitutes a connection point between potentially dissimilar materials and each may provide an opportunity for the BCD to "communicate" with the fabrication ambient.

Wetted surfaces in the BCD include the delivery container, process piping to the storage or "day" tank, pump and valving materials, and the flanges and gaskets used to join the various elements. The construction material of the wetted surface is a particularly important issue in that the solvents used in high-purity chemicals, especially UPW, are excellent extraction agents for any metals and organics present in the wetted surfaces. Therefore, extractables in the process piping or other surfaces will, depending on time and solvents employed, leach into the chemicals flowing through the BCD. Materials selection is thus a critical element in BCD design and construction.

Aqueous chemicals are generally contained using fluorinated polymers such as PVDF, PFA and PTFE. These materials are highly resistant to corrosion by many aqueous chemicals and have low levels of extractables (fluorine may be an exception). Organic and formulated solvents and semi-aqueous blends are generally transported in electropolished stainless steel [KER 93]. It is equally important to recognize that all components of the BCD must be constructed of materials compatible with the chemicals being contained. More than one chemical delivery system has required retrofitting or rebuilding when it was found that one or more of the pumps, valves, flow meters, etc. contained incompatible plastic and/or metal parts that were exposed to the chemicals in the flow path.

BCD design also requires close attention to flow path characteristics. Dead legs are to be rigorously avoided as they can lead to bacterial growth in UPW and can serve as a stagnation point that provides the residence time necessary for chemicals prone to degradation to do so. Such degradation allows the high-purity chemical to become a contaminant to itself. This problem exists in many aqueous and solvent chemistries that employ relatively reactive components such as hydrogen peroxide, ascorbic acid and catechol. Material joints, whether welding, flanges, couplings or other types of connections, are another design characteristic that can lead to a major source of contamination. In this case, contamination can occur because of the presence of voids, crevices and discontinuous surfaces. These irregularities can trap the primary chemical along with other foreign materials that then become contaminants as they are washed out of these features.

Proper flow path design and materials of construction of the BCD are thus critical to prevent the contamination of high-purity fabrication chemicals. Also, it is necessary to design the BCD to prevent the most common type of deadly fabrication contamination – particulates. Particles can contaminate high-purity chemicals during manufacturing, transport, in the BCD, and at the point of use. Filtration is used at all of these locations; however, with proper design and construction, particle addition (and consequently the need for filtration) can be minimized in the BCD. The use of proper materials of construction and rigorous cleaning protocols prior to the introduction of high-purity chemicals into the BCD will minimize particle addition during chemical transport. Eliminating moving parts in the flow path through the use of pressure dispensing rather than pumping may be advantageous for reduced particle operation. However, care must be taken not to introduce bubbles that can also act as contaminants. While in-line filtration removes added particulates in BCDs, it can also be an inadvertent source of particles if the filter is shocked by pulsed or intermittent flow. The use of pulsation dampeners and systems designed to constantly recirculate chemicals greatly reduces this issue [REI 11]. The chemical delivery systems used in semiconductor fabs have matured greatly from the glass- and plastic-bottle, hand-poured systems of the past, but the basic goal of the system has not changed. The steady improvements in chemical delivery are driven largely by the increasing purity and productivity requirements needed for advanced device construction. Arguably, a typical fabrication running 50–100 K 45 nm wafer starts per month would be impossible to operate without an effective high throughput, high-purity chemical delivery system.

4.5. Waste management

Semiconductor device fabrication requires large volumes of UPW and considerable amounts of chemicals; therefore, it generates large volumes of wastewater. The outgoing waste is a blend of water and specialty chemicals that needs to be treated and disposed of appropriately.

The first critical point for effective waste treatment is segregation of the waste directly out of the production equipment, not only for safety reasons (incompatibility between products such as acids and solvents) but also for the ease of treatment. As environmental regulations evolve, segregation is also a way to recycle some products and reduce water consumption. For example, 30 years ago, a plant would have two or three waste streams (acid/base and solvents), whereas today a new fab-line can have up to seven categories of waste:

– organic wastes from lithography and some cleaning steps;

– acids and alkaline wastewater, mainly from the rinsing steps of cleaning and etching processes;

– acid solutions containing high amounts of heavy metals from sources such as the copper baths used in electrochemical deposition processes;

– colloidal siliceous waste arising out of CMP processes;

– wastewater from etching processes that contains fluoride;

– phosphorous blends from etching processes;

– wastewater containing ammonia and other nitrogenous products from cleaning and etching processes.

In addition, facilities must control gas emissions. Gas abatement systems are used to break down dangerous molecules, and gas washing units are used to remove gaseous components at various points in the chemical exhaust network. The water used to wash gases is directed to an appropriate wastewater collector.

This section describes general waste dispositions needed to achieve an environmentally friendly solution for the disposal of these different wastes.

Most plants do not prefer to perform onsite waste treatment. They subcontract the difficult treatments destroying or valorizing most waste streams containing, for example, organics or heavy metals, to a third party. Unfortunately, this mode of operation will no longer support the volume of waste produced daily by semiconductor plants. This requires that simple and economical solutions to this problem be implemented directly onsite. In this section, we discuss conventional treatment principles and advanced solutions that have been developed to meet the need to reuse, recycle and reclaim for sustainable manufacturing according to the waste type (Table 4.11).

Before designing a waste treatment plant, it is recommended to perform a careful diagnostic of all volumes and quantities of waste at the point of use. Because this industry also needs to adapt to the market, it is also important to foresee volume variation. It is a good idea to plan a spare containment to save any unusual spare stream [BOE 99]. This will allow:

– buffering of any volume variation due to production increase;

– extra time for analysis of a particular waste in case of a problem at the production level;

– safe re-direction of a waste stream in case of a problem during treatment.

4.5.1. *Emission guidelines*

The allowable emission levels for liquid and gaseous waste treatment systems are defined by local legislation. For example, the United States will refer to the Resource Conservation and Recovery Act (RCRA), while in France legislation refers to the ICPE legislation [ARR 98].

The guidelines are normally expressed in units of concentration to facilitate monitoring, but flux is also determined for each site in order to control the absolute quantity of each element released into the environment. Dilution of effluents to achieve regulated concentration levels is unacceptable. Frequency of the survey is also a legislative parameter.

Waste type/treatment	Physicochemical	Evapoconcentration	Biochemical reactor	Filtration	Adsorption	Neutralization
Diluted waste (acid/base and rinse water)						***
Fluoride and phosphorous waste	***					*
Nitrogenous waste		***	***			*
Colloids (from CMP processes)	***			**		*
Suspended particles (grinding and sawing processes)	***			**		*
Heavy metals	***					*
Diluted organic wastes	**				**	*

*Technique needs to be used in correlation with another technique (for example neutralization is only a final treatment step);
**Technique can be used depending on the waste stream parameters (flow, concentration, composition, etc.);
***Most adapted technique.

Table 4.11. *The major wastewater streams generated by semiconductor fabrication processes and the associated techniques for waste treatment*

Elements to be surveyed include waste flow, biological oxygen demand (BOD), chemical oxygen demand (COD), total suspended particles, and most of the relevant elements used in the industry. The pH and temperature of the effluents should be continuously monitored. This is the principal waste survey.

Additional surveys are also required to trace other organic micropollutants that are likely to be environmentally hazardous, issued from an industrial plant [DIR 00]. This exhaustive list [CIR 09] includes phenols, chloroform, fluoranthene and other micropollutants. Emission limits are close to the detection limit of most laboratories.

4.5.2. *Liquid waste treatment – main technologies*

4.5.2.1. *Neutralization*

Large amounts of acids and bases are present in the wastewater from semiconductor rinse and chemical bath discharges. These acid and alkaline effluent streams are sent to onsite wastewater treatment facilities for neutralization. The most commonly measured and controlled parameter in the wastewater stream is pH. The neutralization system discharge must lie within a pH range of $6.5 < pH < 8.5$ to ensure that the wastewater is released at a pH close to that of water in the environment. The wastewater pH is adjusted by adding either a strong acid (such as HCl or H_2SO_4) if the incoming pH is basic, or a strong base (such as NaOH) if the incoming pH is acidic.

$$HCl + H_2O + NaOH \rightarrow NaCl + 2H_2O \qquad [4.28]$$

Two types of neutralization systems are common in the semiconductor industry: continuous and batch systems. Continuous systems are the mainstay of the industry and can be used in most applications in the semiconductor industry, in particular when the influent contains only strong acids and strong bases and tends to be within a range of $2.0 < pH < 12.0$. Batch systems can be used for any application regardless of the flow or influent chemistry. They are larger and more complicated than continuous systems; however, this is the safest design since there are no limitations on the load handling capability. It can deal with periodic discharges of highly concentrated acids or bases that may drive the influent pH out of the range of $2.0 < pH < 12.0$. Many fabs have modular neutralization systems designed to be able to switch from a continuous batch system to a batch system.

Neutralization has no effect on toxic element concentrations and certain semiconductor processes generate such hazardous elements at unacceptable concentrations. Rinse water is normally not heavily contaminated by chemical processes; however, some chemical bath effluents have high levels of fluorine, phosphor, nitrogenous compounds or heavy metals that must be removed prior to release. In such cases, appropriate pretreatments are mandatory before the pH adjustment step.

Neutralization is, in fact, the last treatment step before wastewater is released from a plant into a river or city wastewater treatment plant.

4.5.2.2. *Physicochemical treatment*

Physicochemical treatments are standard practice in many industries. They are typically based on the precipitation of metal hydroxides at variable pH. At low pH values, hydroxides are soluble; at higher pH values, metal ions form precipitates that can be separated from the solution.

This method of physicochemical treatment employs three sub-steps: pH adjustment; coagulation/flocculation; and filtration/ decantation [BOE 02]. If several metal ions are present in the solution, one will precipitate preferentially to the others. Thus, in practice, it is not easy to treat all metals at the same time due to such preferential reactions. Also, initial concentrations of different metal ions can vary daily, depending on plant activities. Therefore, if different waste streams can be treated through physicochemical processing, it is preferred to go through different reactors rather than treat all in one. Also, for cost-effectiveness, only concentrated effluents are treated in this way.

The case of hydrofluoric and phosphoric acids:

The industry uses large amounts of hydrofluoric acid in fluoride-based processes, hydrofluoric acid being the main etchant for silicon dioxide. Fluoride is therefore ubiquitous, present at bath concentrations that typically vary between 0.1% and 30%. These processes are spread all over device fabs and segregation at the point of use is required, starting from the installation of new equipment. A dedicated drain network will channel concentrated fluoride waste to a specific treatment.

It is common to address fluoride and phosphorous ions in the same process. The waste is passed through a reactor, and the pH is increased to 10.5–11 with the addition of lime ($Ca(OH)_2$) or whitewash. Process control can be difficult since attaining a pH suitable for precipitation must be done without measuring either the pH or the fluoride concentration. At the concentrations normal for these effluent streams, the fluoride ion will attack any glass electrode. A "Jar test" is commonly done on a sample of the waste to provide a rough idea of the amount of lime to be added to the batch, and the pH target is then slightly exceeded to insure a full reaction of all the ions.

The following reaction takes place:

$$Ca(OH)_2 + 2F^- \rightarrow CaF_2 + 2OH^- \quad [4.29]$$

The CaF_2 is produced as a suspension of fine particles in the reactor. Contact time of the reaction is critical, and batch size must be designed as a carefully dimensioned compromise between the time needed for reaction and the waste flow that must be treated every day.

The addition of commercially available coagulants and flocculants helps to segregate the fine particles as bigger and heavier micelles that are more easily separated from the liquid phase. In this step, it is important not to exceed the pH required for precipitation by too much since higher pH will impede the coagulation reaction, increasing the total suspended solids and the fluoride/phosphorus concentrations in the final waste stream.

The last step of the process is either filtration or sludge settling. Following this step, the liquid waste stream is relatively free of fluoride and phosphorous ions and can be directed to the neutralization treatment.

4.5.2.3. *Biotreatment*

Biotreatment is employed for ammonia and other nitrogenous compounds [BOU 09, DEL 07]. Bacteria are grown on a volcanic-type rock (called "fixed bacteria media"), which has the advantage of combining high surface function with filtration capability. A series of reactors, such as those shown schematically in Figure 4.23, is used, alternating reactors with aerobic bacteria and reactors with anaerobic bacteria to follow the described process.

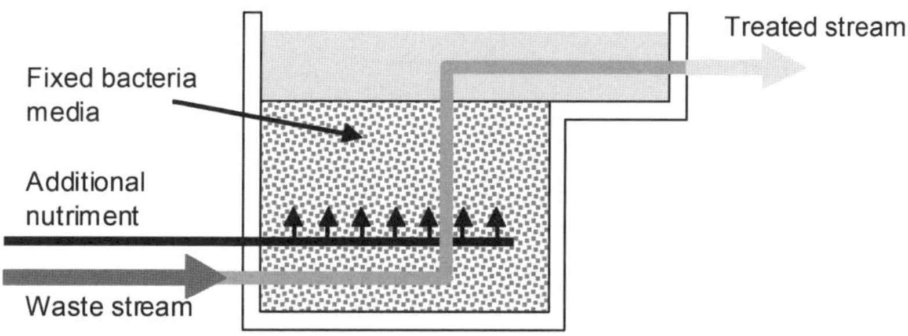

Figure 4.23. *Principle of a bio-reactor*

A three-step process is often used:

1) Reduction of BOD: the reactor is filled with anaerobic bacteria, which nourish themselves on any remaining solvent in the waste stream.

2) Nitrification: the reactor is filled with aerobic bacteria and fed with air or oxygen through a series of nozzles. In these conditions, the oxidation of ammonium subsequently gives nitrite and nitrate anions:

$$NH_4^+ \rightarrow NO_2^- \rightarrow NO_3^- \quad [4.30]$$

3) Denitrification: this last reactor is filled with anoxic bacteria with the controlled addition of an organic solvent (BDO) such as ethanol or methanol. Under these conditions, nitrate anions are transformed into nitrogen gas:

$$NO_3^- \rightarrow N_2 \text{ (gas)} \quad [4.31]$$

The bioprocess is efficient and relatively easy to implement; however, it needs to be carefully controlled:

– It requires temperature stability throughout the year to keep the bacteria active. Bacteria are alive and need to be fed with appropriate amounts of nutriment (either O_2 or an organic solvent). The process also gives poor results on diluted waste.

– As bacteria develop, they progressively seal the porosity of the rock. Redundancy is necessary in each reactor in order to regularly wash the reactor. The setup requires a compromise between the pore sealing and filtration effects. In addition, the use of several small reactors in parallel allows flexibility of the process linked to the waste stream flow.

Again, this treatment is efficient on concentrated effluent as bacteria need to be fed with large quantities of nutriments.

4.5.2.4. *Evapoconcentration*

Evapoconcentration separates water from other components using differences in vapor pressures. This technique is energy-intensive as there is a change of phase, but modern process units have become more efficient. The three main techniques are evaporation under vacuum, multiple effect concentration (Figure 4.24) and mechanical compression.

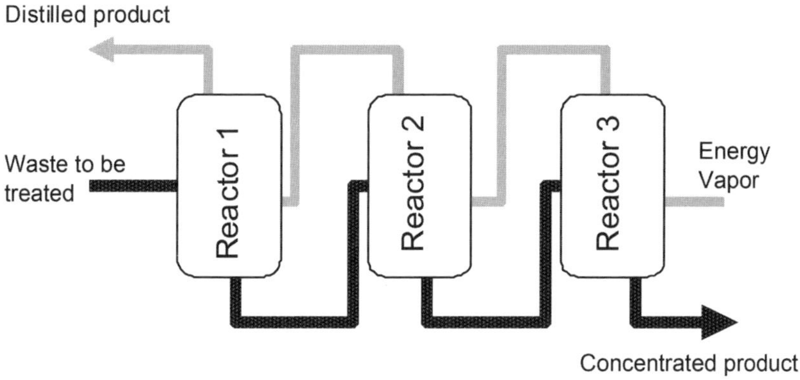

Figure 4.24. *Example of triple-effect counter-current evaporator-concentrator*

The principle is roughly the same for all types of setup: the inputs are the waste to be treated and energy (in the form of electricity, vapor, hot water, etc.). The outputs are the distilled product and the concentrated product. Depending on the case, the product to be valorized will be either the distilled or the concentrated component [ADE 06].

This system is commonly applied in the treatment of wastes containing ammonia or nitrogenous compounds. For example, a waste from NH_4OH-containing processes will be treated in such a system, and gaseous NH_3 (distilled product) will be recovered with high purity permitting valorization of this product. Contrarily, in the case of waste streams containing $(NH_4)_2SO_4$ (arising from gas cleaning systems), water will evaporate at a lower temperature, and the product to be valorized is the concentrated part. In both examples, some of the effluent can be reused in other industries [BIO 06]. Even though evapoconcentration is energy-intensive, it is often a good economical option when the by-product is valuable in other applications.

The downside of this system is the need for a very well-identified incoming waste stream, with two distinguished components and vapor pressures. Once the reactor is designed, very few parameters can be tuned to improve efficiency, except for pressure.

4.5.2.5. *Filtration*

Filtration is a complementary technique to physically remove particles from effluents, discussed in section 4.4. Rough front-end filtration is efficient enough to remove most of the particles or colloids arising from grinding and CMP processes.

Advanced reclaiming can also be set up on grinding effluents, with up to 100% water reuse at the point of use. Some small compact units are commercially available for full water recycling. This has the main advantage of removing hazardous substrate components such as emerging III–V materials from the main waste [WU 05].

4.5.3. *Gas abatement systems*

Waste management must also deal with gas emissions. As seen in section 4.2, many gases are used in the semiconductor industries. All unprocessed gases or by-products from etching or deposition chambers go through the vacuum pumps, then to the plant extraction network [SHE 05].

One gas treatment is required at the final output of the extraction network, as a final treatment before gas is released to the atmosphere. This is generally a standard gas washing unit, in which a range of nozzles sprays basic water (NaOH) with concentrations depending upon pH of the gas. This washing water is then redirected to the neutralization station of the plant, with its amount of solubilized anions and cations.

Regulatory gas analysis is done after this washing step, just before release to the atmosphere.

In specific cases, gas treatment should also be added at the point of use, between the pump equipment and the extraction network. Such a configuration should be carefully analyzed, with a good diagnostic of incoming gases (processes, chamber clean steps, idle mode), in terms of volume and peak flow. Also, the rate of transformation of the feed gas varies from tool-to-tool. For example, the deposition rate of III–V materials is lower than that of nitride deposition. For ease of calculation, the total volume of feed gas is considered to exit the process chamber (i.e. process rate is equivalent to zero), which is the worst-case scenario.

Next, the need for a gas abatement system at the point of use might be motivated by different reasons [BOU 10]:

– Security: is this gas flammable or explosive? What is the risk when mixed with the other gases in the network? (This is typically the case for SiH_4 or H_2.)

– Toxicity and corrosion: is this gas toxic to living organisms? Another point to check is whether it is damaging to the network: gases used for metal etching processes are damaging to the ductwork if not treated outside of the tool. Installing a point of use treatment might be cheaper than changing the ductwork later.

– Environment: does this gas have an environmental impact? If this is a greenhouse gas (GHG) such as NF_3, it might require combustion rather than release into the atmosphere.

Also, mixtures in the network need to be carefully examined. Different gas abatement systems include [GEN 05]:

– Thermal treatment in which residual gases are pyrolyzed to release single elements. Thermal treatment can be achieved either by passing the residual through a torch with CH_4 fuel or through a plasma.

– Chemical treatment, which uses ion exchange to capture contaminant species. In this case, ion-exchange cartridges need to be replaced regularly and disposed of as solid wastes in the appropriate manner.

– Gas washing with soft water: this allows chloride and other reactive ions (such as Cl^-, Br^-, F^-) to solubilize and be evacuated into the "diluted acids" wastewater stream. Very often, this last technique is combined with one of the above.

Emergency release systems (ERS) are also part of gas management. These systems consist of a nitrogen bottle or line able to rapidly release a large quantity of inert gas in order to dilute any dangerous gas. These systems are generally placed near the gas distribution unit or on the by-pass of the abatement system.

The issue of abatement systems is becoming very critical as new gases are used in the fabrication.

4.5.4. *Recycling and reusing*

As worldwide environmental regulations become increasingly stringent, there is a concomitant effort to reduce waste and water consumption.

Recycling UPW consists of re-injecting the used UPW as make up water, while reusing allows UPW to be used in other applications in the fab, where water quality is less stringent [SEM 05].

The first point to note is that equipment manufacturers are seriously tackling the problem of water consumption and wet equipment is now designed in such a way that the recycling UPW loop comes as close as possible to the process chamber, without going through the process chamber. This allows the bacteria issue to be addressed without risk of contamination of the loop by the process in the chamber. In addition, rinse steps are optimized to minimize: to use "just enough" water.

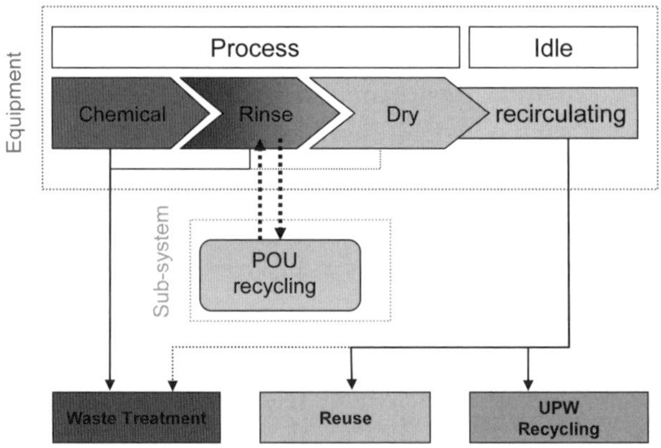

Figure 4.25. *Schematic of a system for recycling or reusing solutions*

General water recycling usually addresses the idle mode, when water passes through the equipment without contacting the process parts. To avoid bacteria growth, a continuous circulation is recommended following the SEMI standards [SEM 09], as discussed in section 4.1.

Depending on the tool configuration, most or all of the water can be returned to the UPW make up. The risk of contamination needs to be carefully studied to determine whether this water can be safely re-injected after the RO unit. If necessary, a separate drain and analysis loop can be designed before re-injecting to the UPW production [DON 98].

"Point of use recycling" must be carefully designed to guarantee water quality through the process. However, some less critical processes can use a dedicated treatment unit. For example, this is the case for grinding activities where segregation of the outgoing water is done and specific treatment is done directly on a closed loop. Systems are available commercially for this specific activity, or can be designed by UPW specialists [PAL 09, DIS 09]. These technologies, first developed for fabs in hydric-stressed countries, are slowly being generalized in other fabs.

Reuse of water for other applications has been widely developed, as water can be reused, particularly for gas washing purposes. This water needs to be re-mineralized to approach soft water characteristics, and injected into the distribution network via a segregated reservoir. This type of reuse only represents 10–20% of the UPW volume, depending on the number of wet scrubbers installed. Other uses of the water need to be considered at the early stages of plant construction.

4.6. List of Abbreviations

AAS:	Atomic Absorption Spectrophotometry
AES:	Atomic Emission Spectroscopy
AISI:	American Iron and Steel Institute
ALD:	Atomic Layer Deposition
ASU:	Air Separation Units
BDS:	Bulk Distribution System
BOD:	Biological Oxygen Demand
BEOL:	Back End Of the Line
CDS:	Chemical Distribution System
CFU:	Colony Forming Unit
COD:	Chemical Oxygen Demand
CMP:	Chemical Mechanical Polishing
CVD:	Chemical Vapor Deposition
DI:	DeIonized water
DBD:	Dielectric Barrier Discharge
DO:	Dissolved Oxygen
EDI:	ElectroDeIonization

ESD: ElectroStatic Discharge
FEOL: Front End Of the Line
FEP: Fluorinated Ethylene Propylene
GF: Graphite Furnace
GOI: Gate Oxide Integrity
IC: Ion Chromatography
ICP: Inductively Coupled Plasma
ITRS: International Technology Roadmap for Semiconductors
LEL: Lower Explosive Level
LFL: Lower Flammable Limit
MB: Mixed Bed
MS: Mass Spectroscopy
MSDS: Material Safety Data Sheet
NIR: Near InfraRed
PE: PolyEthylene
PFA: PerFluoroAlkoxy
PLC: Programmable Logic Controller
POD: Point Of Delivery
POP: Point Of Process
POS: Point Of Supply
POU: Point Of Use
PP: PolyPropylene
ppb: parts-per-billion
ppm: parts-per-million
ppt: parts-per-trillion
PSA: Pressure Swing Absorber
PTFE: PolyTetraFluorEthylene
PVC: PolyVinyl Chloride
PVDF: PolyVinyliDene Fluoride
RO: Reverse Osmosis
SEMI: Semiconductor Equipment and Materials International
SL: SonoLuminescence
SLSI: Super Large Scale Integration
SMR: Steam Methane Reforming
TDS: Total Dissolved Solids
THM: TriHaloMethane
TLV: Threshold Limit Value
TMA: TriMethylAmine
TOC: Total Organic Carbon

TWA:	Total Weight Average
UF:	UltraFiltration
UFL:	Upper Flammability Limit
ULSI:	Ultra Large Scale Integration
UPW:	Ultra Pure Water
UV:	UltraViolet
VMB:	Valve Manifold Box
WGS:	Water-Gas-Shift
XLSI:	Extremely Large Scale Integration
VLSI:	Very Large Scale Integration

4.7. Bibliography

[ADE 06] ADEME, Agence De l'Environnement et de la Maîtrise de l'Energie, L'évapo-concentration, pp. 1–11, 15 June 2006.

[AIM 10] AIMAR P., BACCHIN P., MOREL A., "Filtration membranaire (OI, NF, UF, MFT) aspect théorique", *Techniques de l'ingénieur*, p. J789, 2010.

[AOK 05] AOKI H., NAKABEPPU K., TOMIMORI H., *et al.*, Semiconductor device fabricating method and treating liquid, US patent 6,890,864, 10 May 2005.

[APP 97] APPL M., "The Haber-Bosch heritage: the ammonia production technology", *50th Anniversary of the IFA Technical Conference*, Seville, Spain, 25–26 September 1997.

[ARD 98] ARDEN T., DE DARDEL F., "Echange d'ions", *Dossier Techniques de l'ingénieur*, p. J2860a, 1998.

[ARR 98] Arrêté du 2 Février 1998 relatif aux prélèvements et à la consommation d'eau ainsi qu'aux émissions de toute nature des installations classées pour la protection de l'environnement soumises à autorisation, Journal Officiel, vol. 52, 3 March 1998.

[AVE 05] AVERY C., FEINDEL D., "Static charge control to reduce particle contamination in wafer processing Tools", *SCP 9th International Symposium on Wafer Cleaning and Surface Preparation*, 8–9 June 2005.

[BAK 76] BAKAY C.J., Process for making silane, US patent 3,968,199, 6 July 1976.

[BIL 09] BILOUK S., BROUSSOUS L., NOGUEIRA R.P., *et al.*, "Electrochemical behavior of cobalt in post-via etch cleaning solutions", *Microelectronic Engineering*, vol. 86, no. 10, pp. 2038–2044, 2009.

[BIO 06] BIOCCHI A.L., Réorganisation de la filière de traitement des effluents fluorés, Master degree report, ENGEES, September 2006.

[BOE 99] BOEGLIN J.-C., "Lutte contre la pollution de l'eau", *Techniques de l'ingénieur*, vol. G1, 1250, pp. 1–10, 1999.

[BOE 02] BOEGLIN J.-C., "Traitement physico-chimique de la pollution soluble", *Techniques de l'ingénieur*, vol. G1, 1271, pp. 1–18, 2002.

[BOU 09] BOULENGER P., GALLOUIN Y., "Traitements biologiques anaérobies des effluents industriels", *Techniques de l'ingénieur*, vol. G1305, pp. 1–23, 2009.

[BOU 10] BOUCHARD J., Stratégie rejets gazeux, choix d'un système d'abattement et bilan carbone au LETI, CEA, LETI, Internal report, 2010.

[BUC 03] BUCHANAN D.A., COPEL M.W., FEELY F.R., *et al.*, Nitrogen-rich barrier layer and structures formed, US patent 6,566,281 B1, 20 May 2003.

[CAR 00] CARR G., "The effect of plastic tubing type on oxygen and resistivity measurements in high-purity water", *Ultrapure Water*, pp. 17–21, December 2000.

[CHE 03] CHENG Y.-L., TSAN C.-C., CHENG W.-K., *et al.*, Method for improved cleaning in HDP-CVD process with reduced NF_3 usage, US patent 6,584,987 B1, 1 July 2003.

[CHO 03] CHOI K., LEE C., "Removal efficiency of organic contaminants using ECR H_2 plasma and ECR O_2 plasma", *Journal of the Korean Physical Society*, vol. 42, pp. S702–S707, 2003.

[CHO 06] CHOU C.L., LIN Y.-A., HUAN Y.-A., *et al.*, Wet cleaning method to eliminate copper corrosion, US patent 7,022,610, April 2006.

[CIR 09] Circulaire du 5 janvier 2009 relative à la mise en œuvre de la 2ème phase de l'action RSDE pour les ICPE soumises à autorisation, Annexe 1, Ministère de l'Environnement, 2009.

[DAH 84] DAHM J.C., FRANKA J.G., Process for improving nitride deposition on a semiconductor wafer, US patent 4,428,975, 31 January 1984.

[DEG 89] DEGREMONT S.A., *Mémento, Technique de l'eau*, Lavoisier, Tec et Doc, 1989.

[DEL 07] DELPORTE C., "Traitements biologiques aérobies des effluents industriels", *Techniques de l'ingénieur*, vol. G1300, pp. 1–15, 2007.

[DIR 00] Directive Européenne 2000/60/CE du Parlement Européen et du Conseil du 23 Octobre 2000, Journal Officiel du Conseil Européen, 22 December 2000.

[DIS 09] DISCO CORPORATION, "A super-compact deionized water recycling unit for advanced environmental performance", 2009 available at http://www.dicing-grinding.com.

[DON 98] DONOVAN R.P., MORRISON D.J., TIMON R.P., "Design of recycling system for spent rinse water in Sandia's microelectronics development laboratory", *International SEMATECH*, November 1998.

[EIG 00] EIGA, The hazards of oxygen enriched atmospheres, IGC Doc. 400/00, 2000.

[EIG 06] EIGA, Gaseous hydrogen stations, IGC Doc 15/06/E, 2006.

[EIG 08] EIGA, Safe practices for cryogenic air separation plants, IGC Document 147/08/E, 2008.

[EIG 09] EIGA, Best available techniques for hydrogen production by steam methane reforming, IGC Document 155/09/E, 2009.

[EIG 10] EIGA, Code of practise silane, IGC Doc. 160/10/E, 2010.

[FRA 10] FRANSILLA S., *Introduction to Microfabrication*, John Wiley & Sons Ltd, UK, pp. 153–154, 2010, available at http://www.uread.com/book/introduction-microfabrication-sami-franssila-fransilla/9780470851050.

[GEN 05] GENG H., *Waste Gas Abatement Systems, Semiconductor Manufacturing Handbook*, McGraw-Hill Handbooks, 2005.

[GOT 00] GOTTSCHALK C., LIBRA J.A., SAUPE A., *Ozonation of Water and Waste Water: A Practical Guide to Understanding Ozone and its Application*, Wiley-VCH, Weinheim, Germany, 2000.

[GOT 04] GOTTSCHALK C., SCHWECKENDIEK J., "Using dissolved ozone in semiconductor cleaning equipment", *Micromagazine*, March 2004.

[GOT 10] GOTTSCHALK C., SCHMIDT T., WOLKE K., "Controlled DI water CO_2 gasification to eliminate ESD effects and corrosion: control of contamination, development in equipment", *Proceedings of the 10th Surface Preparation and Cleaning, Conference*, Austin, TX, 2010.

[GOT 11] GOTTSCHALK C., BRAMMER U., LE TIEC Y., "Deoxification of liquids for advanced semiconductors processing", *ECS Transactions, Semiconductor Cleaning Science and Technology 12*, HATTORI T., RUZYLLO J., MERTENS P. and NOVAK R. (eds), vol. 41, no. 5, pp. 287–294, 2011.

[GOT 12] GOTTSCHALK C., "The challenge to reduce dissolved oxygen for advanced processing", *Future Fab*, vol. 41, pp. 60–63, 2012.

[GOV 11] GOVINDARAJAN R., KESWANI M., RAGHAVAN S., "Investigation of galvanic corrosion characteristics between tantalum nitride and poly silicon in dilute HF solution", *ESC Transactions*, vol. 41, no. 5, pp. 15–21, 2011.

[HAL 08] HALLADAY J., TEETER B., NEWCOMB R., *et al.*, "Elimination of ESD defects using $DICO_2$", *Proceedings of the 8th Surface Preparation and Cleaning, Conference*, Austin, TX, 2008.

[HIR 06] HIROMOTO O., OHI T., Process for producing nitrogen trifluoride, US patent 7,018,598 B2, 28 March 2006.

[HOI 83a] HOIGNE J., BADER H., "Rate constants of reactions of ozone with organic and in inorganic compounds in water – I. Non dissociating organic compounds", *Water Research*, vol. 17, pp. 173–183, 1983.

[HOI 83b] HOIGNE J., BADER H., "Rate constants of reactions of ozone with organic and in inorganic compounds in water – II. Dissociating organic compounds and Radicals", *Water Research*, vol. 17, pp. 185–195, 1983.

[HOI 83c] HOIGNE J., BADER H., "Rate constants of reactions of ozone with organic and in inorganic compounds in water – III. Inorganic compounds", *Water Research*, vol. 19, pp. 993–1004, 1983.

[IAE 12] IAEA, International Atomic Energy Agency, "Neutron transmutation doping of silicon at research reactors", *IAETA-TECDOC-1861*, Vienna, pp. 6–8, 2012.

[ITR 11a] ITRS, Environment, Safety and Health, Table ESH6A-B, 2011.

[ITR 11b] ITRS, Yield enhancement, 2011.

[JAN 96] JANSEN H., GARDENIERS H., DE BOER M., *et al.*, "A survey on the reactive ion etching of silicon in microtechnology", *Journal of Micromechanics and Microengineering*, vol. 6, no. 1, pp. 14–28, 1996, available at http://iopscience.iop.org/0960-1317/6/1/002.

[KER 93] KERN W., *Handbook of Semiconductor Wafer Cleaning Technology: Science, Technology, and Applications*, Noyes Publications, pp. 394–400, 1993.

[KUC 10] KUCERA J., *Reverse Osmosis, Design, Processes and Applications,* Wiley, 2010.

[KUH 11] KUHN K., "Variability in nanoscale CMOS technology", *Science China Information Sciences*, vol. 54, no. 5, pp. 936–945, 2011.

[KUM 11] KUMARI S., KESWANI M., SINGH S., *et al.*, "Effect of dissolved CO_2 in de-ionized water in reducing wafer damage during megasonic cleaning in MegPie", *ECS Transaction*, vol. 41, no. 5, pp. 93–99, 2011.

[LAU 09] LAUERHAAS J., "Reduced oxygen cleaning process for advanced Cu/low-k integration", *Proceedings of the 9th Surface Preparation and Cleaning, Conference*, Austin, TX, 2009.

[LEN 00] LENGYEL I., JENSEN K.F., "A chemical mechanism for in situ boron doping during silicon chemical vapor deposition", *Thin Solid Films*, vol. 365, pp. 231–241, 2000.

[LEV 89] LEVY R.A., *Microelectronic Material and Processes*, Kluwer Academic Publishers, Dordrecht, the Netherland, pp. 32–34, 1989, available at http://books.google.fr/books?id=wZPRPU6ne7UC&printsec=frontcover&hl=fr#v=onepage&q&f=false.

[LI 00] LI F., BALAZS M.K., DEAL B.E., "Wafers/substrate: thickness measurement of sub-monolayer native oxide films on silicon wafers", *Solid State Technology*, vol. 43, no. 2, pp. 87, 2000.

[LIE 03] LIEV I., CHYAN H., "Federal standard 2009E for cleanroom: an obsolete document!", *Development in refrigeration, air conditioning, heating and ventilation systems technical seminar*, Kuala Lumpur, Malaysia, 24–25 September 2003.

[MAN 00] MANNING T., "Production of ozone in an electrical discharge using inert gases as catalysts", *Ozone, Science and Engineering*, vol. 22, no. 1, pp. 53–64, 2000.

[MAN 01] MANNING T., HEDDEN J., "Gas mixtures and ozone production in an electrical discharge", *Ozone, Science and Engineering*, vol. 23, no. 2, pp. 95–103, 2001.

[MAT 07] MATZ P., HURD T., CUNNINGHAM K., *et al.*, "Elimination of electrostatic charge-induced 'punch-through' defects using single wafer cleaning processes", *Proceedings of the 212th ECS Meeting*, 7–10 October 2007.

[MAY 64] MAYA W., "Synthesis of trifluorure nitrogen from the elements", *Inorganic Chemistry*, vol. 3, no. 7, pp. 1063, 1964.

[NAG 95] NAGAYAMA T., KADUMURA S., Dry etching method, US patent 5,419,809, 30 May 1995.

[NAK 98] NAKASHIMA H., OMAE K., TAKEBAYASHI T., *et al.*, "Toxicity of silicon compounds in semiconductor industries", *Journal of Occupational Health*, vol. 40, pp. 270–275, 1998.

[NAL 02] NALLAN P.C., KUMAR A., CHININ J.D., Plasma etching of silicon using a chlorine chemistry augmented with sulfur dioxide, US patent 6,415,198 B1, 2 July 2002.

[NEL 97] NELSON S.L., "The effects of oxygen passivation of silicon by cleaning processes on contamination and defects", *Proceedings of the Electrochemical Society,* vols. 97–35, pp. 38–45, 1997.

[OHN 06] OHNO H., OHI T., Process for producing nitrogen trifluorine, US patent 7,018,598 B2, 28 March 2006.

[OSH 89] OSHA, Final Rule on Air Contaminants Project: Carbon Dioxide, comments from the 19 January 1989, available at www.cdc.gov.

[OWE 95] OWENS D.L., *Practical Principle of Ion Exchange*, 2nd ed., Tall Oaks, 210 p., 1995.

[OZO 94] OZONEK J., FIJALKOWSKI S., POLLO I., "A new approach to energy distribution in industrial ozonizers", *Proceedings of the International Ozone Symposium: Application of Ozone in Water and Wastewater Treatment*, Warsaw, Poland, pp. 218–227, 26–27 May 1994.

[PAG 03] PAGLIARO R.H., DOTY M.L., KING D.M., Stable, oxide-free silicon surface preparation, US patent 6,620,743 B2, 16 September 2003.

[PAL 09] PALL, "Water reclaim in silicon processing", *Application Bulletin*, September 2009.

[PEN 11] PENA LOPEZ JORGE A., MANOUSIOUTHAKIS VASILIOS I., "Natural gas based hydrogen production with zero carbon dioxide emissions", *International Journal of Hydrogen Energy*, vol. 36, no. 20, pp. 12853–12868, 2011.

[PER 07] PERNEL C., BILOUK S., NOGUEIRA R.P., *et al.*, "Electrochemical behavior of copper and cobalt", *Advanced Metallization Conference Proceedings*, Albany, NY, 2007.

[POR 89] PORTER M.C., *Handbook of Industrial Membrane Technology*, William Andrew Publishing, 626 p., 1989.

[REI 11] REINHARDT K.A., REIDY R.F., *Handbook of Cleaning for Semiconductor Manufacturing: Fundamentals and Applications*, Scrivener Publishing LLC, pp. 177–180, 2011.

[SAK 09] SAKIMA H., Plasma processing methods and apparatus, US patent 7,481,230 B2, 27 January 2009.

[SEM 05] SEMI, Guideline for treatment of reuse water in semiconductor processing, F98-0305, 2005.

[SEM-11-F63] SEMI, "Guideline for ultrapure water used in semiconductor processing", F63-0211, 2011.

[SEM 12a] SEMI, Specifications for gases, C3-0699, 2012.

[SEM 12b] SEMI, Specification for polymer materials and components used in ultrapure water and liquid chemical distribution systems, F57-0312, 2012.

[SHE 76] SHEWCHUN J., KING F.D., Method of gas doping of vacuum evaporated epitaxial silicon films, US patent 3,949,119, 6 April 1976.

[SHE 05] SHERE J.M., *Semiconductor Industry: Wafer Fab Exhaust Management*, CRC Press, 2005.

[SPI 09] SPICER R., HAWTHOME J., PETERS D., *et al.*, "Controlling process-induced charging heightens productivity", 2009, available at http://www.semiconductor.net/article/CA6641340.html, 2009.

[STA 83] STAEHELIN J., HOIGNE J., "Reaktionsmechanismus und Kinetik des Ozonzerfalls in Wasser in Gegenwart organischer Stoffe", *Vom Wasser*, vol. 61, pp. 337–348, 1983.

[VAN 09] VAN SCHOONEVELD G., GRANT D., CARRIERI D., *et al.*, "Modeling of component lifetime based on accelerated acid gas permeation measurements", *Ultra Pure Fluid and Wafer Cleaning Conference*, February 2009.

[VER 00] VERHAVERBEKE S., PAGLIARO B., "Isothermal Epi-Si deposition at 800°C", *Proceedings of the Electrochemical Society*, vols. 99–36, pp. 445–451, 2000.

[WAL 06] WALTHER S.R., MEHTA S., JEONG U., *et al.*, Shallow-junction fabrication in semiconductor devices via plasma implantation and diffusion, US patent 0,205,192 A1, 14 September 2006.

[WU 05] WU M., SUN D.D., TAY J.H., "Characterisation and ultrafiltration of semiconductor InP waste water for recycling", *Environmental Technology*, vol. 26, no. 1, pp. 111–119, 2005.

[YAW 98] YAWS C., *Chemical Properties Handbook: Physical, Thermodynamics, Environmental Transport, Safety and Health Related Properties for Organic and Inorganic Chemical*, McGraw-Hill, 1998.

[YE 98] YE Y., MA X.D., YIN Z.G., *et al.*, Method and apparatus for cleaning by-product from plasma chamber surfaces, US patent 5,756,400, 26 May 1998.

[ZHA 97] ZHAO J.W., WANG Z., CATABY W.G., Plasma clean with hydrogen gas, US patent 5,660,682, 26 August 1997.

[ZUL 83] ZULEHNER W., "Czochralski growth of silicon", *Journal of Crystal Growth*, vol. 65, nos. 1–3, pp. 189–213, 1983.

Quantity Symbol

C: concentration
H: enthalpy
R: ideal gas constant
T: temperature

Greek Symbols

Π: osmotic pressure

Chapter 5

Surface Functionalization for Micro- and Nanosystems: Application to Biosensors

5.1. Introduction

Nowadays, chemistry plays an important role at the interface between organic and inorganic substrates and the chemical or biological environment to develop hybrid micro- and nanosystems. Effectively, the phenomena in micro- and nanosystems are influenced by the device walls due to a high surface to volume ratio. Thus, there is a great need to chemically tune the properties of a material into a function such as sample conditioning, separation, recognition and detection of species. Chemists have developed a chemistry "tool box", which is focused on the creation of a chemical interface and designed for integrated systems by taking into account the fabrication and application constraints (see Figure 5.1). This approach can cover a very large set of applications such as environment, well-being, medicine, energy and information [PRA 09].

In this chapter, we present the different materials used in micro- and nanosystems, their ability to be modified by different chemical or physical methods, the resulting chemical functions after complete processing and some examples of applications.

Chapter written by Antoine HOANG, Gilles MARCHAND, Guillaume NONGLATON, Isabelle TEXIER-NOGUES and Francoise VINET.

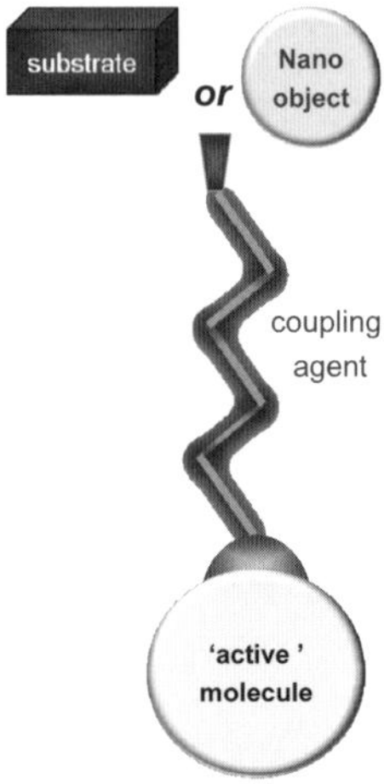

Figure 5.1. *Chemistry "tool box"*

5.2. Materials

The implementation of chemistry in micro- and nanosystems will provide analytical results with high accuracy in small volumes (nL–mL) by operating several functions. Such an approach will eliminate the potential errors induced by the great number of process manipulations. Among the different microsystem developments [BER 10], an example of an ideal point-of-care device [GUB 12] is shown in Figure 5.2. Two main parts constitute the system: a microfluidic network and a cap with an inlet and an outlet; their assembly defines the sample volume to be treated. As several functions are integrated (sample conditioning, separation, recognition and detection of species), this will introduce the management of various materials such as glass, silicon, metals, polymers, tissues and paper.

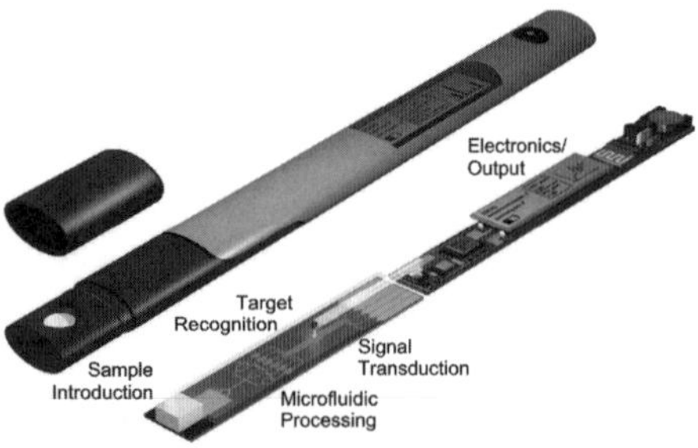

Figure 5.2. *Ideal point-of-care device*

The compounds to be analyzed are quite diverse because they are sampled from soil, water, air or the body. This requires a great adaptability of the chemical properties of the surfaces. Because the ability of these devices to tune the interactions, which take place on the surfaces, depends on the physical properties of the material (optical, electrical, magnetic, etc.), an adequate chemical modification is mandatory. Moreover, the fabrication constraints, including the technological environment or sterilization for portable and *in vivo* applications, have to be taken into account for further chemical implementation on the device materials.

5.2.1. *Metal oxides and semiconductors*

Silicon microfabrication allows cost-effective mass production of micro- and nanostructured surfaces to introduce microfluidic patterns. By using standard deep reactive ion etching, a high surface to volume ratio is obtained in silicon material [GUB 12]. By tuning the parameters during the lithographic mask (length, space) and the depth during the etching step, a broad range of microstructures can be designed as a function of the application needs. By designing the space between the pillars, a microstructured silicon substrate can be applied to either analytical systems for proteomics [SAR 05] or liquid–liquid extraction for chemical detection [BER 09]. Figure 5.3 shows the general pattern of the fluid flow and a cross-section of the micropillars. This is a generic approach for either the hydrolysis of proteins or chemical and biological (red cells, white cells, peptides, DNA, RNA, etc.) separation. In the case of chromatography, a minimal space of up to 1 μm is required between the pillars to enhance the interaction between molecules and surfaces.

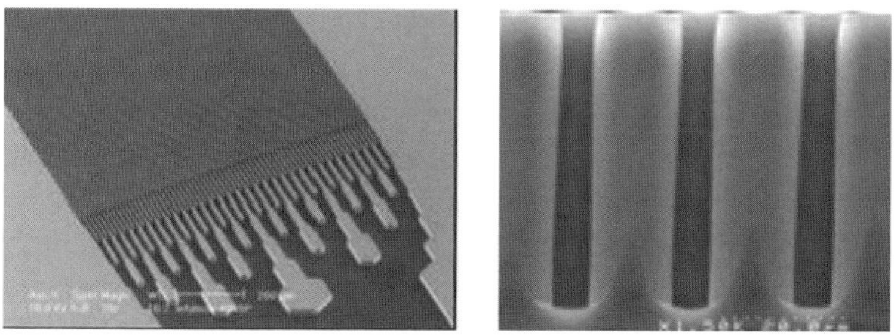

Figure 5.3. *View of a proteomics reactor (left); micropillar cross-section, SEM image before oxidation: a good etching homogeneity is achieved all over the thickness of the layer (right)*

In the field of environmental analysis, it is essential to control the quality of liquids, especially water, and to detect and identify traces of contaminants. Heavy metals such as lead, mercury and chrome are toxic even at very low concentration, and water companies need to check their concentration in the water they distribute. Figure 5.4 shows the design of a microsystem for water analysis in which the micropillars allow fluid stabilization.

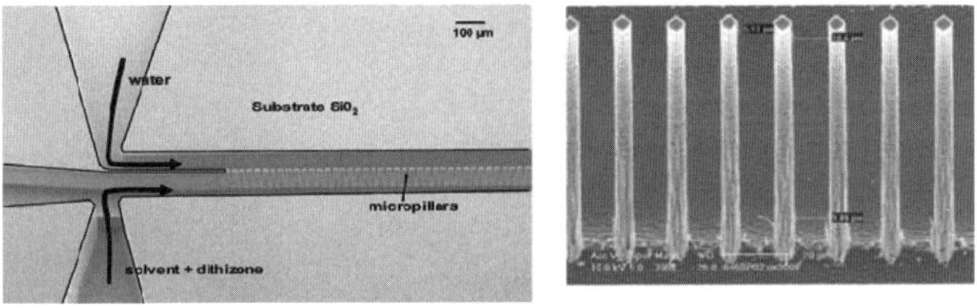

Figure 5.4. *Left: water and solvent are co-flowing through the system separated by micropillars. Right: close up of the micropillars*

The perfectly ordered surface enhances the capability of the fluids to go straight through the microsystem compared with a disordered packed bed system [BIL 05] as shown in Figure 5.5. This is of particular interest in a species separation mechanism.

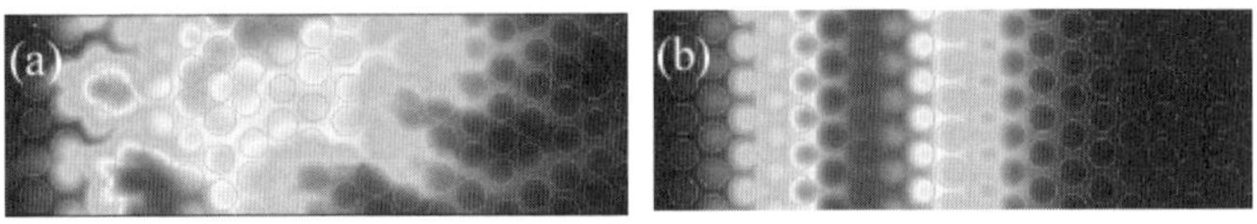

Figure 5.5. *(a) Computational fluid dynamic study of the flow through disordered pack bed and (b) perfectly ordered packed bed*

These structures can undergo further oxidation to be easily chemically modified, due to a hydroxylated chemical function on the surface (see section 5.3), and to greatly enhance the liquid handling performance of complex fluidic microsystems or to elaborate a specific function. Other metal oxides, such as indium, nitride and hafnium, are also materials of importance. The conductivity and transparency of indium oxide offer specific properties for biosensing [BER 06]. Silicon oxinitride or hafnium oxide are also widely used in conjunction with silicon oxide in the production of planar waveguides [WU 06]. Glass and quartz afford the advantage of ultraviolet (UV) transparency compared to other cited materials, particularly when optical detection is required.

The thermal conductivity [CHE 96] of silicon is another important property when thermally assisted chemical or biochemical reactions have to be performed in one of the modules of the microsystem.

Apart from the microstructuration, silicon process technology is also very promising in the development of nanostructures for fluidics and detection purposes [CRA 00, LI 07, AGA 11]. The recent improvements in fabrication create freestanding objects, channels and pores with nanometric dimensions. On a molecular scale, these nanostructures provide new applications for actuating and sensing molecular interactions which can be integrated into microsystems. In comparison, nanoparticles are "stand-alone" nanosystems.

Semiconductor nanocrystals (NCs), also called quantum dots (QDs), are of great importance to biomedical labeling, light-emitting diodes (LEDs), solar cells, lasers and sensors. II–VI and IV–VI semiconductors such as CdSe, CdTe and PbSe as well as their alloys have been intensively studied; however, despite their appealing optical properties, the intrinsic toxicity of these materials requires new developments for compatible environmental regulation [REI 09].

5.2.2. *Carbon nanotubes and silicon nanowires*

Semiconductor nanowires and nanotubes exhibit novel electronic and optical properties owing to their unique structural one-dimensionality and possible quantum confinement effects in two dimensions. In parallel to the fabrication of these novel materials, the improvement of the nanocharacterization methods ensures the availability of reproducible and well-characterized structures. Silicon and carbon belong to the same chemical family due to their hydrogenated surface, which can be modified chemically for large surface area improvement [FON 11], detection [WAN 11] or brain–computer interface [LIU 11] purposes. As shown in Figure 5.6, the technology ensures the *in situ* growth of carbon nanotubes in liquid chromatography microcolumns. The resulting property is an increase of the developed surface and, thus, of the retention capacity of the column [FON 11]. The carbon nanotubes can be used with or without chemical modification. Without modification, their graphite-like structure causes hydrophobic interactions.

Figure 5.6. *(Left) SEM picture of 1.2 µm thickness on micropillars and (right) filled channel in situ grown carbon nanotubes*

Semiconductor nanowires and nanotubes exhibit novel electronic and optical properties owing to their unique structural one-dimensionality and possible quantum confinement. In a field effect transistor (FET) configuration, label-free detection of biological species has been demonstrated [PAT 06, LIE 11, POG 07, WON 08]. As shown in Figure 5.7, a silicon nanowire of thickness 25 nm and width 30 nm is achieved by using microelectronics technology [HAL 09].

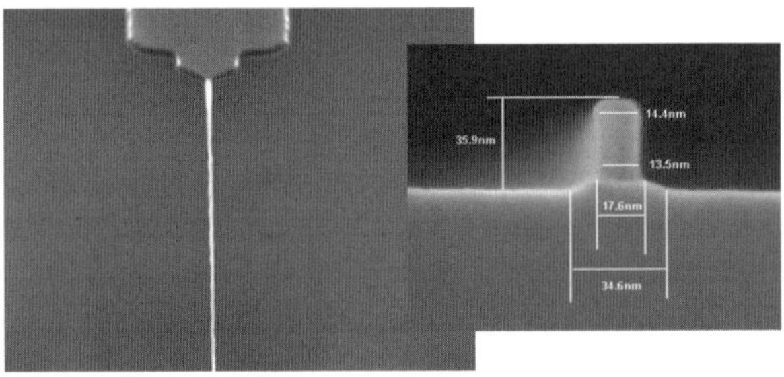

Figure 5.7. *(Left) SEM top view picture of a silicon nanowire and (right) SEM cross-section of a silicon nanowire*

5.2.3. *Metals*

Noble metal substrates are used in microsystems because they are stable, inert and easy to functionalize in mild conditions. They offer the capability of both implementing the chemistry in the devices and detecting events occurring inside them [MER 08] due to the optical property of the metal. To develop immunodetection through impedance measurements, well-controlled electrodes can be obtained with gold material [AME 09]. Several studies show that neural electrodes are essential tools for the study of the nervous system and related diseases for sensitive detection of neural electrical activity. Gold is one of the best candidates due to the low impedance of this metal [SEK 10].

The literature on noble metal nanoparticles is already very dense, with gold nanoparticles [HUA 07] being particularly well investigated. Their optical properties as well as their scalability down to nanometric size have enhanced their use in bioapplications.

5.2.4. *Polymers*

Over the last decade, various polymers have emerged as possible materials for MEMS structures using high-throughput fabrication techniques like hot embossing or injection molding [HEC 04]. Polymers are complementary to previously described materials (silicon and derivative metals) and are mainly used for disposable liquid handling systems. The main issue is the stability of their surface modification. Effectively, the chemical composition of silicon, oxides and metals are quite reproducible compared to commercially available polymers.

Among these materials, cyclo olefin copolymer (COC) presents an excellent UV transparency, a low autofluorescence and a high mechanical strength. However, due to its natural hydrophobicity, without any treatment, the processed compounds can be adsorbed, resulting in sample loss and degradation of the analytical performances of the system [FRE 07].

Another easy way to process polymer is polydimethylsiloxane (PDMS) [ANA 00]. PDMS has been widely used in microfluidics for rapid prototyping. It bonds strongly to glass, has good optical transparency and elastomeric properties, and is permeable to gases. However, the hydrophilic surface modification performance of PDMS is known to recover its hydrophobicity shortly after oxidation modification. A reproducible and long-term surface treatment is a key issue [YAO 12]. Recent results have been published and concern the performances of polyurethane (PU)-based microfluidic devices as illustrated in Figure 5.8 [WU 12].

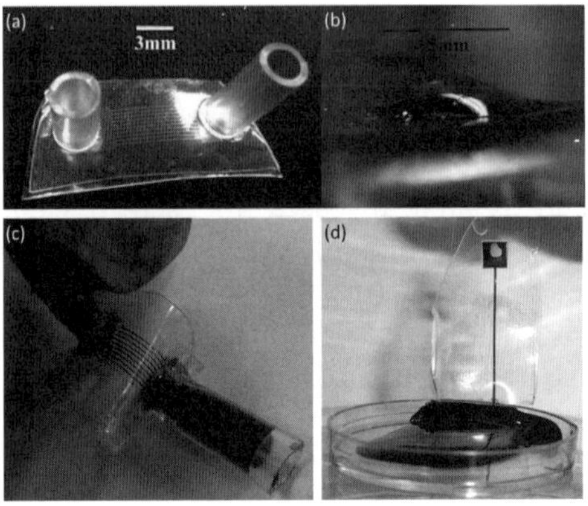

Figure 5.8. *Photos of PU-based microfluidic devices: (a) fluidic connections, (b) surface characterization, (c) flow through, and (d) capillarity experiments*

Polyimides are widely used in microelectronics as a dielectric support for electrode processing. These polymers are thermally stable (> 300°C), present minimal swelling in the presence of water, can be processed in thin and conformable films, and up to now are biocompatible and do not induce cytotoxicity, which make them suitable for implantable conductive systems [BEU 00].

5.2.5. *Paper and tissues*

As new applications are emerging, such as early diagnosis in remote underdeveloped area or home tests, low-cost, portable, and easy-to-use systems need to be developed. Paper [MAR 08b] appears to be a well-adapted material because it is thin, lightweight, and easy to stack, store and transport. Because it is cellulose based, it can be easily chemically modified and its white color improves the contrast of colorimetric detection. Moreover, the printing techniques are quite compatible with this flexible material for patterning. All these factors will contribute to the development of paper-based devices [MAR 10].

In Figure 5.9, we can observe preliminary experiments with 3D channels in paper support [MAR 10]. Wax has been printed to define the channels. The intrinsic hydrophilic property of the cellulosic material is compatible with the flow-through, water-based solutions. Two different dyed solutions have been applied as a proof of the concept of complex microfluidic devices.

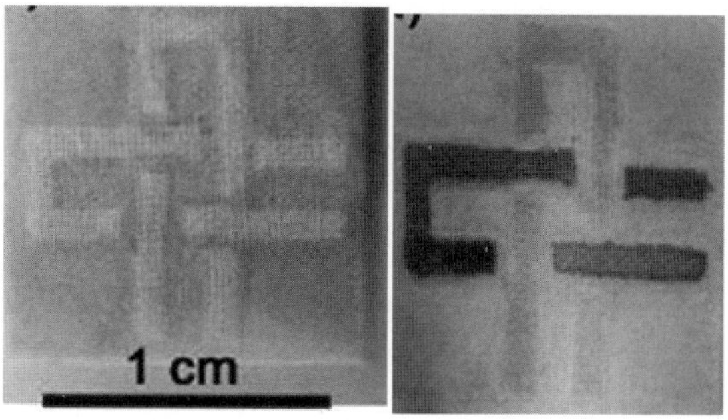

Figure 5.9. *Microfluidic channel wax treated without (left) and with flow-through solutions with two different dyes (right)*

This paper-based technology may also find applications in developed countries for home tests or in specific environments for quick tests.

The integration of micro- and nanotechnologies on textiles [LUP 06, COY 10] or wound dressing is ongoing for sensing, actuating, communicating, processing, and power sourcing. The wound dressing area [ZHO 11] will also benefit from these technologies.

5.3. Functionalization process

To tune the surface properties of the previously described materials, an activation process followed by various processes has been developed and presented here.

5.3.1. *Activation and cleaning treatments*

A pretreatment is often required to activate surface materials before subsequent functionalization and to clean contaminants from substrates. Indeed, contaminants such as particles (dust, bacteria, pollen and clothing particles), inorganic contaminants (salts, ions and heavy metal) and organic contaminants (smog, skin oil, fluxes, lubricants and solvent vapors monomers from plastic) are commonly adsorbed on surfaces. It is necessary to remove them in order to ensure the best functionalization. In the case of silicon-based substrates (glass, quartz, silicon dioxide, silicon nitride and PDMS), a surface hydroxylation [DSO 99, DUG 03] is required to create silanols (Si-OH) before the silanization process (see Figure 5.10).

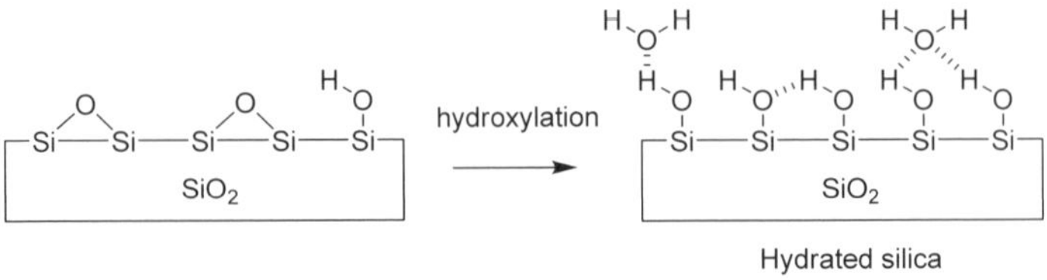

Figure 5.10. *Schematic representation of the hydroxylation of a silicon oxide surface*

In the same manner, cleaning treatments are used to modify chemical properties of a polymer to activate the surface and allow further functionalization. Activation and cleaning treatments can be performed in the liquid phase or can use plasma depending on the type and sensitivity of the substrate.

5.3.1.1. *Wet cleaning treatments*

Wet treatments are widely used to clean and activate material surfaces. They are extensively used in the semiconductor industry [MOY 09, ISH 86]. Depending on the substrate, acidic or basic solutions can be used. Hydrofluoric acid or diluted hydrofluoric acid removes oxides from the area of interest, etches silicon oxides and dioxides, and reduces metal contamination of the surface, whereas phosphoric acid is dedicated to silicon nitride surface. Hydrochloric, nitric and sulfuric acids act as oxidizing agents and react with metallic contaminants. Potassium hydroxide etches silicon, labeling and silicon oxides to a lesser extent. Ammonium hydroxide is less aggressive and reacts with silicon only at 80°C. Hydrogen peroxide in combination with acidic or basic solution is strongly reactive. Piranha clean, also called "Caro" solution, is composed of sulfuric acid and peroxide hydrogen. This mixture, heated at 140°C, oxidizes and dissolves organic contaminants adsorbed on wafers. Ammonium hydroxide with hydrogen peroxide oxidizes and slightly etches to undercut and remove particles from the surface.

The most widespread wet processing in the semiconductor industry is called RCA [KER 70]. It mainly consists of two steps: the first step is the oxidation step of the bare wafer surface in an alkaline medium made of an ammonia and hydrogen peroxide mixture called SC1 (standard clean #1). This first treatment results in the rapid growth of a Si oxide layer whose surface charge contributes to eject the adherent nanoparticles. But the alkaline properties of this solution present the inconvenience of a substrate etching at the nanoscopic scale and induce an unfavorable effect on surface microroughness. In the second step, an acidic reactant called SC2 (standard clean #2), a mixture of HCl with H_2O_2, is used to eliminate the residual traces of metal contaminants. An additional step with ultrapure water dilutes chemicals and rinses substrates after chemical cleaning.

Using a large volume of water and chemicals, wet cleaning is a major polluting source, and the optimization of wet cleaning steps is necessary. New environmentally friendly technologies have been developed using fewer chemicals, recycling water, or processing in dry conditions. Mechanical cleaning technologies such as centrifugal spraying, water jets, brush scrubbing and pressurized fluids can also be used. Furthermore, many advances are based on the use of the diluted chemistries and ozonated ultrapure water as a replacement for hydrogen peroxide or even sulfuric-based mixtures. Chapters 3 and 4 describe all these wet treatments in greater depth.

5.3.1.2. *Plasma treatments*

Plasma treatment is probably the most versatile surface treatment technique and is widely used to clean and activate a variety of materials such as silicon, steel, PDMS and polymers before a functionalization or bonding step [LIN 97, ZHO 10, PRA 09]. Different types of gas, such as argon, oxygen, nitrogen, ammonia, fluorine, carbon dioxide and water can be used. For example, oxygen plasma treatment can increase the surface energy and thus the wettability of the materials' surface, whereas fluorine plasma treatment can decrease the surface energy and improve the chemical inertness. The plasma treatment is uniformly spread on all the surfaces of the materials. Moreover, modification is usually confined to the top several hundred Ångstroms and does not affect the bulk properties. This technique requires a vacuum system which increases the cost of operation [PRA 09, CHA 96]. Surface modification is mainly due to the reaction between active gas-phase species and surface species. Organic materials are removed from the surface by chemical reactions and physical etching on the surface to form volatile products. Oxygen-containing plasmas are used to remove organic contaminants from surfaces. Etching differs from cleaning only in the amount of materials removed from the surface. Oxygen plasmas and oxygen- and fluorine-containing plasmas are frequently used for etching polymers.

Plasma is a state of matter similar to gas containing charged and neutral species, including some of the following: electrons, positive ions, negative ions, radicals, atoms, and molecules. Ionization is generally induced by a strong electromagnetic field applied with a laser or microwave generator. Thus, the generation and sustenance of plasma requires an energy source to produce ionization. The plasma process is extremely complex depending on the material properties, RF volume, RF frequency, power level, gas flow rate, gas composition, gas pressure, sample temperature and reactor geometry. All these parameters render a good understanding of the interactions between the plasma and the surface that is difficult to achieve. Typical conditions of oxygen plasma cleaning are 20–500 sccm O_2 flow and 200–600 W for 2 to 5 min.

5.3.2. *Silanization processes*

Silanization is a functionalization process through a chemical reaction between an organofunctional alkoxysilane molecule and hydroxyl groups of a substrate to form stable covalent bonds. Silanization is currently applied to inorganic substrates such as glass and silicon oxide to modify their physical and chemical properties. Silanes are silicon chemicals with hydrolyzable groups and organic substitution allowing permanent property modification (see Figure 5.11). Property modifications include hydrophobicity, release, absorption, orientation, hydrophilicity, and charge conduction. Silanization is applied for anti-stiction, water repellents, dielectric or optical coatings and to the creation of self-assembled monolayers. A large variety of silanes substituted with hydrophobic groups, such as methyl, aryl, linear, branched or fluorinated alkyl, or with hydrophilic groups such as hydroxylic, ionic, or polar function, are available. Stability of the self-assembled monolayers of silane has been verified up to 250°C for amino silane and up to 350°C for fluorosilane [CHA 10].

Depending on the materials and applications, different techniques of silanization are possible: wet processes or dry processes.

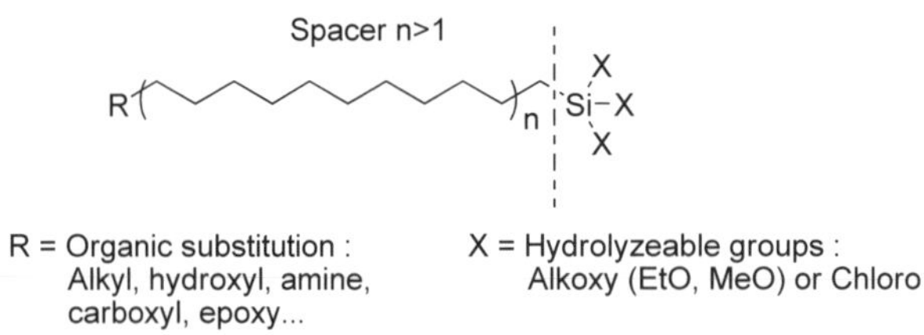

Figure 5.11. *Chemical structure of organosilane*

5.3.2.1. *Wet processes*

Silanization using wet processes implies the use of solvents such as water, alcohol, toluene and hydrocarbon. Silanization is often catalyzed under acidic or basic conditions or by the addition of amine. Deposition from aqueous alcohol solutions is the easiest method for preparing silylated surfaces [ARK 08]. Hydrolysis of alkoxysilanes occurs after silane addition. Silanols can then rapidly react with inorganic surfaces (see Figure 5.12). After several minutes of reaction, the substrates are removed from the solution and rinsed with ethanol. Deposition in aqueous solutions is used in commercial systems but is limited to water-soluble silanes. Long-chain alkyl and aromatic silanes are excluded. The main disadvantage of the aqueous method is the difficulty to control the deposition process, leading to bad reproducibility, an uncontrollable surface density and a heterogeneous deposit [JON 85].

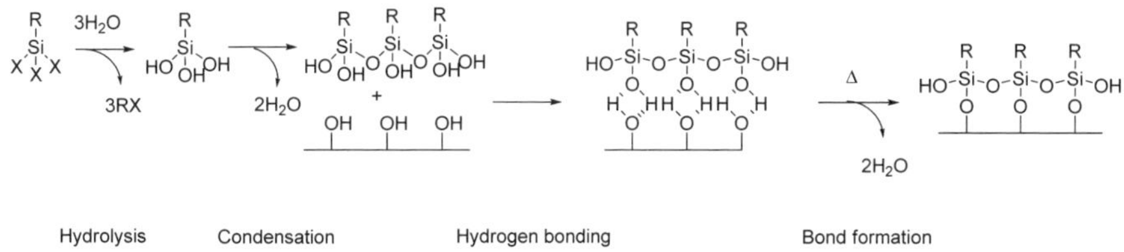

Figure 5.12. *Hydrolytic deposition of silanes [ARK 77]*

Anhydrous liquid phase deposition is preferred for small particles and micro- and nanostructured substrates. Highly reactive silanes, such as chlorosilanes or methoxysilanes, are required for this method. Anhydrous solvents such as toluene, tetrahydrofurane (THF) or hydrocarbon solutions are prepared containing 2–5% silane. The mixture is generally refluxed for 12–24 h with the substrate to be used. After washing with solvents, that is ethanol and acetone, the substrate is dried with air or using explosion-proof oven. This cumbersome protocol gains in reproducibility compared to aqueous solutions. This reaction involves a direct nucleophilic displacement of the silane chlorines by the surface silanol (see Figure 5.13). If water is present on the substrate, polymerization of the silanes occurs and bulk deposition results. Thus, pre-drying at 130–150°C for 3–4 h is mandatory for self-assembled monolayer (SAM) deposition.

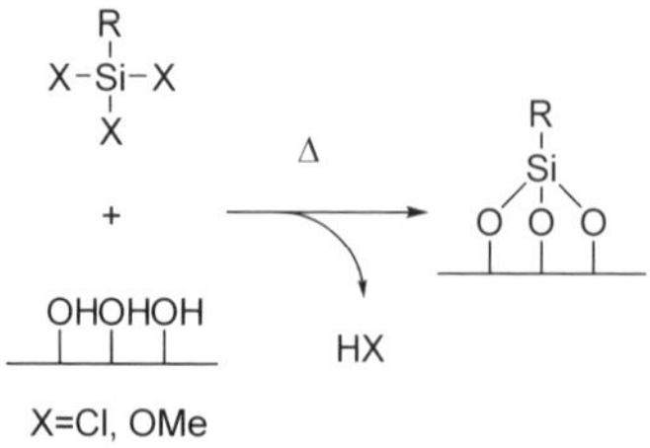

Figure 5.13. *Anhydrous deposition of silanes*

To create SAMs, long-chain aliphatic molecules are fully extended to maximize packing density and Van der Waals interactions [LOV 05, XIA 98]. Thus, it is possible to introduce almost any functional group. Formation of SAMs is facilitated on planar and/or curved surfaces. The main disadvantage is that defects are practically inevitable. Furthermore, SAMs are mechanically and chemically fragile. Then, the functional groups are located only on the surface.

5.3.2.2. *Chemical Vapor Deposition*

Silanes can be applied to substrates under dry conditions by chemical vapor deposition methods. These methods use no or low amount of solvents. Silane molecules are vaporized, diffused in a closed chamber and allowed to react with the hydroxyl group present on the substrate surface [SUG 02, JON 85, ZHA 10]. This deposition process is highly conformal and in favor of a monolayer [SEV 09, BOE 05].

Moreover, silane diffusion, and hence silanization reaction, rates are generally higher in the vapor phase. Thus, reaction conditions can be more easily controlled, resulting in more reproducible surface modifications. The main disadvantage of vapor phase deposition is the restriction to silanes with a vapor pressure under 5 Torr at 100°C. Indeed, silanes poorly vaporizable and temperature sensitive cannot be deposited using this method.

In MVD equipment from Applied Microstructures, substrates are supported in a closed process chamber next to a silane reservoir which is heated to sufficient temperature to achieve 5 Torr vapor pressure. Alternatively, a vacuum can be used until silane evaporation is observed (see Figure 5.14). In general, the temperature is maintained above 35°C and below 50°C. Chlorosilanes and methoxysilanes usually deposit within 5–60 min using water as a catalyst [ZHA 10, BOE 05].

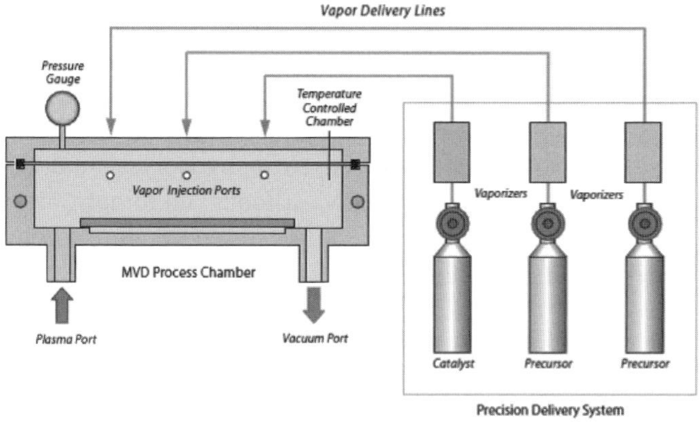

Figure 5.14. *Apparatus for vapor phase silylation [KOB 07]*

In addition, vapor phase deposition can be processed using a solution of silane in toluene, and the toluene is brought to reflux allowing sufficient silane to enter the vapor phase through partial pressure contribution. In this way, the substrate temperature should be maintained above 50°C and below 120°C to promote reaction. Amine-functional silanes usually deposit rapidly (within 30–45 min) without a catalyst [ZHA 10, SEV 03, BOE 05]. Reaction of other silanes requires extended reaction times, usually 4–24 h.

5.3.2.3. *Casting method*

A third method by direct casting of neat silanes can be performed to functionalize silicon oxide surfaces. This simple method was originally performed on hot substrates [HUS 03], but casting neat reagents in an ambient atmosphere was also studied [KOB 07]. A pure liquid silane is deposited on a freshly pretreated and dried substrate positioned in a Teflon microreactor. The microreactor is placed in an oven at 60–120°C for 10 min to induce reaction. After reaction, the substrates are sequentially rinsed with acetone and heated in a Soxhlet extractor with xylene to remove excess silanes. This method can be used with a number of different non-volatile compounds to form SAMs. The advantage of this method is the rapidity of the functionalization process. The disadvantages are that this procedure uses more compounds than other methods that use dilute solutions of silanes or vapor phase deposition. Thus, this casting method cannot be used for large-scale silanization process. Moreover, the monolayers formed by this method are generally poorly packed and ordered in particular with short carbon chains or branched silanes.

5.3.3. *Sol-gel and polymer deposition techniques*

5.3.3.1. *Sol-gel processes*

The sol-gel process is a widely used technique to fabricate glass and ceramic materials. Sol-gel is prepared from hydrolysis and condensation polymerization of a metal (currently silicon, aluminum, zirconium or titanium) alkoxide solution, followed by a temperature program which controls the densification process (see Figure 5.15).

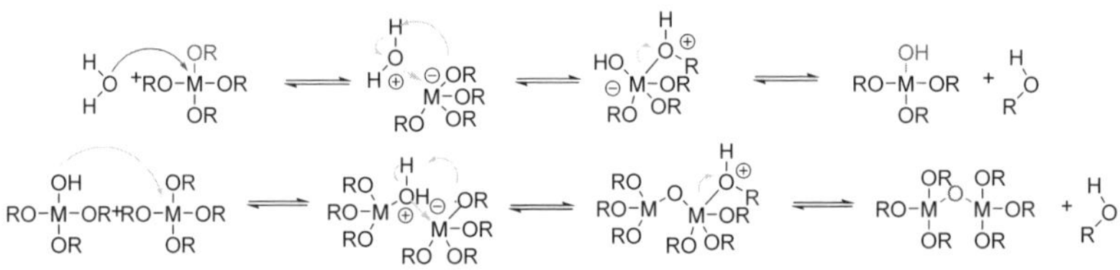

Figure 5.15. *Sol-gel polymerization process*

Sol-gel process has been used for optical chemical and biosensors [MAC 95, AVA 08, DES 10], and for anti-reflection coating [CHE 01]. Several methods such as dip coating, spin coating, spraying and electrodeposition have been developed to deposit sol-gel on a substrate [SHA 07].

5.3.3.2. *Dip coating*

Dip coating is an immersion method where substrates are immersed into a solution containing the "sol" (a, b) and then carefully removed (c) under stable conditions to obtain (d) regular thickness film (see Figure 5.16). Thus, when substrates are removed, solutions drains onto the substrate. Finally, substrates are covered by a uniform and porous sol-gel type of film.

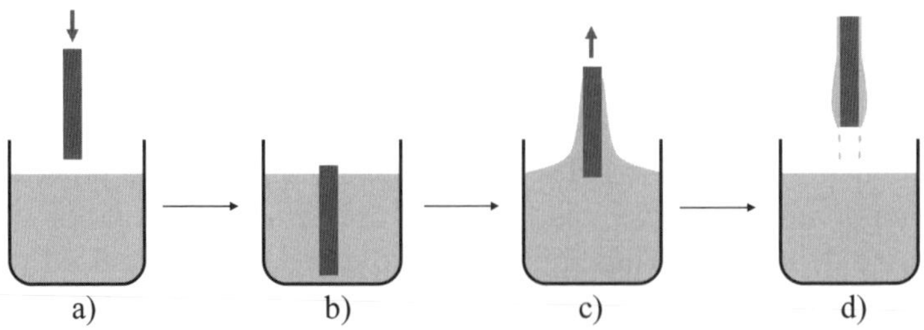

Figure 5.16. *Dip coating principle*

5.3.3.3. *Spin coating*

Spin coating is a common centrifugation technique to deposit thin films on a flat substrate [VOR 95, BIR 04, ZHA 04]. Substrates are covered with excess solution and then rotated at a high speed. Fluid is spread over the surface. Then, the solvent evaporates, and a homogeneous thin film is obtained (see Figure 5.17).

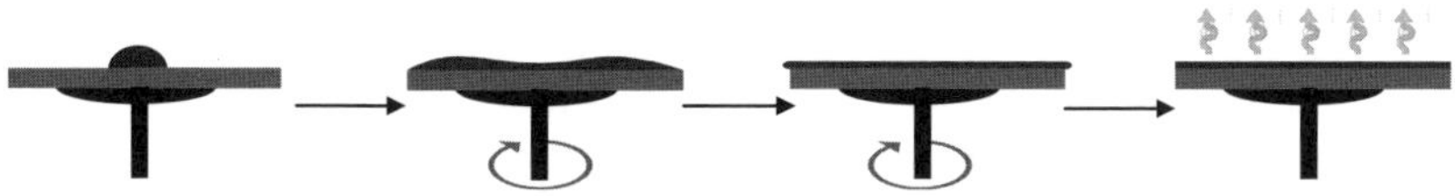

Figure 5.17. *Spin coating principle*

Spin coating methods can be applied to solutions of silanes, sol-gel, polymer, alginates, etc. For the spin coating of silanes, processing can occur under hydrolytic or dry conditions. Hydrolytic conditions will lead to maximum functionalization and polylayer deposition. On the contrary, dry conditions will form monolayers.

5.3.3.4. *Spray coating*

Spray coating technology is an alternative to spin coating to deposit sol-gel or polymers such as photoresists [FIS 04, WET, BAS 10]. It consists of mixing a liquid with a gas flow and guiding the mixture through a nozzle to generate a spray of droplets. The droplets have a diameter of approximately 1 mm and are composed of polymer and solvent. Solvent should evaporate quickly in order to obtain a dry material when the droplets reach the sample surface. Several parameters can modify the viscosity of the sprayed solution and thus the quality of the coating: solvent quantity, flow, gas pressure, substrate holder temperature, distance between nozzle and substrate, speed out of the nozzle and the way the substrate is scanned. This technique is fast and simple but reproducibility is difficult without an automatic spray coater.

5.3.4. *Localization processes*

Localization or patterning of the functionalization coating is of interest for applications such as lab-on-chip or micro- and nanoelectromechanical systems (MEMS and NEMS). Patterning can be processed by several methods depending on the class of material. Organics such as silanes, sol-gel, polymers, or biological materials will be deposited in a different manner according to their vulnerability. Conventional micropatterning tools are photolithography and microcontact printing. However, nanolithographic techniques, such as pip-pen nanolithography, are rapidly rising [ZHO 11].

5.3.4.1. *Spotting*

The simplest way to localize is to deposit a drop of solution onto a substrate (see Figure 1.18). This method called "spotting" or "printing" can be performed using contact or non-contact printer [PIR 02].

Figure 5.18. *Non-contact printing on microstructured wafer*

For contact printing, the solution is deposited by bringing a pin or a needle in contact with the surface. Commonly used for DNA microarray fabrication, this deposition method is highly dependent on the surface properties, and some substrates cannot tolerate contact printing due to their surface structure or patterning. Thus, non-contact techniques, such as ink-jet printing, were developed. Liquid droplets are formed and ejected by thermal, acoustic, or piezoelectric phenomenon through nozzle apertures [PAR 07]. The functional resolution is 20–30 μm. Indeed, the droplet diameters are usually no smaller than 10–20 μm, and placement errors are typically ±10 μm.

To get better resolution, water-based ink droplets can also be confined on a hydrophilic substrate with a pattern of narrow, repelling, hydrophobic surface regions [SIR 00, STV 03, WAN 06] (see Figure 5.19). These methods require separate patterning systems and processing steps to define assist features.

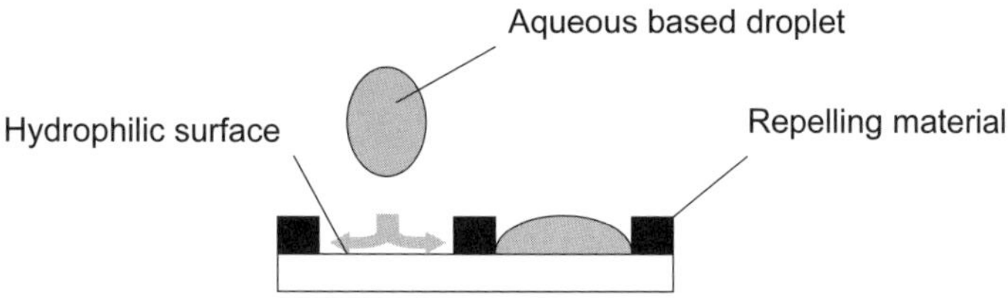

Figure 5.19. *Droplet confinement on a patterned surface*

The spotting method was originally developed for biological applications, but polymers can be deposited using this method.

5.3.4.2. *Photolithography*

Over the past three decades, photolithography has been one of the main methods used for the patterning of polymers [NIE 08, MEN 07, REV 03]. Photolithography is a process that selectively exposes a polymer-coated surface to irradiation and then removes selected areas of the film via dissolution with an appropriate solvent (see Figure 5.20). The polymer film called photoresist must be sensitive to light, and the substrate is generally flat. Photolithography is a cost-effective, high-throughput technique, which generates high-resolution patterns less than few tens of nanometers in size. This technique is, in particular, adequate to create patterns with exact control over the shape and size over an entire, large and flat substrate. It requires clean operating conditions and is currently used in microfabrication in the semiconductor industry.

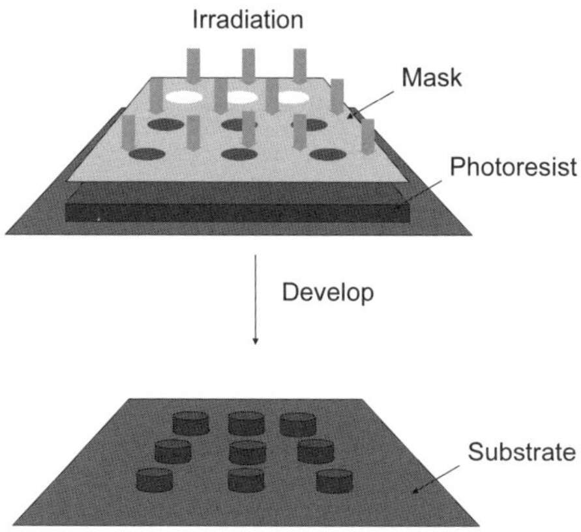

Figure 5.20. *Patterning of polymers by photolithography*

Microarrays of DNA, cells and proteins [REV 03, KOH 02, SEO 11, FOD 91] as well as sensors and actuators [HOF 99] have been produced using photolithography. Combination of photolithography and surface silanization has been recently considered for selective surface attachment of proteins and cells (see Figure 5.21) [SEO 11, SWE 11, CHI 09, LEN 11].

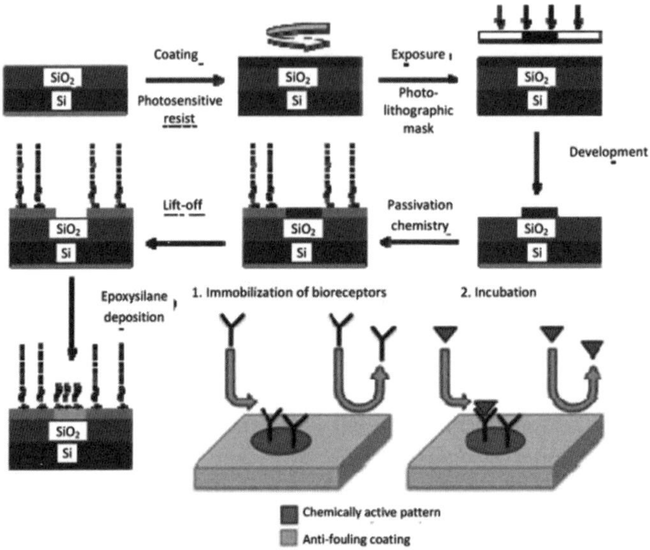

Figure 5.21. *Combination of photolithography and surface silanization for antibody patterning*

Besides, optical techniques can been used to selectively unprotect aminosilane films protected by a modified 2-nitrophenylethoxycarbonyl (NPEOC) photoreactive group [ALA 11]. Thus, a polyethylene-glycol (PEG)-modified NPEOC aminosilane was deposited on silicon wafers. The PEG function is used to prevent non-specific adsorption. The substrate was irradiated at 325 nm through a mask to cleave the NPEOC group, liberate the amino function and immobilize the protein. These simple photocleavable protecting group strategies combined with near-field exposure lead to the control of surface reactivity at nanometer resolution.

5.3.4.3. *Scanning beam lithography*

The principle of the scanning beam lithography technique is to write patterns into a polymer, such as PMMA, a photoresist or directly onto a substrate with a focused ion beam. Thus, the material is removed from the surface through ion bombardment, creating a pattern at sub-50 nm resolution [GAT 05]. This technique (see Figure 5.22) was applied to form aminopropyl triethoxysilane (APTES) patterns to graft DNA rafts and proteins [SAR 06, ZHA 05].

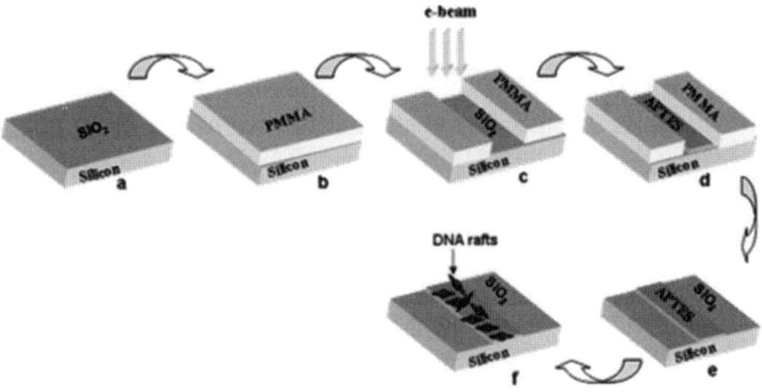

Figure 5.22. *Schematic diagram of the molecular lift-off process [SAR 06]*

5.3.4.4. *Dip-pen nanolithography*

Dip-pen nanolithography (DPN) is a versatile method used to precisely position molecules onto a surface [GAT 05]. An Atomic Force Microscopy (AFM) tip is inked with a solution of the material to be transferred to the surface (see Figure 5.22). The majority of DPN experiments have used gold surfaces as the substrates and thiol molecules as the inks. However, dip-pen nanolithography of mercaptoalkoxysilanes on glass under specific humidity conditions has been reported [JUN 03].

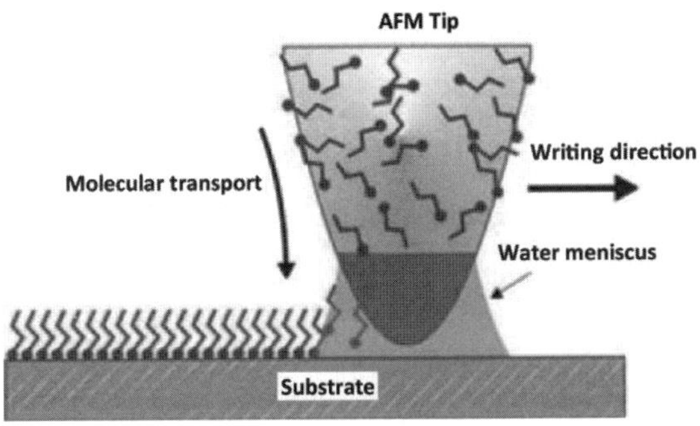

Figure 5.23. *Schematic representation of dip-pen nanolithography*

This method still remains challenging to generate reproducible structures between scans because of variations in the surface topography of the substrate and differences in the shape of the tip.

5.3.4.5. *Microcontact printing*

Microcontact printing is a widely used technique which consists of transferring a material called ink from a topographically patterned stamp onto a substrate (see Figure 5.24). Originally, stamps were made from PDMS and the inks composed of alkanethiols brought onto gold, silver, or palladium substrates [GAT 05, KUM 93]. This technique has been extended to pattern several materials such as biomolecules, particles and polymers. The patterning of silanes is also feasible using microcontact printing [LEE 09].

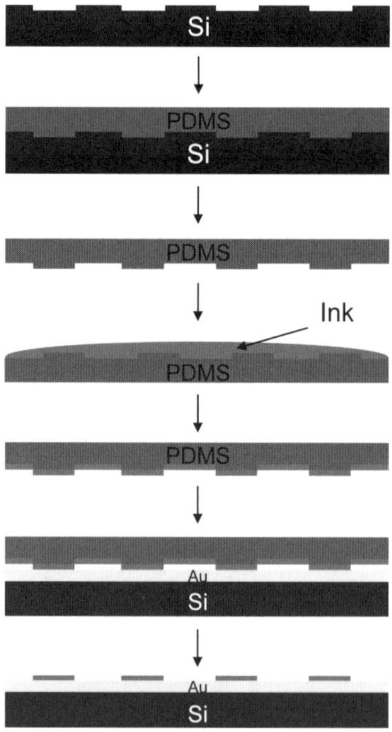

Figure 5.24. *Microcontact printing*

Recently, an enzymatic soft lithography based on microcontact printing and enzymatic degradation has been developed to create the pattern of poly-L-lysine [GUY 08]. Furthermore, soft lithography using PDMS stamps has been proposed to selectively block the outer gold membrane surfaces of nanopores for label-free biosensing [GYU 03, GYU 08] (see Figure 5.25).

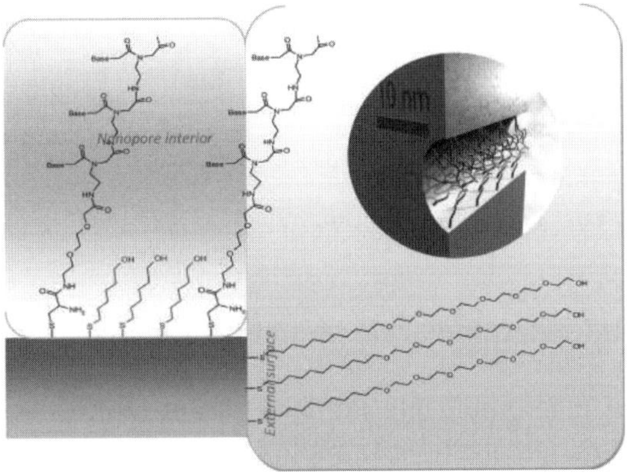

Figure 5.25. *Selective chemical modification of the inner and outer surfaces of Au nanopore membranes [GYU 08]*

5.3.4.6. *Electrochemical functionalization*

Electro-induced grafting is a powerful and versatile technique to chemically coat conducting surfaces such as gold film, doped silicon substrate, stainless steel grids or carbon surfaces [PAL 04]. This grafting method leads to thin, transparent, and insulating coatings chemically linked on conducting surfaces. Vinylic monomers, such as acrylonitrile and methyl methacrylate or diazonium salts, can be grafted by polymerization onto the surface with various coating thicknesses.

Using this technique, some diazonium-modified organosilyl/-germyl polyoxotungstate have been covalently grafted onto silicon surfaces in order to realize POM-based multilevel memory devices [JOO 10] (see Figure 5.26).

The electrografting can be localized on designed areas of a composite conducting surface. The method relies on the local work function of electrons, which can easily be tuned by patterning different conducting materials on the surface [CHA 04, FLA 10].

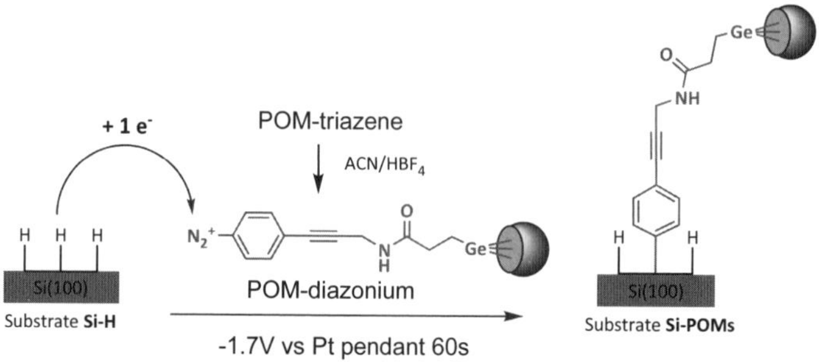

Figure 5.26. *Electrografting of POM-diazonium on silicon substrate*

5.4. Molecule and macromolecule immobilization

Molecule and macromolecule immobilization is most often the ultimate step in surface functionalization. Immobilization is the outcome of strong or weak attachment of probes (molecules or biomolecules) on the surface, according to the use of a chemical coupling method [HER 08, JON 08, NEB 09]. The probes to be immobilized represent a large variety of synthetic or natural molecules and macromolecules. Among them, biomolecules such as proteins, nucleic acids, saccharides and their templates are widely studied because of their numerous applications, mainly in the biotechnology field. Moreover, these biomolecules often require difficult-to-achieve immobilization conditions in terms of solvents, temperatures and reactants in order to keep their physicochemical nature [BIN 08, JON 08, NEB 09, WAN 03]. The methods most currently described in the literature for their attachment to surfaces will be discussed in this section.

5.4.1. *Adsorption and covalent grafting*

Two strategies can be used for immobilizing a molecule or biomolecule on a surface [JON 08]. The first strategy is based on the physical adsorption phenomenon. The second strategy relies on the creation of covalent bonds.

Molecular adsorption generally occurs due to electrostatic interactions and/or Van der Waals forces, including hydrogen bonding and/or hydrophobic interactions, between the probe compound and the surface substrate. As an example, one of the most popular DNA array constructions consists of depositing natural negatively charged DNA chains onto positively charged polylysine coated surfaces to obtain efficient grafting. If no further chemical modifications of DNA are required in this case, the chemical bonds created are generally weak and will not be stable, for instance, in high ionic buffers, such as physiological conditions (sodium saline 8.5 g/L).

On the other hand, covalent grafting involves specific chemical functions present on both the probe and the surface, or implies the use of a cross-linker reagent, which possesses functional extremities for strongly binding the two partners (probe and surface).

Figure 5.27 shows the reaction between a chemically modified surface substrate (with function X) and a probe (with function Y) leading to the covalent grafting without (a) or with (b) the use of a cross-linker. A cross-linker is a chemical molecule or macromolecule with an A-L-B-like structure in which L is a chemical linker, and A and B are reactive functional groups, able to form covalent bound with X and Y, respectively. In the case of no cross-linker use, chemical functions X and Y are specific and have to be branched on the probe and the surface before the immobilization step.

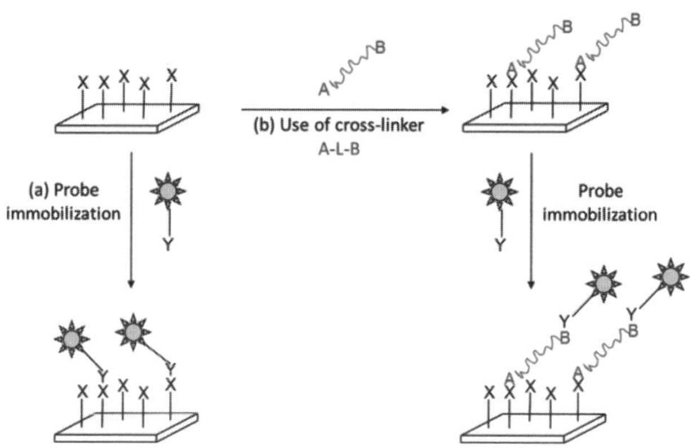

Figure 5.27. *Covalent grafting of probes on the surface without (a) and with (b) the use of a cross-linker*

Among the multitude of chemical reactions, we have selected the most usual X and Y functional groups for introducing molecule or macromolecule immobilization on different surface substrates. These chemical groups and the corresponding grafting mechanisms that can be used to establish covalent bonds are detailed in the following section.

5.4.2. *Chemical functions for covalent grafting*

The list of chemical functions that can be present on the probe (Y) or surface (X) and the associated coupling methods detailed below is not exhaustive, and the reader can refer to more complete books for further insight [HER 08]. Those that are described here are the most frequently used for covalent grafting, especially in the case of biomolecules. The chemical functions listed below can be positioned either on the probe (Y function, see Figure 5.27) or on the surface substrate (X function, see Figure 5.27). They can be naturally present (such as the thiol or amine functions in proteins) or be introduced by previous modifications of the molecules/biomolecules/ surfaces and/or by the use of cross-linkers. It is noticeably the case for chemical groups used in "click chemistry", which are chemical functions not present in "natural products", such as nucleotides, amino acids and saccharides.

5.4.2.1. *Amines*

The amine group ($-NH_2$) reacts with electrophilic moieties such as N-hydroxy succinimide(NHS)-activated esters, aldehydes, iso(thio)cyanates, epoxides, and anhydrides, leading to stable amine or amide bonds. The choice of the electrophilic group for modifying the surface (or the probe) depends on its stability in the grafting conditions. For the immobilization step, primary amines are strongly preferred because they are the most reactive among the amine compound class. Reaction 5.1 shows the general reaction between a primary amine and an NHS ester derivative to yield the coupling product assembled by a stable amide bond.

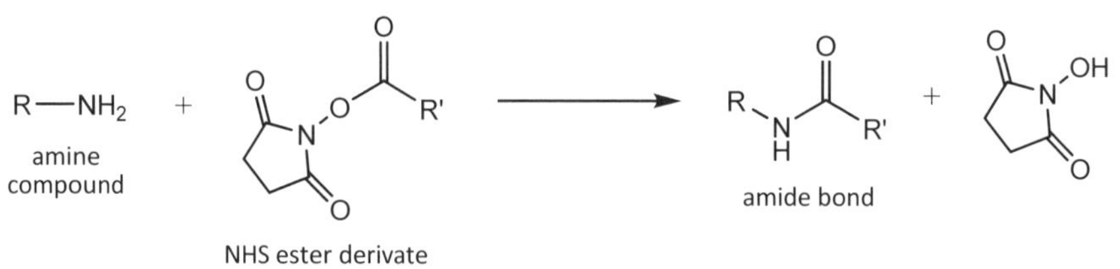

Reaction 5.1. *Reaction between amine and NHS functions*

5.4.2.2. *Thiols*

Sulfhydryl-containing molecules or surfaces (–SH) are known to be modified by alkylation reaction or disulfide interchange. Most of these reactions are rapid and occur in high yield to give stable thioether or disulfide bonds. The most commonly used alkylation reactions occur between a sulfhydryl compound and an iodoacetyl derivative, as shown in reaction 5.2, or between a sulfhydryl compound and a maleimide (maleic acid imide) group, to give a thioether bond.

Reaction 5.2. *Reaction between thiol and iodoacetate functions*

Iodoacetate is known to have a very high reactivity toward sulfhydryl cysteine residues; therefore, it may be used specifically for cysteine-containing protein immobilization without preliminary chemical modification.

Reaction 5.3 illustrates the disulfide interchange reaction, for instance, in the presence of an activated pyridyl disulfide derivative, which gives a new mixed disulfide link comprising a portion of the original disulfide compound.

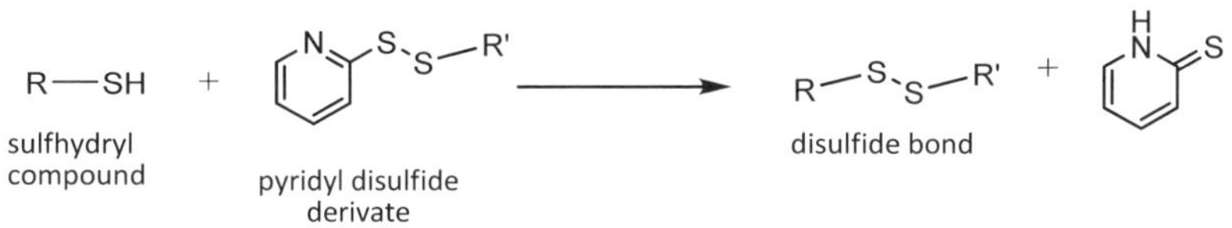

Reaction 5.3. *Disulfides interchange from thiol function and disulfide derivative reaction*

The obtained S–S bond is reversible, and the reaction can be reversed using disulfide reducing agents such as dithiotreitol (DTT) as shown in reaction 5.4.

Reaction 5.4. *Reduction of the S–S bond by DTT agent*

5.4.2.3. *Hydroxyls*

Similar to the amine group, the hydroxyl group (–OH) can react with electrophilic moieties (i.e. isocyanate or activated ester NHS), but with lower reactivity. Another strategy for the conjugation of hydroxyl-bearing moieties consists of transforming the –OH group into an electrophilic function by the use of an activated hydroxylic such as the carbonyldiimidazole (see reaction 5.5). The obtained intermediate is reactive toward the amine compounds, which allows the stable carbamate linkage formation.

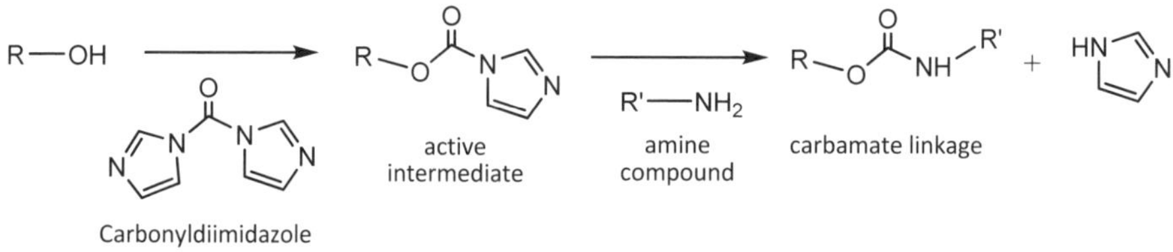

Reaction 5.5. *Hydroxyl group activation by carbonyldiimidazole for reacting with amine compound*

5.4.2.4. *"Click chemistry"*

"Click chemistry" is a concept and not a specific reaction [KOL 03, MOS 07]. It includes reactions which mainly give high chemical yields, generate no or inoffensive by-products and require a mild experimental process (low temperature, free of toxic solvents). One of the most popular reactions within the click chemistry concept is the azide-alkyne Huisgen cycloaddition; it can be performed in aqueous buffer at room temperature in the presence of a copper (I) catalyst [MEL 08]. As shown in reaction 5.6, no by-product is formed, only the desired compound is obtained with a triazole linkage which can have two stereoisomer forms.

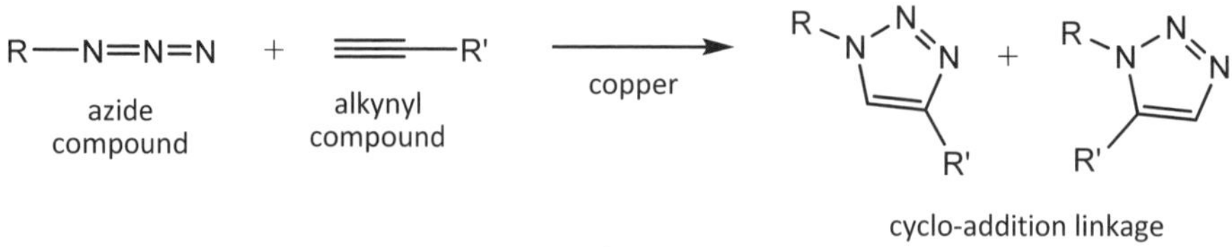

Reaction 5.6. *Click chemistry using the Huisgen cyclo-addition*

The advantages of using click chemistry in the probe/substrate conjugation step lie in the lack of by-products and the control of the probe grafting orientation on the surface due to the reaction specificity between the implemented chemical functions. On the other hand, the use of click chemistry necessitates the preliminary introduction of the exogenous alkyne and azide functions on both the probe molecule and substrate surface. However, numerous cross-linkers are nowadays commercially available for material (polymers, silica, gold, carbon nanotubes, etc.) functionalization by click chemistry whose field of application is rapidly growing [BIN 08, BOI 08, FLE 06, IHA 09, NEB 09, DIJ 09, WAN 03].

5.4.3. *Applications and examples*

Generally, to immobilize biomolecules without preliminary chemical modifications, the surface first has to be modified by a function chosen to be specific with the chemical groups naturally present on the molecule. The available biomolecules of low molecular weight (peptides, oligonucleotides and oligosaccharides) and high molecular weight (proteins, DNA, polysaccharides, etc.) combined with their surface immobilization strategies are summarized in Table 5.1.

Appropriate immobilization strategy	Proteins		Nucleic acids		Sugars	
	Peptides	Proteins	Oligonucleotides	DNA	Oligosaccharides	Polysaccharides
Adsorption	No	Yes	No	Yes	No	Yes
Covalent grafting	Yes	Yes	Yes	No	Yes	Yes

Table 5.1. *Biomolecules and their immobilization strategies*

Adsorption as an immobilization strategy is not applicable to low molecular weight compounds even if they are charged. Indeed, the attachment with the surface using physical adsorption becomes significant when multiple bonds are established, as in the case of polymers. Proteins and sugar polymers (polysaccharides) can also be immobilized on different surface substrates by covalent grafting without any chemical modifications, contrary to DNA, because they have reactive functions. Non-modified DNA only displays poorly reactive phosphate groups in its structure.

5.4.3.1. *Peptide and protein grafting*

Peptides and proteins present amine functions on the lysine residues. For this reason, they will preferentially be grafted onto surfaces presenting electrophilic functional groups. In this case, the aldehyde group can be preferred to the NHS ester function because it is more stable during the immobilization step realized in aqueous media (see reaction 5.7). The amine reacts with the aldehyde to yield an unstable imine intermediate, which is subsequently reduced for stabilizing the desired product. Practically, the substrate is directly immersed in the reducing agent solution, typically aqueous sodium borohydride, without washing after the formation of the imine intermediate.

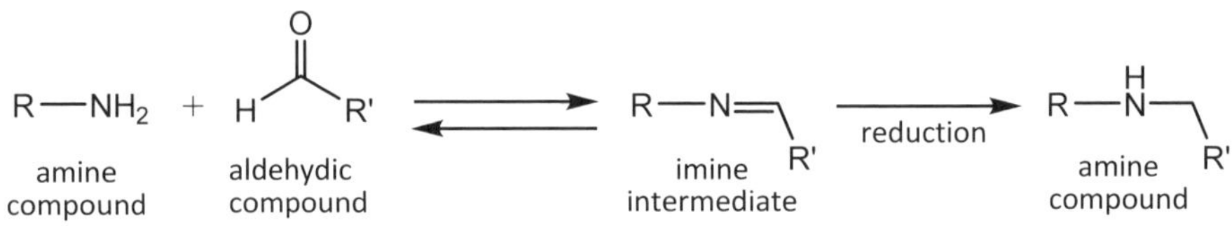

Reaction 5.7. *Reaction between amine and aldehyde functions.*

5.4.3.2. *Nucleic acid and DNA grafting*

For natural nucleic acids, such as DNA, which do not have modified chemical functions, the use of non-covalent grafting by electrostatic interactions is commonly adopted. For shorter nucleic acids, such as synthetic oligonucleotides (10–100 base pairs), covalent grafting is usually implemented, and a preliminary chemical modification is required before the immobilization step. In most cases, amine or thiol groups are used to modify the oligonucleotides, which are automatically spotted on the covered corresponding electrophilic function surface (see section 5.3.3). Figure 5.28 shows the ejection heads from spotting machinery and the fluorescence results of the hybridization step between the dye-containing DNA target and different oligonucleotide probes spotted on the surface. The substrate is a glass plate and the drop size is approximately 120–150 μm.

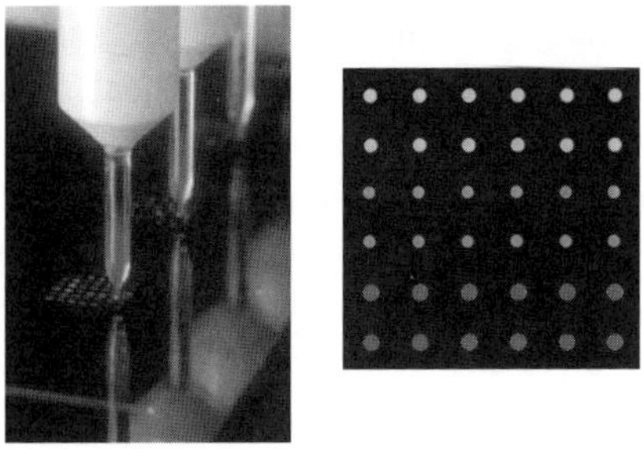

Figure 5.28. *Spotter picture and fluorescence results after hybridization step*

5.4.3.3. *Saccharide grafting*

For the covalent grafting of saccharides on surfaces, their treatment with sodium periodate ($NaIO_4$) prior to the immobilization step leads to the activated probes with aldehyde groups, which can then be spotted on the oxyamine ($-ONH_2$) or hydrazine ($-NH-NH_2$) functionalized surfaces, using reaction 5.8.

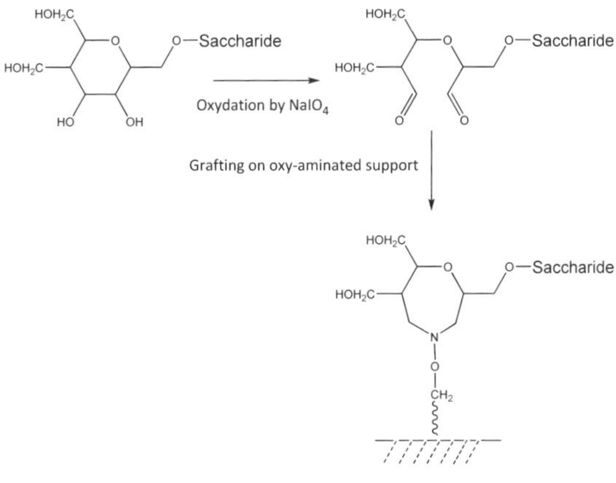

Reaction 5.8. *Reaction between oxyamine and aldehyde functions.*

5.4.4. *The chemical tool box*

A wide range of chemical reactions exist for grafting on functionalized surface substrate molecules and sensitive biomolecules, which often require mild coupling conditions (limited pH and temperature range, aqueous buffer). The selection of the suitable reaction should take into account the chemical functions already available in the molecule or surface to be grafted, the possibility/difficulty of introducing other chemical groups, the sensitivity of the chemical groups to reaction conditions, and the coupling efficiencies. However, some parameters are difficult to master. For instance, different antibodies (included in protein category) could display different coupling efficiencies on the same substrate, despite the similarity of their structure, because they present different hydrophobicity or present differently to the surface [JON 08]. This example demonstrates that it is important to have available and continue to develop different coupling strategies in the chemical tool box.

5.4.5. *Applications*

As previously described, a chemical "tool box" has been developed in order to tune the surface properties of various materials. We describe hereafter a selection of applications labeling by functionalized microsystems.

5.5. Analytes capture

5.5.1. *Example of in vitro application*

The capturing of bacteria has numerous applications in the rapidly evolving area of lab-on-chip technology [GRO 03, FLO 05, HWA 08, JAC 11, HWA 11]. The capture and release of bacteria can be used to concentrate bacteria in a large volume of water. It could also be used as a filter to separate bacteria from a fluid sample. In general, bacterial cells have negatively charged surfaces [TER 06]. Thus, electrostatic interactions are predominantly involved in bacterial adhesion, and cationic surface functionalization can be used to electrostatically trap bacteria with high efficiency. CEA-Léti has developed a polyethylenimine-coated silicon micropillar array chip to capture bacteria. Polythylenmine (PEI) is a simple cationic polymer which easily adsorbs on negatively charged surfaces (see Figure 5.29).

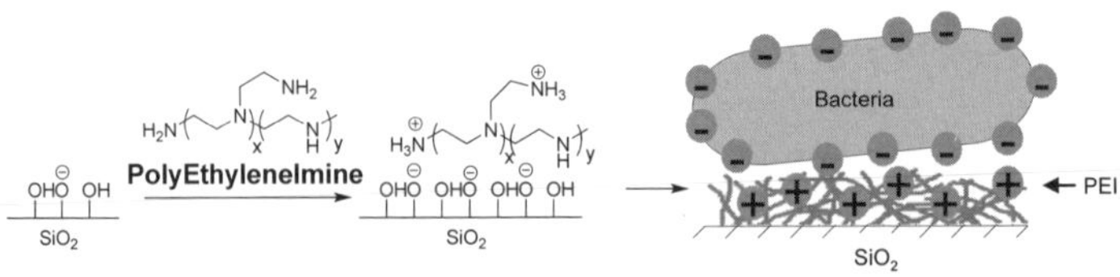

Figure 5.29. *Adsorption of PEI on a SiO_2 surface followed by bacteria capture*

This branched polymer has an exceptionally high positive charge at low pH and it can react with negatively charged cells (see Figure 5.30).

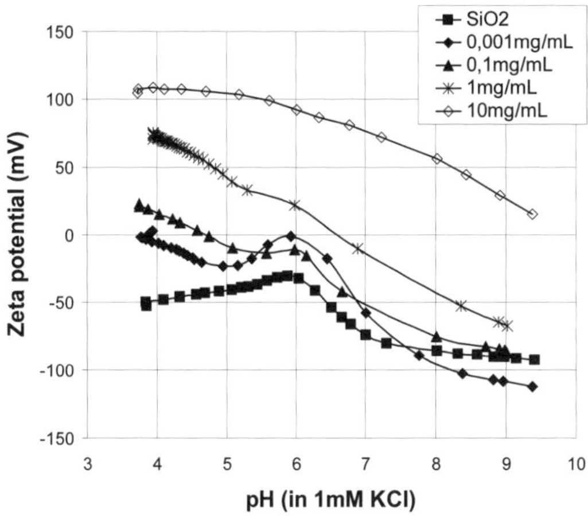

Figure 5.30. *Effect of PEI concentration on the zeta potential-pH behavior of SiO_2 planar samples*

The silicon micropillar array chips consist of a flow channel of etched silicon (see Figure 5.31). The channel contains a pyramid array of flow dividers used to distribute the bacteria along the flow channel as uniformly as possible followed by an array of silicon pillars. The purpose of these pillars is to greatly increase the functionalized surface area which the bacteria are exposed to, and to decrease the average distance between the bacteria and the nearest surface. This will improve the probability that bacteria will come close enough to a surface for the adhesion forces to be significant, and the increased surface area improves the binding capacity.

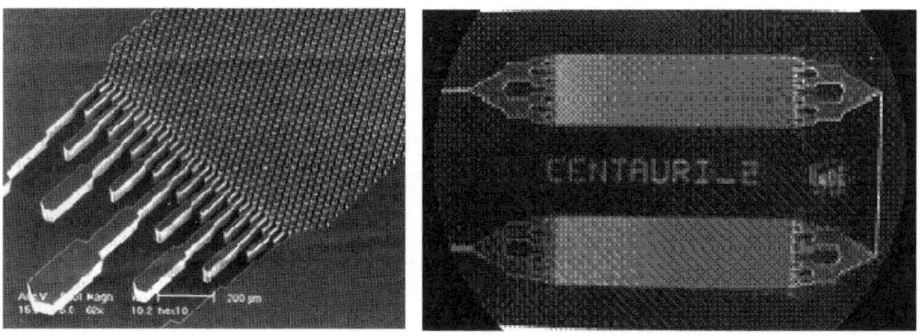

Figure 5.31. *SEM image of a silicon micropillar array chip and top view of chips*

Binding experiments were performed with 200 µL of *Escherichia coli* at the concentration of 10^6 bacteria per mL in an aqueous buffer (pH 8) with a flow rate of 100 µL/min. The pillar chip functionalized with PEI demonstrated a capture efficiency of 97.9%. The control tests showed no bacterial capture in the control chip as expected and further confirmed that the adhesion is a direct result of the surface functionalization.

A PEI chip infused with fluorescent bacteria was cut in half and placed under a microscope to obtain a side profile view of the bacterial adhesion on the pillars. Fluorescence microscopy images of the pillars with adhered bacteria can be seen in Figure 5.32.

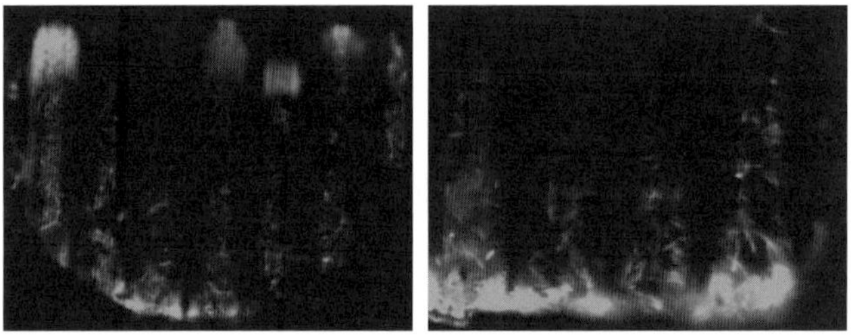

Figure 5.32. *Side profiles of pillars with adhered bacteria in green (pillars are 20 µm in diameter)*

The release of bacteria was investigated in order to demonstrate the proof of the concept of a chip concentrating and releasing bacteria. Preliminary results using basic conditions were highly promising, showing a 25.8% bacteria release without optimization.

5.5.2. *Example of in vivo application*

Clinical diagnostic and therapeutic monitoring techniques for brain tumors are evolving toward miniaturization not only to provide new information but also to minimize side effects. The first diagnosis is generally based on scanner and magnetic resonance imaging (MRI) results; however, these techniques detect the presence and localization of a tumor, but not its nature and type. To obtain this kind of information, tumor analysis is required. Recently, molecular profiling has been validated as an important complement to this anatomopathological examination [SIK 07, JOH 06] because this approach provides a more precise diagnosis and prognostic analysis. Biopsies are needed for relevant biomarker discovery allowing this profiling. However, biopsies can be difficult to obtain, in particular, for brain tissue. Thus, various developments have been made to miniaturize biopsy tools in order to reach inaccessible areas and minimize the amount of tissue sampled. Microgrippers [LEE 96], microneedles [MOO 05] or different types of microspikes provided with barbs [PAR 05] have been developed and validated on animal tissue. Even if these systems are miniaturized, the principle is still to extract a piece of tissue, rendering some areas and organs unexplorable.

A miniaturized minimally invasive device particularly adapted to human brain tissue investigation was developed by CEA-LETI [COS 06]. This tool consists of a metallic rod of 1 mm in diameter, which is the standard size for a brain biopsy tool, with a functionalized silicon chip. Each chip is designed individually and manufactured collectively on 200 mm silicon wafers. The design includes micropillars of 20 or 80 µm diameter, with a pitch of 60 or 240 µm (see Figure 5.33). The geometry and topography of the surface has a double function: overall surface increase and matrix/protein co-crystallization control because samples are analyzed by laser desorption/ionization [COS 06].

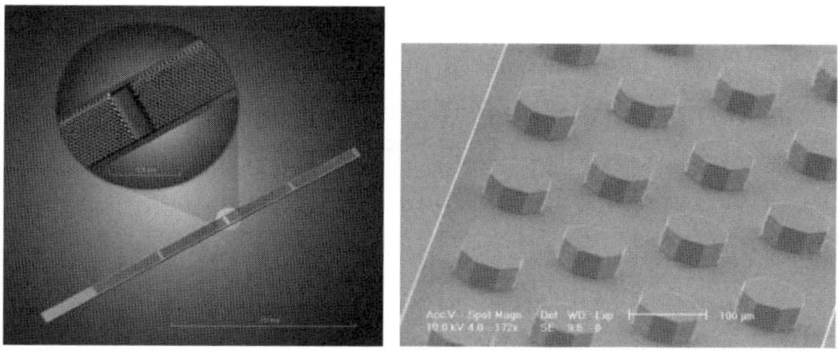

Figure 5.33. *Processed silicon chip (left) and detail of the microstructuration (right)*

To capture proteins by a simple apposition of the silicon chip on the tumor, the silicon chip was functionalized by silanization [COS 06, GRA 07] to obtain different surface behaviors: hydrophobic, hydrophilic, cationic and anionic surfaces.

For its surgical use, the functionalized chip is assembled on a specific surgical metal rod made of surgical stainless steel. Its surface is specifically designed for receiving the silicon chip that is attached with biocompatible glue. Next, the chip is placed in contact with fresh extracted tissue for approximately 30 sec to 1 min and then directly analyzed in a laser desorption mass spectrometer-TOF analysis (SELDI-TOF). Several mass spectra (each peak is a protein that could be identified and potentially a new biomarker) were obtained depending on the type of chemical functionalization. These results clearly show the significant role of chemical functionalization and microstructuration on the protein adsorption yield, thus enhancing the sensitivity and specificity of the mass spectra analysis drastically (see Figure 5.34).

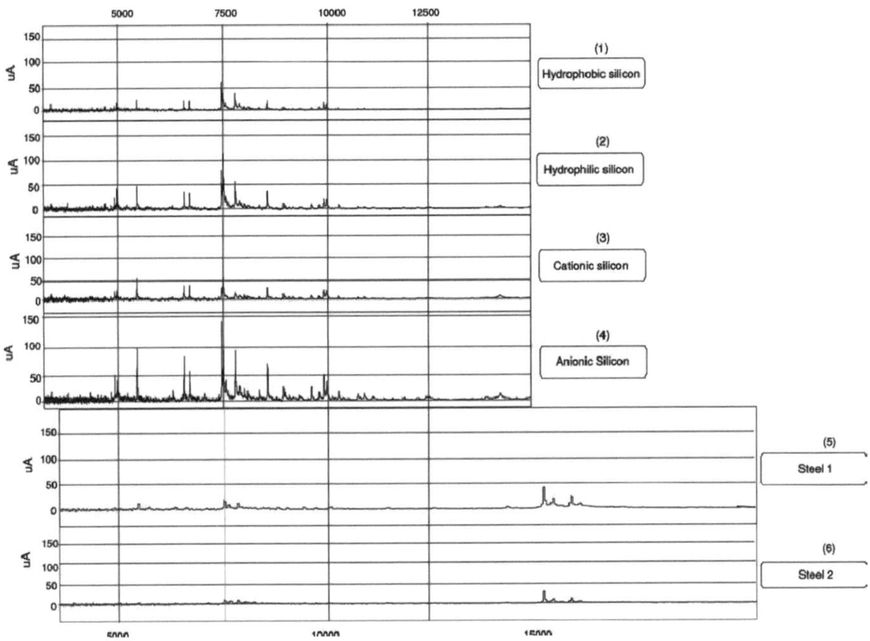

Figure 5.34. *Mass spectra of a sample harvested by: (1) hydrophobic functionalized silicon surface, (2) hydrophilic functionalized silicon surface, (3) cationic functionalized silicon surface, (4) anionic functionalized silicon, (5) surgical stainless steel surface, (6) another surgical stainless steel surface*

5.5.3. *Detection of analytes*

The research area of biosensors [MAR 07a] is very wide, both with respect to the analytes/molecules that are sensed and the sensing technology used. Biosensors are analytical tools for the analysis of biomaterial samples to gain an understanding of their biocomposition, structure and function by converting a biological response into an electrical or optical signal. The analytical devices are composed of a biological/chemical recognition element by a sensing part, directly interfacing with a signal transducer which together, relates the concentration of the analyte to a measurable response. A common feature of all chemical sensors is the fact that they use a responsive thin film/layer that selectively interacts with the analyte and causes a measurable change in optical/electric/resonance property that is evidenced by the optical and electronic signals. Many different transducer technologies are used for the interrogation of the sensor signal such as electrochemical, electrical (impedance), optical (fluorescence, surface plasmon resonance (SPR), surface enhanced Raman scattering (SERS)), piezoelectric quartz crystal microbalance (QCM) and thermal. These biosensors are commonly incorporated into lab-on-a-chip or microfluidic devices to facilitate sample handling, sensing and data processing.

For *in vitro* applications, the requirements on materials and procedures are less stringent than for *in vivo* biosensors. Labeling the *in vivo* detection of (small) biomolecules requires direct sensing, using smart responsive materials in combination with the appropriate transducer technology that has sufficient biocompatibility which remains a key issue.

5.5.3.1. *Example of in vitro applications*

Clinical diagnostics is one of the most promising applications for microfluidic lab-on-a-chip or lab-on-card systems. DNA chips, which provide multiparametric data, are privileged tools for genomic analysis. However, automation of molecular biology protocol and use of these DNA chips in fully integrated systems remain a great challenge. The simplicity of chip and/or card/ instrument interfaces is among the most critical issues to be addressed. Indeed, current detection systems for DNA chip reading are often complex, expensive, bulky, and even limited in terms of sensitivity or accuracy [AHN 04, AND 00, AUR 04]. Furthermore, for liquid handling in lab-on-cards, many devices use complex and bulky systems either to directly manipulate fluids or to ensure pneumatic or mechanical control of integrated valves. All these drawbacks prevent or limit the use of DNA-chip-based integrated systems for point-of-care testing or as a routine diagnostic tool. In response to clinical diagnostic applications, an optoelectronic DNA chip integrated into a plastic card was reported [MAR 08a] (see Figure 5.35).

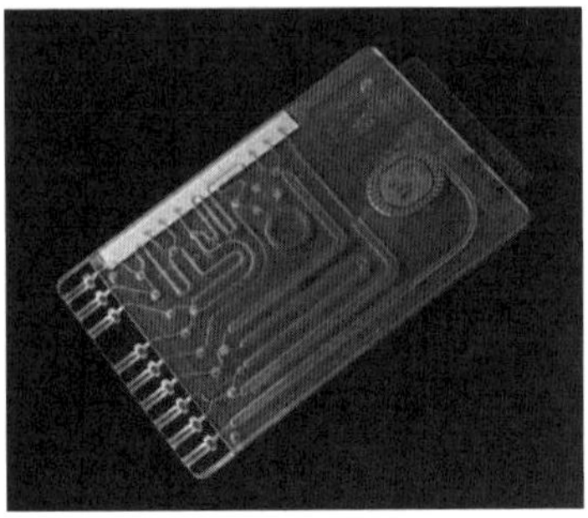

Figure 5.35. *Pictures of the lab-on-chip*

Before integration, the active pixel sensor (APS) was functionalized [MAL 05] by successive steps comprising a rehydration step with oxygen plasma in order to generate silanol groups on the chip surface, a silanization with 5,6-epoxyhexyltriethoxysilane, a hydrolysis and an oxidative cleavage to obtain aldehyde functions, and finally, the immobilization of the oligonucleotide probes. The detection of the targets was performed by an indirect approach including a chemiluminescence reaction based on the oxidation of luminol by horseradish peroxidase (HRP) enzyme. This reaction is triggered by the addition of a commercial substrate solution. Photons generated by the reaction are detected by the underlying photodiodes and converted into electrons. The total charge accumulated in each pixel over the acquisition period is then interpreted as a gray level. Considering the whole detector is based on APS technology, a quantitative picture of the integrated light intensity generated on the chip is produced (see Figure 5.36).

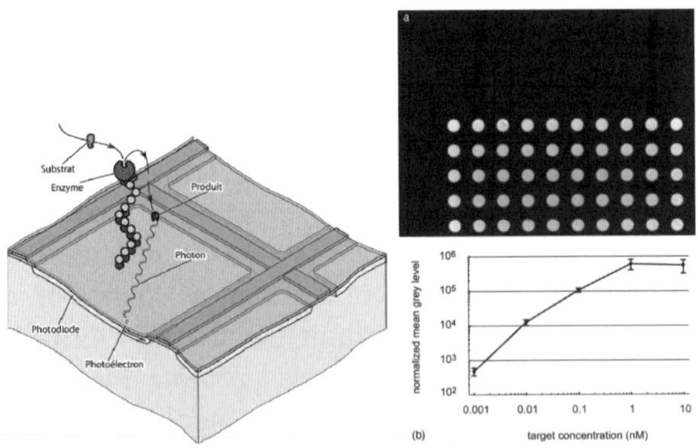

Figure 5.36. *Schematic representation of the optoelectronic DNA chip at the level of 1 pixel (left); dynamic detection range of the optoelectronic DNA chip (right)*

5.5.3.2. *Example of in vivo application*

The ProeTEX project was a European integrated project (FP6) aiming at developing a new generation of equipment for the market of emergency operators, like civil protection rescuers and firefighters. Among the numerous sensors integrated in the inner garment, the Na^+ sensor appeared as one of most essential. Indeed, during interventions in disasters, emergency personnel are exposed to extreme physical conditions, and important regulation disorders can be observed such as an abnormal concentration of sodium in the blood plasma [DAV 01, SHA 05] which eventually causes a cerebral edema and ultimately results in death. Thus, monitoring the sodium concentration seems to be an important parameter to ensure the good health of emergency personnel. However, this application requires not only a portable and not very bulky device in order to prevent hindering the firefighter during intervention, but also a non-invasive measurement. Thus, the monitoring of Na^+ concentration in sweat can be an interesting alternative because sweat is an ultra-filtrate of the blood plasma [NAT 04].

An electrochemical sensor (see Figure 5.37) was developed [MAR 09] to selectively detect and quantify Na^+ ions in natural sweat. This sensor is constituted of several sensing parts integrated into a textile pump used to direct the flow of sweat and a wearable electronic board driving the sensing parts and a temperature sensor and converting electrical measurements data into ionic concentration.

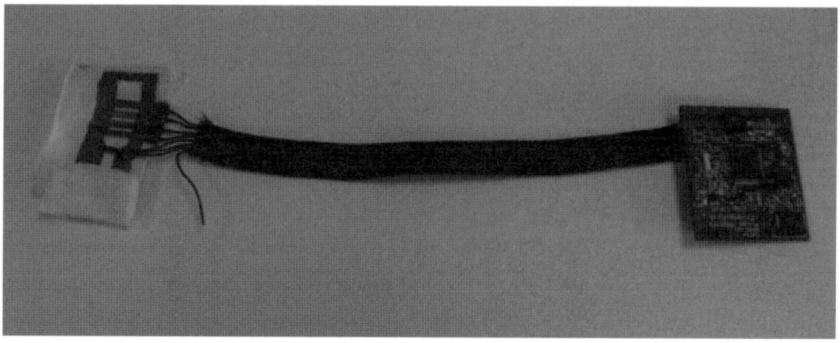

Figure 5.37. *Electrochemical sensor with the electronic board and the sensing parts integrated into a textile pump*

The developed electrochemical sensing part is an electrochemical cell composed of four gold electrodes, three working electrodes to ensure the reproducibility of the measurement, and one reference electrode. This electrochemical cell is coupled with a temperature probe to enhance the reliability of the measurement because electrochemical measurement depends on the temperature. A polypyrrole layer was deposited on the working electrodes by electropolymerization at 1.2 V versus an Ag/AgCl electrode in trapping a 4-tert-butylcalix[4]arene tetraacetic acid tetraethyl ester [MAR 07b]. This host molecule is able to complex Na^+ ions selectively in a solution containing competitors such as K^+, Mg^{2+}. Thus, the complexation of the targeted ion (with positive charges) by the host molecule leads to a modification of the sensing part charge, and this modification is followed by the evolution of the open circuit potential (OCP). These functionalized electrodes, called ionic selective electrodes [MAL 00, PHI 07], were also integrated into the inner garment of the equipment and then, their functionality was validated with success in terms of response time, sensitivity, reversibility and accuracy in real conditions (see Figure 5.38).

Figure 5.38. *Validation of the sensors in real conditions*

5.6. Conclusion

The development of robust and multifunctional micro- and nanosystems is strongly related to surface properties. The grafting mechanisms and the resulting characterized surfaces are a key issue for the development and industrialization of these systems. This has been carried out not only in the classical silicon-dominated domains but also outside the conventional MEMS and NEMS technologies, including plastics and, more recently, paper.

As the average world population ages, healthcare and well-being will become a great societal challenge and, thus, new designs, easy-to-use portable or implantable micro- and nanosystems will be developed. A versatile chemistry "tool box" for surface monitoring will be of great importance for these new developments.

5.7. List of Abbreviations

NC:	NanoCrystals
QD:	Quantum Dots
MEMS:	Micro Electro Mechanical System
NEMS:	Nano Electro Mechanical System
COC:	Cyclo Olefin Copolymer
PDMS:	PolyDiMethylSiloxane
PU:	PolyUrethane
RF:	Radio Frequency
SAM:	Self-Assembled Monolayer
THF:	TetraHydroFurane
MVD:	Molecular Vapor Deposition
NPEOC:	NitroPhenylEthOxyCarbonyl
PEG:	PolyEthylene Glycol
PMMA:	PolyMethylMetAcrylate
DPN:	Dip-Pen Nanolithography
APTES:	AminoPropylTriEthoxySilane
POM:	PolyOxoMetalate
NHS:	N-Hydroxy Succinimide
DTT:	DiThioTreitol
PEI:	PolyEthylenimine
MRI:	Magnetic Resonance Imaging
SELDI-TOF MS:	Surface Enhanced Laser Desorption Ionization–Time Of Flight Mass Spectrometry
SPR:	Surface Plasmon Resonance
SERS:	Surface Enhanced Raman Scattering
APS:	Active Pixel Sensor
HRP:	HorseRadish Peroxidase
OCP:	Open Circuit Potential

5.8. Bibliography

[AGA 11] AGACHE V., BLANCO-GOMEZ G., BALERAS F., *et al.*, "An embedded microchannel in a MEMS plate resonator for ultrasensitive mass sensing in liquid", *Lab on a Chip*, vol. 11, no. 15, pp. 2598–2603, 2011.

[AHN 04] AHN C.H., JIN-WOO C., BEAUCAGE G., *et al.*, "Disposable smart lab on a chip for point-of-care clinical diagnostics", *Proceedings of the IEEE*, vol. 92, no. 1, pp. 154–173, 2004.

[ALA 11] ALANG AHMAD S.A., WONG L.S., UL-AQ E., *et al.*, "Protein micro- and nanopatterning using aminosilanes with protein-resistant photolabile protecting groups", *Journal of the American Chemical Society*, vol. 133, no. 8, pp. 2749–2759, 2011.

[AME 09] AMEUR S., MARTELET C., JAFFREZIC-RENAULT N., *et al.*, "Sensitive immunodetection through impedance measurements onto gold functionalized electrodes", *Applied Biochemistry and Biotechnology*, vol. 89, no. 2–3, pp. 1621–170, 2009.

[ANA 08] ANASTASOVA S., "Morphology of sol-gel produced composite films for optical oxygen sensors", *Applied Surface Science*, vol. 254, no. 6, pp. 1545–1558, 2008.

[AND 00] ANDERSON R.C., SU X., BOGDAN G.J., *et al.*, "A miniature integrated device for automated multistep genetic assays", *Nucleic Acids Research*, vol. 28, no. 12, p. 60, 2000.

[ARK 77] ARKLES B.C., "Tailoring surfaces with silanes", *Chemtech*, vol. 7, no. 12, pp. 766–778, 1977.

[ARK 08] ARKLES B.C., *Silicon Compounds: Silanes and Silicones: A Survey of Properties and Chemistry*, 2nd ed., Gelest, Inc., Morrisville, PA, 2008.

[AUR 04] AUROUX P.A., KOC Y., DE MELLO A., *et al.*, "Miniaturised nucleic acid analysis", *Lab on a Chip*, vol. 4, pp. 534–546, 2004.

[BAS 10] BASU B.J., HARIPRAKASH V., ARUNA S.T., *et al.*, "Effect of microstructure and surface roughness on the wettability of superhydrophobic sol–gel nanocomposite coatings", *Journal of Sol-Gel Science and Technology*, vol. 56, no. 3, pp. 278–286, 2010.

[BER 06] BERMUDEZ V.M., BERRY A.D., KIM H., *et al.*, "Functionalization of indium tin oxide", *Langmuir,* vol. 22, pp. 11113–11125, 2006.

[BER 09] BERTHIER J., TRAN V.-M., MITTLER F., *et al.*, "The physics of a co-flow micro-extractor: interface stability and optimal extraction length", *Sensors and Actuators A,* vol. 149, no. 1, pp. 56–64, 2009.

[BER 10] BERTHIER J., MOURIER V., SARRUT N., *et al.*, "Some examples of micro-devices for biotechnology developed at the Department of Technologies for Life Sciences and Healthcare of the LETI", *International Journal of Nanotechnology,* vol. 7, no. 4–8, pp. 802–818, 2010.

[BEU 00] BEUTEL S.T., "Micromachined, polyimide-based devices for flexible neural interfaces", *Biomedical Microdevices*, vol 2, no. 4, pp. 283–294, 2000.

[BIL 05] BILLEN J., GZIL P., VERVOORT N., *et al.*, "Influence of the packing heterogeneity on the performance of liquid chromatography supports", *Journal of Chromatography A,* vol. 1073, no. 1–2, pp. 53–61, 2005.

[BIN 08] BINDER W.H., SACHSENHOFFER R., "'Click' chemistry in polymer and material science: an update", *Macromolecular Rapid Communications,* vol. 29, no. 12–13, pp. 952–981, 2008.

[BIR 04] BIRNIE D.P., "Surface skin development and rupture during sol-gel spin-coating", *Journal of Sol-Gel Science and Technology,* vol. 31, no. 1–3, pp. 225–228, 2004.

[BOE 05] BOERIO F.J., WAGH V.H., DILLINGHAM R.G., "Vapor-phase silanation of plasma-polymerized silica-like films by gamma-aminopropyltriethoxysilane", *Journal of Adhesion,* vol. 81, no. 2, pp. 115–142, 2005.

[BOI 08] BOISSELIER E., SALMON L., RUIZ J., *et al.*, "How to very efficiently functionalize gold nanoparticles by 'click' chemistry", *Chemical Communications,* pp. 5788–5790, 2008.

[CHA 96] CHAN C.M., KO T.M., HIRAOKA H., "Surface modification by plasmas and photons", *Surface Science Reports,* vol. 24, no. 1–2, pp. 1–541, 1996.

[CHA 10] CHANDEKAR A., SENGUPTA S.K., WHITTEN J.E., "Thermal stability of thiol and silane monolayers: a comparative study", *Applied Surface Science,* vol. 256, no. 9, pp. 2742–2749, 2010.

[CHA 04] CHARLIER J., AMEUR S., BOURGOIN J.P., *et al.*, "Mask-free localized grafting of organic polymers at the micrometer or submicrometer scale on composite conductors or semiconductor substrates", *Advanced Functional Materials*, vol. 14, no. 2, pp. 125–132, 2004.

[CHE 08] CHEE R.E., CHUA J.H., AGARWAL A., *et al.*, "Label free detection of proteins with surface-functionalized silicon nanowires", in LIM C. T., GOH J.C.H. (eds), *Proceedings of the 13th International Conference on Biomedical Engineering (ICBME)*, vol. 33, pp. 838–841, 2008.

[CHE 01] CHEN D., "Anti-reflection coatings made by sol–gel processes: a review", *Solar Energy Materials and Solar Cells,* vol. 68, no. 3–4, pp. 313–336, 2001.

[CHE 96] CHENG J., SHOFFNER M.A., HVICHIA R.E., *et al.*, "Chip PCR. II: investigation of different PCR amplification systems in microfabricated silicon–glass chips", *Nucleic Acids Research,* vol. 24, no. 2, pp. 380–385, 1996.

[CHI 09] CHIU C.-S., LEE H.M., GWO S., "Site-selective biofunctionalization of aluminum nitride surfaces using patterned organosilane self-assembled monolayers", *Langmuir,* vol. 26, no. 4, pp. 2969–2974, 2009.

[COS 06] COSNIER M.L., MARTIN F., BOUAMRANI A., *et al.*, "A new micro minimally invasive biopsy tool for molecular analysis", *Proceedings of the 28th IEEE Engineering in Medicine and Biology Society (EMBS) Annual International Conference*, New York City, vol. 1, pp. 2820–2823, 2006.

[COY 10] COYLE J., LAU K.-T., MOYNA N., *et al.*, "BIOTEX-biosensing textiles for personalized healthcare management", *IEEE Transactions on Information Technology in Biomedicine*, vol. 14, no. 2, pp. 364–370, 2010.

[CRA 00] CRAIGHEAD H.G., "Nanomechanical systems", *Science,* vol. 290, pp. 1532–1535, 2000.

[DAV 01] DAVIS D.P., VIDEEN J.S., MARINO A., *et al.*, "Exercise-associated hyponatremia in marathon runners: a two-year experience", *The Journal of Emergency Medicine*, vol. 21, no. 1, pp. 47–57, 2001.

[DES 10] DESCAMPS M.N., BORDY T., HUE J., *et al.*, "Real-time detection of formaldehyde by a fluorescence-based sensor", in JAKOBY B., VELLEKOOP M.J. (eds), *Proceedings of the Eurosensors XXIV Conference*, Elsevier Science Bv, Amsterdam, pp. 1009–1012, 2010.

[DSU 99] D'SOUZA A.S., PANTANO C.G., "Mechanisms for silanol formation on amorphous silica fracture surfaces", *Journal of the American Ceramic Society*, vol. 82, no. 5, pp. 1289–1293, 1999.

[DUG 03] DUGAS V., CHEVALIER Y., "Surface hydroxylation and silane grafting on fumed and thermal silica", *Journal of Colloid and Interface Science*, vol. 264, no. 2, pp. 354–361, 2003.

[FIS 04] FISCHER K., SUSS R., "Spray coating: a solution for resist film deposition across severe topography", *Proceedings of the 29th IEEE Electronics Manufacturing Technology Symposium*, pp. 338–341, 2004.

[FLA 10] FLAVEL B.S., GROSS A.J., GARRETT D.J., *et al.*, "A simple approach to patterned protein immobilization on silicon via electrografting from diazonium salt solutions", *Applied Materials and Interfaces*, vol. 2, no. 4, pp. 1184–1190, 2010.

[FLE 06] FLEMING D.A., THODE C.J., WILLIAMS M.E., "Triazole cycloaddition as a general route for functionalization of Au nanoparticles", *Chemistry of Materials*, vol. 18, no. 9, pp. 2327–2334, 2006.

[FLO 05] FLORIANO P.N., CHRISTODOULIDES N., ROMANOVICZ D., *et al.*, "Membrane-based on-line optical analysis system for rapid detection of bacteria and spores", *Biosensors and Bioelectronics,* vol. 20, no. 10, pp. 2079–2088, 2005.

[FOD 91] FODOR S.P.A., READ J.L., PIRRUNG M.C., *et al.*, "Light-directed, spatially addressable parallel chemical synthesis", *Science*, vol. 251, no. 4996, pp. 767–773, 1991.

[FON 11] FONVERNE A., RICOUL F., DEMESMAY C., *et al.*, "New carbon nanotubes growth process in a closed microfabricated channel for liquid chromatography application", *Sensors and Actuators A: Physical*, vol. 67, no. 2, pp. 517–523, 2011.

[FRE 07] FRÉCHET J.M.J., "Hydrophilic surface modification of cyclic olefin copolymer microfluidic chips using sequential photografting", *Journal of Separation Science*, vol. 30, pp. 1088–1093, 2007.

[GAT 05] GATES B.D., XU Q., STEWART M., *et al.*, "New approaches to nanofabrication: molding, printing, and other techniques", *Chemical Reviews*, vol. 105, no. 4, pp. 1171–1196, 2005.

[GRA 07] GRANIER M., Novel silane compounds and use thereof in order to functionalize solid media and immobilize biological molecules on said media, W.O. Patent 2 007 088 186, 9 August 2007.

[GRO 03] GROW A.E., WOOD L.L., CLAYCOMB J.L., *et al.*, "New biochip technology for label-free detection of pathogens and their toxins", *Journal of Microbiological Methods*, vol. 53, no. 2, pp. 221–233, 2003.

[GUB 12] GUBALA V., HARRIS L.F., RICCO A.J., *et al.*, "Point of care diagnostics: status and future", *Analytical Chemistry*, vol. 84, pp. 487–515, 2012.

[GUY 11] GUYOMARD-LACK A., DELORME N., MOREAU C., *et al.*, "Site-selective surface modification using enzymatic soft lithography", *Langmuir*, vol. 27, no. 12, pp. 7629–7634, 2011.

[GYU 03] GYURCSANYI R.E., VIGASSY T., PRETSCH E., "Biorecognition-modulated ion fluxes through functionalized gold nanotubules as a novel label-free biosensing approach", *Chemical Communications*, vol. 20, pp. 2560–2561, 2003.

[GYU 08] GYURCSANYI R.E., "Chemically-modified nanopores for sensing", *TrAC Trends in Analytical Chemistry*, vol. 27, no. 7, pp. 627–639, 2008.

[HAL 09] HALTE C., Détection directe d'événements chimiques et biologiques en milieu liquide sur nanostructures de silicium, PhD Thesis, Joseph Fourier University, Grenoble I, France, 2009.

[HEC 04] HECKELE M., SCHOMBURG W.K., "Review on micro molding of thermoplastic polymers", *Journal of Micromechanics and Microengineering*, vol. 14, no. 3, pp. 1–14, 2004.

[HER 08] HERMANSON G.T., *Bioconjugate Techniques*, Academic Press, 2008.

[HOF 99] HOFFMANN J., PLOTNER M., KUCKLING D., *et al.*, "Photopatterning of thermally sensitive hydrogels useful for microactuators", *Sensors and Actuators A-Physical*, vol. 77, no. 2, pp. 139–144, 1999.

[HUA 07] HUANG X., JAIN P.K., EL-SAYED I.H., *et al.*, "Special focus: nanoparticles for cancer diagnosis and therapeutics: review", *Nanomedicine*, vol. 2, no. 5, pp. 681–693, 2007.

[HUS 03] HUSSEINI G.A., PEACOCK J., SATHYAPALAN A., *et al.*, "Alkyl monolayers on silica surfaces prepared using neat, heated dimethylmonochlorosilanes with low vapor pressures", *Langmuir*, vol. 19, no. 12, pp. 5169–5171, 2003.

[HWA 08] HWANG K.-Y., LIM H.-K., JUNG S.-Y., *et al.*, "Bacterial DNA sample preparation from whole blood using surface-modified Si pillar arrays", *Analytical Chemistry*, vol. 80, no. 20, pp. 7786–7791, 2008.

[HWA 11] HWANG K.-Y., KWON S.H., JUNG S.-O., *et al.*, "Miniaturized bead-beating device to automate full DNA sample preparation processes for Gram-positive bacteria", *Lab on a Chip*, vol. 11, no. 21, pp. 3649–3655, 2011.

[IHA 09] IHA R.K., WOOLEY K.L., NYSTROM A.M., *et al.*, "Applications of orthogonal click chemistries in the synthesis of functional soft materials", *Chemical Reviews*, vol. 109, no. 11, pp. 5620–5686, 2009.

[ING 07] INGEBRANDT S., HAN Y., NAKAMURA F., *et al.*, "Label-free detection of single nucleotide polymorphisms utilizing the differential transfer function of field-effect transistors", *Biosensors and Bioelectronics*, vol. 22, pp. 2834–2840, 2007.

[ISH 86] ISHIZAKA A., SHIRAKI Y., "Low-temperature surface cleaning of silicon and its application to silicon MBE", *Journal of the Electrochemical Society*, vol. 133, no. 4, pp. 666–671, 1986.

[JAC 11] JACKERAY R., ZAINUL ABID C.K.V., SINGH G., *et al.*, "Selective capturing and detection of Salmonella typhi on polycarbonate membrane using bioconjugated quantum dots", *Talanta*, vol. 84, no. 3, pp. 952–962, 2011.

[JOH 06] JOHNSON M.D., FLOYD J.L., CAPRIOLI R.M., "Proteomics in diagnostic neuropathology", *Journal of Neuropathology and Experimental Neurology*, vol. 65, no. 9, pp. 837–845, 2006.

[JON 08] JONKHEIJM P., WEINRICH D., SCHRODER H., *et al.*, "Chemical strategies for generating protein biochip", *Angewandte Chemie International Edition*, vol. 47, no. 50, pp. 9618–9647, 2008.

[JON 85] JONSSON U., OLOFSSON G., "Chemical vapor-deposition of silanes", *Thin Solid Films*, vol. 124, no. 2, pp. 117–123, 1985.

[JOO 10] JOO N., RENAUDINEAU S., DELAPIERRE G., *et al.*, "Organosilyl/-germyl polyoxotungstate hybrids for covalent grafting onto silicon surfaces: towards molecular memories", *Chemistry – A European Journal*, vol. 16, no. 17, pp. 5043–5051, 2010.

[JUN 03] JUNG H., KULKARNI R., COLLIER C.P., "Dip-Pen nanolithography of reactive alkoxysilanes on glass", *Journal of the American Chemical Society*, vol. 125, no. 40, pp. 12096–12097, 2003.

[KER 70] KERN W., POUTINEN D.A., "Cleaning solutions based on hydrogen peroxide for use in silicon semiconductor technology", *RCA Review*, vol. 31, no. 2, pp. 187–191, 1970.

[KOB 07] KOBRIN B., CHIN J., ASHURST W.R., "Vapor deposition of composite organic-Inorganic films", *Proceedings of the World Tribology Congress III (WTC)*, vol. 2, 2007, available at http://dx.doi.org/10.1115/WTC2005-64335.

[KOH 02] KOH W.G., REVZIN A., PISHKO M.V., "Poly(ethylene glycol) hydrogel microstructures encapsulating living cells", *Langmuir*, vol. 18, no. 7, pp. 2459–2462, 2002.

[KOL 03] KOLB H.C., SHARPLESS K.B., "The growing impact of click chemistry on drug discover", *Drug Discovery Today*, vol. 8, no. 24, pp. 1128–1137, 2003.

[KUM 93] KUMAR A., WHITESIDES G.M., "Features of gold having micrometer to centimeter dimensions can be formed through a combination of stamping with an elastomeric stamp and an alkanethiol ink followed by chemical etching", *Applied Physics Letters*, vol. 63, no.14, pp. 2002–2004, 1993.

[LEE 96] LEE A.P., CIARLO D.R., KRULEVITCH P.A., *et al.*, "A practical microgripper by fine alignment, eutectic bonding and SMA actuation", *Sensors and Actuators A: Physical*, vol. 54, no. 1–3, pp. 755–759, 1996.

[LEE 09] LEE J.-H., HON M.-H., CHUNG Y.-W., *et al.*, "Microcontact printing of organic self-assembled monolayers for patterned growth of well-aligned ZnO nanorod arrays and their field-emission properties", *Journal of the American Ceramic Society*, vol. 92, no. 10, pp. 2192–2196, 2009.

[LEN 11] LENCI S., TEDESCHI L., PIERI F., *et al.*, "UV lithography-based protein patterning on silicon: towards the integration of bioactive surfaces and CMOS electronics", *Applied Surface Science*, vol. 257, no. 20, pp. 8413–8419, 2011.

[LI 07] LI M., TANG H.X., ROUKES M.L., "Ultra-sensitive NEMS-based cantilevers for sensing, scanned probe and very high-frequency applications", *Nature Nanotechnology*, vol. 2, pp. 114–120, 2007.

[LI 09] LI L., DAOU T.J., TEXIER I., *et al.*, "Highly luminescent $CuInS_2$/ZnS core/shell nanocrystals: cadmium-free quantum dots for in vivo imaging", *Chemistry of Materials*, vol. 21, no. 12, pp. 2422–2429, 2009.

[LIE 11] LIEBER C.M., "Semiconductor nanowires: a platform for nanoscience and nanotechnology", *MRS Bulletin*, vol. 36, no. 12, pp. 1052–1063, 2011.

[LIN 97] LIN T.J., ANTONELLI J.A., YANG D.J., *et al.*, "Plasma treatment of automotive steel for corrosion protection: a dry energetic process for coatings", *Progress in Organic Coatings*, vol. 31, no. 4, pp. 351–361, 1997.

[LIU 11] LIU J., APPAIX F., BIBARI O., *et al.*, "Control of neuronal network organization by chemical surface functionalization of multi-walled carbon nanotube arrays", *Nanotechnology*, vol. 22, no. 19, pp. 195101, 2011.

[LOV 05] LOVE J.C., ESTROFF L.A., KRIEBEL J.K., *et al.*, "Self-assembled monolayers of thiolates on metals as a form of nanotechnology", *Chemical Reviews*, vol. 105, no. 4, pp. 1103–1170, 2005.

[LUP 06] LUPRANO J., "European projects in smart fabrics, interactive textiles: sharing opportunities and challenges", *Workshop on Wearable Technology*, IST, Helsinki, 2006.

[MAC 95] MAC CRAITH B.D., MC DONAGH C., O'KEEFFE G., *et al.*, "Sol-gel coatings for optical chemical sensors and biosensors", *Sensors and Actuators B: Chemical*, vol. 29, no. 1–3, pp. 51–57, 1995.

[MAL 00] MALINOWSKA E., GAWART L., PARZUCHOWSKI P., *et al.*, "Novel approach of immobilization of calix[4]arene type ionophore in 'self-plasticized' polymeric membrane", *Analytica Chimica Acta*, vol. 421, no. 1, pp. 93–101, 2000.

[MAL 05] MALLARD F., MARCHAND G., GINOT F., *et al.*, "Opto-electronic DNA chip: high performance chip reading with an all-electric interface", *Biosensors and Bioelectronics*, vol. 20, no. 9, pp. 1813–1820, 2005.

[MAR 07a] MARCHAND G., COSNIER M.-L., DAVID N., *et al.*, "An electrochemical sensor integrated on fabric", *Proceedings of the 11th International Conference on Miniaturized Systems for Chemistry and Life Sciences (μTAS)*, Paris, France, October 2007.

[MAR 07b] MARTIN F., GRANIER M., LANNEAU G., Novel silane compounds and use thereof in order to functionalize solid media and immobilize biological molecules on said media, W.O. Patent 2 007 088 186, 2007.

[MAR 08a] MARCHAND G., BROYER P., LANET V., *et al.*, "Optoelectronic DNA chip-based integrated card for clinical diagnostics", *Biomedical Microdevices*, vol. 10, no. 1, pp. 35–45, 2008.

[MAR 08b] MARTINEZ A.W., PHILLIPS S.T., WILEY B.J., *et al.*, "FLASH: a rapid method for prototyping paper-based microfluidic devices", *Lab on a Chip*, vol. 8, no. 12, pp. 2146–2150, 2008.

[MAR 09] MARCHAND G., BOURGERETTE A., ANTONAKIOS M., *et al.*, "Development of a hydration sensor integrated on fabric", *Proceedings of the 6th IEEE International Workshop on Wearable Micro and Nano Technologies for Personalized Health*, Oslo, 24–26 June 2009.

[MAR 10] MARTINEZ A.W., PHILLIPS S.T., WHITESIDES G.M., "Diagnostics for the developing world: microfluidic paper-based analytical devices", *Analytical Chemistry*, vol. 82, no. 1, pp. 3–10, 2010.

[MEL 08] MELDAL M., TORNOE C.W., "Cu-catalyzed azide-alkyne cycloaddition", *Chemical Reviews*, vol. 108, no. 8, pp. 2952–3015, 2008.

[MEN 07] MENARD E., MEITL M.A., SUN Y., *et al.*, "Micro- and nanopatterning techniques for organic electronic and optoelectronic systems", *Chemical Reviews*, vol. 107, no. 4, pp. 1117–1160, 2007.

[MER 08] MERCEY E., SADIR R., MAILLART E., *et al.*, "Polypyrrole oligosaccharide array and surface plasmon resonance imaging for the measurement of glycosaminoglycan binding interaction", *Analytical Chemistry*, vol. 80, no. 9, pp. 3476–3482, 2008.

[MOO 05] MOON S.J., CHOI Y.H., LEE S.S., "A micro machined biopsy tool for a capsule type endoscope", *Proceedings of the 13th International Conference on Solid-State Sensors*, Actuators Microsystems, Seoul, Korea, pp. 368–371, June 2005.

[MOS 07] MOSES J.E., MOORHOUSE A.D., "The growing applications of click chemistry", *Chemical Society Reviews*, vol. 36, pp. 1249–1262, 2007.

[MOY 09] MOYNIHAN M., "An overview of wet chemistry processing for the manufacture of silicon solar cells", *CircuiTree*, vol. 22, no. 7, pp. 16–21, 2009.

[NEB 09] NEBHANI L., BARNER-KOWOLLIK C., "Orthogonal transformations on solid substrates: efficient avenues to surface modification", *Advanced Materials*, vol. 21, no. 34, pp. 3442–3468, 2009.

[NIE 08] NIE Z., KUMACHEVA E., "Patterning surfaces with functional polymers", *Nature Materials*, vol. 7, 84, p. 277–290, 2008.

[NRC 04] NATIONAL RESEARCH COUNCIL, *Monitoring Metabolic Status: Predicting Decrements in Physiological and Cognitive Performance*, The National Academies Press, Washington, DC, 2004.

[PAL 04] PALACIN S., BUREAU C., CHARLIER J., *et al.*, "Molecule-to-metal bonds: electrografting polymers on conducting surfaces", *ChemPhysChem*, vol. 5, no. 10, pp. 1468–1481, 2004.

[PAR 05] PARK S., LEE A., JEONG M.-J., *et al.*, "A disposable MEMS-based micro-biopsy catheter for the minimally invasive tissue sampling", *Proceedings of the IEEE International Conference on "Intelligent Robots and Systems" (IROS)*, pp. 2768–2773, 2005.

[PAR 07] PARK J.U., HARDY M., KANG S.J., *et al.*, "High-resolution electrohydrodynamic jet printing", *Nature Materials*, vol. 6, pp. 782–789, 2007.

[PAT 06] PATOLSKY F., ZHENG G., LIEBER C.M., "Fabrication of silicon nanowire devices for ultrasensitive, label-free, real-time detection of biological and chemical species", *Nature Protocols*, vol. 1, pp. 1711–1724, 2006.

[PHI 07] PHILLIPS F., KACZOR K., GANDHI N., *et al.*, "Measurement of sodium ion concentration in undiluted urine with cation-selective polymeric membrane electrodes after the removal of interfering compounds", *Talanta*, vol. 74, no. 2, pp. 255–264, 2007.

[PIR 02] PIRRUNG M.C., "How to make a DNA chip", *Angewandte Chemie-International Edition*, vol. 41, no. 8, pp. 1276–1289, 2002.

[POG 07] POGHOSSIAN A., "Label-free detection of single nucleotide polymorphisms utilizing the differential transfer function of field-effect transistors", *Biosensors and Bioelectronics*, vol. 22, pp. 2834–2840, 2007.

[PRA 09] PRAKASH S., KARACOR M.B., BANERJEE S., "Surface modification in microsystems and nanosystems", *Surface Science Reports*, vol. 64, no. 7, pp. 233–254, 2009.

[REI 09] REISS P., "Highly luminescent CuInS2/ZnS core/shell nanocrystals: cadmium-free quantum dots for in vivo imaging", *Chemistry of Materials*, vol. 21, no. 12, pp. 2422–2429, 2009.

[REV 03] REVZIN A., TOMPKINS R.G., TONER M., "Surface engineering with poly(ethylene glycol) photolithography to create high-density cell arrays on glass", *Langmuir*, vol. 19, no. 23, pp. 9855–9862, 2003.

[SAR 05] SARRUT N., BOUFFET S., MITTLER F., *et al.*, "Enzymatic digestion and liquid chromatography in micro-pillar reactors: hydrodynamic versus electro-osmotic flow", *Microfluidics, BioMEMS, and Medical Microsystems III*, vol. 5718, pp. 99–109, 2005.

[SAR 06] SARVESWARAN K., HU W., HUBER P.W., *et al.*, "Deposition of DNA rafts on cationic SAMs on silicon [100]", *Langmuir*, vol. 22, no. 26, pp. 11279–11283, 2006.

[SEK 10] SEKER E., BERDICHEVSKY Y., BEGLEY M.R., *et al.*, "The fabrication of low-impedance nanoporous gold multiple-electrode arrays for neural electrophysiology studies", *Nanotechnology*, vol. 21, no. 12, p. 125504, 2010.

[SEO 11] SEO J.H., CHEN L.-J., VERKHOTUROV S.V., *et al.*, "The use of glass substrates with bi-functional silanes for designing micropatterned cell-secreted cytokine immunoassays", *Biomaterials*, vol. 32, no. 23, pp. 5478–5488, 2011.

[SEV 03] SEVER R.R., ALCALA R., DUMESIC J.A., *et al.*, "Vapor-phase silylation of MCM-41 and Ti-MCM-41", *Microporous and Mesoporous Materials*, vol. 66, no. 1, pp. 53–67, 2003.

[SHA 05] SHAMSUDDIN A.K.M., YANAGIMOTO S., KUWAHARA T., *et al.*, "Changes in the index of sweat ion concentration with increasing sweat during passive heat stress in humans", *European Journal of Applied Physiology*, vol. 94, no. 3, pp. 292–297, 2005.

[SHA 07] SHAPIRO L., MARX S., MANDLER D., "Preparation and characterization of ultra-thin sol–gel films", *Thin Solid Films*, vol. 515, no. 11, pp. 4624–4628, 2007.

[SIK 07] SIKORA K., "Personalized medicine for cancer: from molecular signature to therapeutic choice", *Advances in Cancer Research*, vol. 96, pp. 345–369, 2007.

[SIR 00] SIRRINGHAUS H., KAWASE T., FRIEND R.H., *et al.*, "High-resolution inkjet printing of all-polymer transistor circuits", *Science*, vol. 290, no. 5499, pp. 2123–2126, 2000.

[STA 07] STACHOWIAK T.B., MAIR T.A., HOLDEN T.G., *et al.*, "Hydrophilic surface modification of cyclic olefin copolymer microfluidic chips using sequential photografting", *Journal of Separation Science*, vol. 30, no. 7, pp. 1088–1093, 2007.

[STI 00] STIEGLITZ T., BEUTEL H., SCHUETTLER M., *et al.*, "Micromachined, polyimide-based devices for flexible neural interfaces", *Biomedical Microdevices*, vol. 2, no. 4, pp. 283–294, 2000.

[STU 03] STUTZMANN N., FRIEND R.H., SIRRINGHAUS H., "Self-aligned, vertical-channel, polymer field-effect transistors", *Science*, vol. 299, no. 5614, pp. 1881–1884, 2003.

[SUG 02] SUGIMURA H., HOZUMI A., KAMEYAMA T., *et al.*, "Organosilane self-assembled monolayers formed at the vapor/solid interface", *Surface and Interface Analysis*, vol. 34, no. 1, pp. 550–554, 2002.

[SWE 11] SWEETMAN M.J., SHEARER C.J., SHAPTER J.G., *et al.*, "Dual silane surface functionalization for the selective attachment of human neuronal cells to porous silicon", *Langmuir*, vol. 27, no. 15, pp. 9497–9503, 2011.

[TER 06] TERADA A., YUASA A., KUSHIMOTO T., *et al.*, "Bacterial adhesion to and viability on positively charged polymer surfaces", *Microbiology*, vol. 152, no. 12, pp. 3575–3583, 2006.

[VAN 09] VAN DIJK M., RIJKERS D.T.S., LISKAMP R.M.J., *et al.*, "Synthesis and applications of biomedical and pharmaceutical polymers via click chemistry methodologies", *Bioconjugate Chemistry*, vol. 20, no. 11, pp. 2001–2016, 2009.

[VOR 95] VOROTILOV K., PETROVSKY V., VASILJEV V., "Spin coating process of solgel silicate films deposition: effect of spin speed and processing temperature", *Journal of Sol-Gel Science and Technology*, vol. 5, no. 3, pp. 173–183, 1995.

[WAN 03] WANG Q., CHAN T.R., HILGRAF R., *et al.*, "Bioconjugation by copper(I)-catalyzed azide-alkyne [3+2] cycloaddition", *Journal of the American Chemical Society*, vol. 125, no. 11, pp. 3192–3193, 2003.

[WAN 11] WANG Y., "A review of carbon nanotubes-based gas sensors", *Journal of Sensors*, Article ID 493904, 2011.

[WAN 06] WANG J.Z., GU J., ZENHAUSERN F., *et al.*, "Low-cost fabrication of submicron all polymer field effect transistors", *Applied Physics Letters*, vol. 88, no. 13, pp. 133502–133502_3, 2006.

[WAN 09] WANG Y., YEOW J.T.W., "A review of carbon nanotubes-based gas sensors", *Journal of Sensors*, vol. 2009, p. 24, 2009.

[WET 10] WETHERHOLD R.C., HARRY Z., "A rapid chemical method for improving peel strength at Cu-epoxy interfaces", *Theoretical and Applied Fracture Mechanics*, vol. 53, no. 1, pp. 42–46, 2010.

[WU 06] WU P., HOGREBE P., GRAINGER D.W., "DNA and protein microarray printing on silicon nitride waveguide surfaces", *Biosensors and Bioelectronics*, vol. 21, no. 7, pp. 1252–1263, 2006.

[WU 12] WU W.-I., SASK K.N., BRASH J.L., *et al.*, "Polyurethane-based microfluidic devices for blood contacting applications", *Lab on a Chip*, vol. 12, no. 5, pp. 960–970, 2012.

[XIA 98] XIA Y., WHITESIDES G.M., "Soft lithography", *Annual Review of Materials Science*, vol. 28, pp. 153–184, 1998.

[YAN 08] YANG Y., BITTNER A.M., BALDELLI S., *et al.*, "Study of self-assembled triethoxysilane thin films made by casting neat reagents in ambient atmosphere", *Thin Solid Films*, vol. 516, no. 12, pp. 3948–3956, 2008.

[YAO 12] YAO M., FANG J., "Hydrophilic PEO-PDMS for microfluidic applications", *Journal of Micromechanics and Microengineering*, vol. 22, no. 2, p. 025012, 2012.

[ZHA 04] ZHANG X., LU H., SOUTAR A.M., *et al.*, "Thick UV-patternable hybrid solgel films prepared by spin coating", *Journal of Materials Chemistry*, vol. 14, no. 3, pp. 357–361, 2004.

[ZHA 05] ZHANG G.-J., TANII T., ZAKO T., *et al.*, "Nanoscale patterning of protein using electron beam lithography of organosilane self-assembled monolayers", *Small*, vol. 1, no. 8–9, pp. 833–837, 2005.

[ZHA 10] ZHANG F., SAUTTER K., LARSEN A.M., *et al.*, "Chemical vapor deposition of three aminosilanes on silicon dioxide: surface characterization, stability, effects of silane concentration, and cyanine dye adsorption", *Langmuir*, vol. 26, no. 18, pp. 14648–14654, 2010.

[ZHO 10] ZHOU J., ELLIS A.V., VOELCKER N.H., "Recent developments in PDMS surface modification for microfluidic devices", *Electrophoresis*, vol. 31, no. 1, pp. 2–16, 2010.

[ZHO 11a] ZHOU J., Study and development of a smart wound dressing technology which can detect/kill the colonisation of pathogenic bacteria, PhD Thesis, ID 28349, University of Bath, United Kingdom, 2011.

[ZHO 11b] ZHOU X., BOEY F., HUO F., *et al.*, "Chemically functionalized surface patterning", *Small*, vol. 7, no. 16, pp. 2273–2289, 2011.

List of Authors

Jeffery BUTTERBAUGH
TEL FSI, Inc.
Chaska, MN
USA

Véronique CARRON
CEA LETI
Grenoble
France

Julie CREN
CEA LETI
Grenoble
France

Christiane GOTTSCHALK
MKS Instruments Inc.
Wilmington, MA
USA

Paul-Henri HAUMESSER
CEA LETI
Grenoble
France

Jean-Michel HARTMANN
CEA LETI
Grenoble
France

Antoine HOANG
CEA LETI
Grenoble
France

Vincent JOUSSEAUME
CEA LETI
Grenoble
France

Martin KNOTTER
NXP Semiconductors
AE Nijmegen
The Netherlands

Yannick LE TIEC
CEA LETI
Grenoble
France

Didier LOUIS
CEA LETI
Grenoble
France

Sylvain MAÎTREJEAN
CEA LETI
Grenoble
France

Gilles MARCHAND
CEA LETI
Grenoble
France

François MARTIN
CEA LETI
Grenoble
France

Kevin MCLAUGHLIN
Sachem Inc.
Austin, TX
USA

Guillaume NONGLATON
CEA LETI
Grenoble
France

Carole PERNEL
CEA LITEN
Grenoble
France

Catherine PEYNE
CEA LETI
Grenoble
France

Isabelle TEXIER-NOGUES
CEA LETI
Grenoble
France

Patrick VALENTI
Air Liquide
Centre de Recherche
Claude - Delorme
Jouy en Josas
France

Françoise VINET
CEA LETI
Grenoble
France

Index

C

D

E

F

G

H

N

O

P

Q

R

S

T

U

V

W

X

Y

Z